MARKOV CHAINS

HOLDEN-DAY SERIES IN PROBABILITY AND STATISTICS

E. L. Lehmann, Editor

MARKOV CHAINS

DAVID FREEDMAN

University of California, Berkeley

HOLDEN-DAY

San Francisco, Cambridge, London, Amsterdam

ISBN 0-8162-3004-8

Library of Congress Catalog Card Number: 73-111613

Printed in the United States of America.

TO WILLIAM FELLER

PREFACE

A long time ago I started writing a book about Markov chains, Brownian motion, and diffusion. I soon had two hundred pages of manuscript and my publisher was enthusiastic. Some years and several drafts later, I had a thousand pages of manuscript, and my publisher was less enthusiastic. So we made it a trilogy:

Markov Chains
Brownian Motion and Diffusion
Approximating Countable Markov Chains

familiarly — *MC, B & D,* and *ACM.*

I wrote the first two books for beginning graduate students with some knowledge of probability; if you can follow Sections 10.4 to 10.9 of *Markov Chains* you're in. The first two books are quite independent of one another, and completely independent of the third. This last book is a monograph which explains one way to think about chains with instantaneous states. The results in it are supposed to be new, except where there are specific disclaimers. It's written in the framework of *Markov Chains,* and contains the four *MC* chapters you will need for reference.

Most of the proofs in the trilogy are new, and I tried hard to make them explicit. The old ones were often elegant, but I seldom saw what made them go. With my own, I can sometimes show you why things work. And, as I will argue in a minute, my demonstrations are easier technically. If I wrote them down well enough, you may come to agree.

The approach in all three books is constructive: I did not use the notion of separability for stochastic processes and in general avoided the uncountable axiom of choice. Separability is a great idea for dealing with any really large class of processes. For Markov chains I find it less satisfactory. To begin with, a theorem on Markov chains typically amounts to a statement about a probability on a Borel σ-field. It's a shame to have the proof depend on the existence of an unnamable set. Also, separability proofs usually have two parts. There is an abstract part which establishes the existence of a separable version. And there is a combinatorial argument, which establishes some property of the separable version by looking at the behavior of the process on a countable set of times. If you take the constructive approach, the combinatorial argument alone is enough proof.

When I started writing, I believed in regular conditional distributions. To me they're natural and intuitive objects, and the first draft was full of them. I told it like it was, and if the details were a little hard to supply, that was the reader's problem. Eventually I got tired of writing a book intelligible only to me. And I came to believe that in most proofs, the main point is estimating a probability number: the fewer complicated intermediaries, the better. So I switchd to computing integrals by Fubini. This is a more powerful technique than you might think and it makes for proofs that can be checked. Virtually all the conditional distributions were banished to the Appendix. The major exception is Chapter 4 of *Markov Chains,* where the vividness of the conditional distribution language compensates for its technical difficulty.

In *Markov Chains,* Chapters 3 to 6 and 8 cover material not usually available in textbooks — for instance: invariance principles for functionals of a Markov chain; Kolmogorov's inequality on the concentration function; the boundary, with examples; and the construction of a variety of continuous-time chains from their jump processes and holding times. Some of these constructions are part of the folklore, but I think this is the first careful public treatment.

Brownian Motion and Diffusion dispenses with most of the customary transform apparatus, again for the sake of computing probability numbers more directly. The chapter on Brownian motion emphasizes topics which haven't had much textbook coverage, like square variation, the reflection principle, and the invariance principle. The chapter on diffusion shows how to obtain the process from Brownian motion by changing time.

I studied with the great men for a time, and saw what they did. The trilogy is what I learned. All I can add is my recommendation that you buy at least one copy of each book.

User's guide to *Markov Chains*

In one semester, you can cover Sections 1.1-9, 5.1-3, 7.1-3 and 9.1-3. This gets you the basic results for both discrete and continuous time. In one year you could do the whole book, provided you handle Chapters 4, 6, and 8 lightly. Chapters 2-4, 6 and 8 are largely independent of one another, treat specialized topics, and are more difficult; Section 8.5 is particularly hard. I do recommend looking at Section 6.6 for some extra grip on Markov times.

Sections 10.1-3 explain the cruel and unusual notation, and the reference system; 10.4-9 review probability theory quickly; 10.10-17 do the more exotic analyses which I've found useful at various places in the trilogy; and a few things are in 10.10-17 just because I like them.

Chapter 10 is repeated in *B & D;* Chapters 1, 5, 7 and 10 are repeated in *ACM*. The three books have a common preface and bibliography. Each has its own index and symbol finder.

Acknowledgments

Much of the trilogy is an exposition of the work of other mathematicians, who sometimes get explicit credit for their ideas. Writing *Markov Chains* would have been impossible without constant reference to Chung (1960). Doob (1953) and Feller (1968) were also heavy involuntary contributors. The diffusion part of *Brownian Motion and Diffusion* is a peasant's version of Itô and McKean (1965).

The influence of David Blackwell, Lester Dubins and Roger Purves will be found on many pages, as will that of my honored teacher, William Feller. Ronald Pyke and Harry Reuter read large parts of the manuscript and made an uncomfortably large number of excellent suggestions, many of which I was forced to accept. I also tested drafts on several generations of graduate students, who were patient, encouraging and helpful. These drafts were faithfully typed from the cuneiform by Gail Salo.

The Sloan Foundation and the US Air Force Office of Scientific Research supported me for various periods, always generously, while I did the writing. I finished two drafts while visiting the Hebrew University in Jerusalem, Imperial College in London, and the University of Tel Aviv. I am grateful to the firm of Cohen, Leithman, Kaufman, Yarosky and Fish, criminal lawyers and xerographers in Montreal. And I am still nostalgic for Cohen's Bar in Jerusalem, the caravansary where I wrote the first final draft of *Approximating Countable Markov Chains*.

David Freedman

Berkeley, California
July, 1970

TABLE OF CONTENTS

Part I.   Discrete time

Part II.   Continuous time

MARKOV CHAINS

1

INTRODUCTION TO DISCRETE TIME

1. FOREWORD

Consider a stochastic process which moves through a countable set I of states. At stage n, the process decides where to go next by a random mechanism which depends only on the current state, and not on the previous history or even on the time n. These processes are called *Markov chains with stationary transitions and countable state space*. They are the object of study in the first part of this book. More formally, there is a countable set of states I, and a stochastic process $X_0, X_1, \ldots$ on some probability triple $(\mathscr{X}, \mathscr{F}, \mathscr{P})$, with $X_n(x) \in I$ for all nonnegative integer n and $x \in \mathscr{X}$. Moreover, there is a function P on $I \times I$ such that

$$\mathscr{P}\{X_{n+1} = j \mid X_0, \ldots, X_n\} = P(X_n, j).$$

That is, the conditional distribution of X_{n+1} given $X_0, \ldots, X_n$ depends on X_n, but not on n or on $X_0, \ldots, X_{n-1}$. The process X is said to be *Markov with stationary transitions P*, or to have *transitions P*. Suppose I is reduced to the *essential range*, namely the set of j with $\mathscr{P}\{X_n = j\} > 0$ for some n. Then the transitions P are unique, and form a stochastic matrix. Here is an equivalent characterization: X is Markov with stationary transitions P iff

$$\mathscr{P}\{X_n = j_n \text{ for } n = 0, \ldots, N\} = \mathscr{P}\{X_0 = j_0\} \, \Pi_{n=0}^{N-1} \, P(j_n, j_{n+1})$$

for all N and $j_n \in I$. If $\mathscr{P}\{X_0 = j\} = 1$ for some $j \in I$, then X is said to *start*

I want to thank Richard Olshen for checking the final draft of this chapter.

from j or to have *starting state j*. This involves no real loss in generality, as one sees by conditioning on X_0.

(1) Definition. *A stochastic matrix P on I is a function on* $I \times I$, *such that:*

$$P(i,j) \geqq 0 \quad \text{for all } i \text{ and } j \text{ in } I;$$

and

$$\Sigma_{j \in I} P(i,j) = 1 \quad \text{for all } i \text{ in } I.$$

If P and Q are stochastic matrices on I, so is PQ, where

$$(PQ)(i,k) = \Sigma_{j \in I} P(i,j)Q(j,k).$$

And so are P^n, where $P^1 = P$ and $P^{n+1} = PP^n$.

Here are three examples: let Y_n be independent and identically distributed, taking the values 1 and -1 with equal probability $\frac{1}{2}$.

(2) Example. Let $X_0 = 1$. For $n = 1, 2, \ldots$, let $X_n = Y_n$. Then $\{X_n\}$ is a Markov chain with state space $I = \{-1, 1\}$ and stationary transitions P, where $P(i,j) = \frac{1}{2}$ for all i and j in I. The starting state is 1.

(3) Example. Let $X_0 = 0$. For $n = 1, 2, \ldots$, let $X_n = X_{n-1} + Y_n$. Then $\{X_n\}$ is a Markov chain with the integers for state space and stationary transitions P, where

$$P(n, n+1) = P(n, n-1) = \frac{1}{2}$$

$$P(n, m) = 0 \quad \text{when } |n - m| \neq 1.$$

The starting state is 0.

(4) Example. Let $X_n = (Y_n, Y_{n+1})$ for $n = 0, 1, \ldots$. Then $\{X_n\}$ is a Markov chain with state space I and stationary transitions P, where I is the set of pairs (a, b) with $a = \pm 1$ and $b = \pm 1$, and

$$P[(a, b), (c, d)] = 0 \quad \text{when } b \neq c$$

$$= \frac{1}{2} \quad \text{when } b = c.$$

By contrast, let $\hat{X}_n = Y_n + Y_{n+1}$. Now $\hat{X}_n$ is a function of X_n. But $\{\hat{X}_n\}$ is not Markov.

Return to the general Markov chain X with stationary transitions. For technical reasons, it is convenient to study the distribution of X rather than X itself. The formal exposition begins in Section 3 by describing these distributions. This will be repeated here, with a brief explanation of how to translate the results back into statements about X. Introduce the space I^∞ of I-sequences. That is, I^∞ is the set of functions ω from the nonnegative integers

1.1] FOREWORD 3

to I. For $n = 0, 1, \ldots$, define the coordinate function ξ_n on I^∞ by

$$\xi_n(\omega) = \omega(n) \quad \text{for } \omega \in \Omega.$$

Then $\xi_0, \xi_1, \ldots$ is the coordinate process. Give I^∞ the smallest σ-field $\sigma(I^\infty)$ over which each coordinate function is measurable. Thus, $\sigma(I^\infty)$ is generated by the cylinders

$$\{\xi_0 = i_0, \ldots, \xi_n = i_n\}.$$

For any $i \in I$ and stochastic matrix P on I, there is one and only one probability P_i on I^∞ making the coordinate process Markov with stationary transitions P and starting state i. In other terms:

$$P_i\{\xi_n = i_n \text{ for } n = 0, \ldots, N\} = \Pi_{n=0}^{N-1} P(i_n, i_{n+1}),$$

for all N and $i_n \in I$ with $i_0 = i$. The probability P_i really does depend only on P and i.

Now I^∞ is the sample space for X, namely the space of all realizations. More formally, there is a mapping M from $\mathscr{X}$ to I^∞, uniquely defined by the relation

$$\xi_n(Mx) = X_n(x) \quad \text{for all } n = 0, 1, \ldots \text{ and } x \in \mathscr{X}.$$

That is, the nth coordinate of Mx is $X_n(x)$, and Mx is the sequence of states X passes through at x, namely: $(X_0(x), X_1(x), X_2(x), \ldots)$. Check that M is measurable. Fix $i \in I$ and a stochastic matrix P on I. Suppose X is Markov with stationary transitions P and starting state i. With respect to the *distribution* of X, namely $\mathscr{P}M^{-1}$, the coordinate process is Markov with stationary transitions P and starting state i. Therefore $\mathscr{P}M^{-1} = P_i$. Conversely, $\mathscr{P}M^{-1} = P_i$ implies that X is Markov with stationary transitions P and starting state i. Now probability statements about X can be translated into statements about P_i. For example, the following three assertions are all equivalent:

(5a) $\qquad\qquad P_i\{\xi_n = i \text{ for infinitely many } n\} = 1.$

(5b)   For some Markov chain X with stationary transitions P and starting state i,
$$\mathscr{P}\{X_n = i \text{ for infinitely many } n\} = 1.$$

(5c)   For all Markov chains X with stationary transitions P and starting state i,
$$\mathscr{P}\{X_n = i \text{ for infinitely many } n\} = 1.$$

Indeed, the set talked about in (5b) is the M-inverse image of the set talked about in (5a); and $P_i = \mathscr{P}M^{-1}$.

The basic theory of these processes is developed in a rapid but complete way in Sections 3–9; Sections 10, 12, and 14 present some examples, while Sections 11 and 13 cover special topics. Readers who want a more leisurely discussion of the intuitive background should look at (Feller, 1968, XV) or (Kemeny and Snell, 1960). Here is a summary of Sections 3–9.

2. SUMMARY

The main result in Section 3 is the strong Markov property. To state the best case of it, let the random variable τ on I^∞ take only the values $0, 1, \ldots, \infty$. Suppose the set $\{\tau = n\}$ is in the σ-field spanned by $\xi_0, \ldots, \xi_n$ for $n = 0, 1, \ldots,$ and suppose

$$P_i\{\tau < \infty \text{ and } \xi_\tau = j\} = 1 \quad \text{for some } j \in I.$$

Then the fragment

$$(\xi_0, \ldots, \xi_\tau)$$

and the process

$$\xi_\tau, \xi_{\tau+1}, \xi_{\tau+2}, \ldots$$

are P_i-independent; the P_i-distribution of the process is P_j. This is a special case of the *strong Markov property*.

(6) Illustration.   Let τ be the least n with $\xi_n = j$, and $\tau = \infty$ if there is no such n; the assumption above is $P_i\{\tau < \infty\} = 1$.

To state the results of Section 4, write:

$$i \to j \quad \text{iff} \quad P^n(i,j) > 0 \quad \text{for some } n = 1, 2, \ldots ;$$

$$i \leftrightarrow j \quad \text{iff} \quad i \to j \text{ and } j \to i$$

$$i \text{ is } essential \quad \text{iff} \quad i \to j \quad \text{implies } j \to i.$$

(7) Illustration.   Suppose $I = \{1, 2, 3, 4\}$ and P is this matrix:

$$\begin{pmatrix} 1 & 0 & 0 & 0 \\ 0 & \frac{1}{2} & \frac{1}{2} & 0 \\ 0 & \frac{1}{2} & \frac{1}{2} & 0 \\ \frac{1}{4} & \frac{1}{4} & \frac{1}{4} & \frac{1}{4} \end{pmatrix}.$$

Then 1,2,3 are essential and 4 is inessential. Moreover, $1 \leftrightarrow 1$ while $2 \leftrightarrow 3$.

For the rest of this summary,

suppose   all $i \in I$ are essential.

Then $\leftrightarrow$ is an equivalence relation. For the rest of this summary, suppose that I consists of one equivalence class, namely,

$$\text{suppose}\quad i \to j \text{ and } j \to i \quad \text{for all } i \text{ and } j \text{ in } I.$$

Let *period* i be the greatest common denominator (g.c.d.) of the set of $n > 0$ with $P^n(i, i) > 0$. Then period i does not depend on i; say it is d. And I is the disjoint union of sets $C_0, C_1, \ldots, C_{d-1}$, such that

$$i \in C_n \quad \text{and} \quad P(i, j) > 0 \quad \text{imply } i \in C_{n \oplus 1},$$

where $\oplus$ means addition modulo d.

(8) Illustration. Suppose $I = \{1, 2, 3, 4\}$ and P is this matrix:

$$\begin{pmatrix} 0 & 0 & \frac{1}{4} & \frac{3}{4} \\ 0 & 0 & \frac{1}{2} & \frac{1}{2} \\ \frac{2}{3} & \frac{1}{3} & 0 & 0 \\ \frac{1}{2} & \frac{1}{2} & 0 & 0 \end{pmatrix}.$$

Then I has period 2, and $C_0 = \{1, 2\}$ and $C_1 = \{3, 4\}$.

For the rest of the summary,

$$\text{suppose}\quad \text{period } i = 1 \quad \text{for all } i \in I.$$

To state the result of Section 5, say

$$i \text{ is } recurrent \quad \text{iff}\quad P_i\{\xi_n = i \text{ for infinitely many } n\} = 1$$
$$i \text{ is } transient \quad \text{iff}\quad P_i\{\xi_n = i \text{ for infinitely many } n\} = 0.$$

This classification is exhaustive. Namely, the state i is either recurrent or transient, according as $\Sigma_n P^n(i, i)$ is infinite or finite. And all $i \in I$ are recurrent or transient together. These results follow from the strong Markov property. Parenthetically, under present assumptions: if I is finite, all $i \in I$ are recurrent.

(9) Example. Suppose $I = \{0, 1, 2, \ldots\}$. Let $0 < p_n < 1$. Suppose $P(0, 1) = 1$ and for $n = 1, 2, \ldots$ suppose $P(n, n + 1) = p_n$ and $P(n, 0) = 1 - p_n$. Suppose all other entries in P vanish; see Figure 1. The states are recurrent or transient according as $\Pi\, p_n$ is zero or positive.

HINT.   See (16) below.   ★

For the rest of this summary,

$$\text{suppose}\quad \text{all } i \in I \text{ are recurrent.}$$

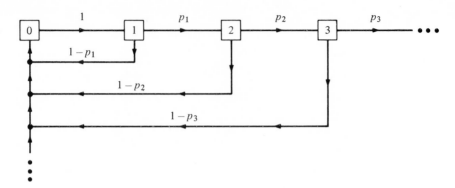

Figure 1

To state the result of Section 6, let $Y_1, Y_2, \ldots$ be a sequence of independent, identically distributed random variables, taking only the values $1, 2, 3, \ldots$ with probabilities $p_1, p_2, p_3, \ldots$. Let $\mu = \Sigma\, np_n$, and suppose

$$\text{g.c.d. } \{n : p_n > 0\} = 1.$$

Let $U(m)$ be the probability that

$$Y_1 + \cdots + Y_n = m \quad \text{for some } n = 0, 1, 2, \ldots.$$

Then

$$\lim_{m \to \infty} U(m) = 1/\mu.$$

This result is called the *renewal theorem*. It is used in Section 7, together with strong Markov, to show that

$$\lim_{n \to \infty} P^n(i, j) = \pi(j),$$

where $1/\pi(j)$ is the P_j-expectation of the least $m > 0$ with $\xi_m = j$.

To state the result of Section 8, say

$$j \text{ is } \textit{positive recurrent} \quad \text{iff} \quad \pi(j) > 0$$
$$j \text{ is } \textit{null recurrent} \quad \text{iff} \quad \pi(j) = 0.$$

Then all $i \in I$ are either positive recurrent or null recurrent together.

(10) Example. Let $I = \{0, 1, 2, \ldots\}$. Let $p_n > 0$ and $\Sigma_{n=1}^{\infty}\, p_n = 1$. Let $P(0, n) = p_n$ and $P(n, n - 1) = 1$ for $n = 1, 2, \ldots$. See Figure 2. The states are positive recurrent or null recurrent according as $\Sigma_{n=1}^{\infty}\, np_n$ is finite or infinite.

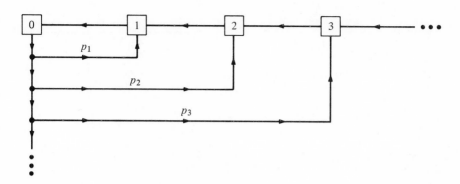

Figure 2

HINT.   See (16) below.
For the rest of this summary,

suppose   all $i \in I$ are positive recurrent.

To state the result of Section 9, say a measure m on I is *invariant* iff

$$m(j) = \Sigma_{i \in I}\, m(i)P(i,j) \quad \text{for all } j \in I.$$

Recall that $\pi(j) = \lim_{n \to \infty} P^n(i,j)$. Then π is an invariant probability. And any invariant signed measure which has finite total mass is a scalar multiple of π. A signed measure m has *finite mass* if $\Sigma_{i \in I} |m(i)| < \infty$.

The results in Sections 3–9 are standard, so references are sparse. The basic results for finite I are usually due to Markov himself. The extension to countable I is usually due to Kolmogorov or Lévy.

3.   THE MARKOV AND STRONG MARKOV PROPERTIES

Let I be a finite or countably infinite set. Give I the discrete σ-field, that is, the σ-field of all its subsets. Let I^∞ be the space of all I-sequences, namely, functions from the nonnegative integers to I. For $\omega \in I^\infty$ and $n = 0, 1, \ldots$ let $\xi_n(\omega) = \omega(n)$. Call ξ the *coordinate process*. Give I^∞ the *product σ-field* $\sigma(I^\infty)$, namely, the smallest σ-field such that $\xi_0, \xi_1, \ldots$ are measurable. A *matrix P on I* is a function $(i,j) \to P(i,j)$ from $I \times I$ to the real line. Say P is *stochastic* iff $P(i,j) \geqq 0$ for all i, j and $\Sigma_j P(i,j) = 1$ for all i. Say P is *substochastic* iff $P(i,j) \geqq 0$ for all i, j and $\Sigma_j P(i,j) \leqq 1$ for all i.   Let P be a stochastic matrix on I, and p a probability on I. There is a unique probability P_p on $(I^\infty, \sigma(I^\infty))$ such that for all n and all $i_0, \ldots, i_n$ in I,

(11) $P_p\{\xi_m = i_m \text{ for } m = 0, 1, \ldots, n\} = p(i_0) \; \Pi_{m=0}^{n-1} P(i_m, i_{m+1})$;

by convention, an empty product is 1. For example, use the Kolmogorov consistency theorem (10.53). If $p\{i\} = 1$, write P_i for P_p.

Sometimes, it is convenient to define P_p even for substochastic P and subprobabilities p. To do this, let I^* be the set of all finite I-sequences, including the empty sequence. Give $I^\infty \cup I^*$ the smallest σ-field which contains all subsets of I^*, and all sets in $\sigma(I^\infty)$. Then $\xi_0, \xi_1, \ldots$ are partially defined and measurable on $I^\infty \cup I^*$; namely, $\xi_m(\omega) = \omega(m)$ is defined provided $\omega \in I^\infty$ or $\omega \in I^*$ has length at least $m + 1$. For this purpose, a sequence of length $m + 1$ is a function from $\{0, 1, \ldots, m\}$ to I. Then there is still one and only one probability P_p on $I^\infty \cup I^*$ satisfying (11). Of course, P_p may assign quite a lot of mass to the empty sequence.

Let $X_0, X_1, \ldots$ be an I-valued stochastic process on a probability triple $(\mathcal{X}, \mathcal{F}, \mathcal{P})$. Then X is a measurable mapping from $(\mathcal{X}, \mathcal{F})$ to $(I^\infty, \sigma(I^\infty))$:

$$[X(x)](n) = X_n(x) \quad \text{for } x \in \mathcal{X} \text{ and } n = 0, 1, \ldots.$$

The *distribution* of X is $\mathcal{P}X^{-1}$, a probability on $\sigma(I^\infty)$. More generally, let $X_0, X_1, \ldots$ be a partially defined, I-valued process on $(\mathcal{X}, \mathcal{F}, \mathcal{P})$. That is, X_n is a function from part of $\mathcal{X}$ to I; and domain $X_{n+1} \subset$ domain X_n; and $\{X_n = i\} \in \mathcal{F}$. Then X is a measurable mapping from $\mathcal{X}$ to $I^\infty \cup I^*$. And $\mathcal{P}X^{-1}$ resides in $I^\infty \cup I^*$.

(12) Definition. $X_0, X_1, \ldots$ *is a* Markov chain with stationary transitions P and starting distribution p *iff the distribution of X is P_p. If $p\{i\} = 1$, say the chain starts from i, or has starting state i. In particular, for stochastic P and probabilities p, the coordinate process $\xi_0, \xi_1, \ldots$ is a Markov chain with stationary transitions P and starting probability p, on the probability triple* $(I^\infty, \sigma(I^\infty), P_p)$.

From now on, unless otherwise noted,

(13) Convention. P is a stochastic matrix on the finite or countably infinite set I. And p is a probability on I.

For (14) and later use, define the *shift* T as this mapping from I^∞ to I^∞:

$$(T\omega)(m) = \omega(m + 1) \quad \text{for } m = 0, 1, \ldots \text{ and } \omega \in I^\infty.$$

For $n = 1, 2, \ldots$,

$$(T^n\omega)(m) = \omega(m + n) \quad \text{for } m = 0, 1, \ldots \text{ and } \omega \in I^\infty.$$

It is convenient to adopt this definition even for $n = 0$, so T^0 is the identity function. Thus, $T^n\omega$ is ω shifted n times to the left; the first n terms of ω disappear during this transaction. In slightly ambiguous notation,

$$T^n\omega = (\omega(n), \omega(n+1), \dots).$$

Formally,

$$\xi_m \circ T^n = \xi_{m+n};$$

where $\circ$ is composition. Write

$$T^{-n}B = (T^n)^{-1}B = \{T^n \in B\} = \{\omega : \omega \in I^\infty \text{ and } T^n\omega \in B\}.$$

Then

$$T^{-n}\{\xi_0 = j_0, \dots, \xi_m = j_m\} = \{\xi_n = j_0, \dots, \xi_{n+m} = j_m\}.$$

So T^n is measurable.

Theorem (14) makes an assertion about regular conditional distributions: these objects are discussed in the Appendix. And (14) uses the symbol P_{ξ_n}. This is an abbreviation for a function Q of pairs (ω, B), with $\omega \in I^\infty$ and $B \in \sigma(I^\infty)$, namely:

$$Q(\omega, B) = P_{\xi_n(\omega)}(B).$$

(14) Theorem. (*Markov property*). P_ξ *is a regular conditional P_p-distribution for T^n given $\xi_0, \dots, \xi_n$.*

PROOF.   For $\omega \in I^\infty$ and $B \in \sigma(I^\infty)$, let

$$Q(\omega, B) = P_{\xi_n(\omega)}(B).$$

For each ω, the function $B \to Q(\omega, B)$ is a probability. For each B, the function $\omega \to Q(\omega, B)$ is measurable on $\xi_0, \dots, \xi_n$, because ξ_n is. What I need is

$$P_p\{A \text{ and } T^n \in B\} = \int_A Q(\omega, B) \, P_p(d\omega)$$

for all A measurable on $\xi_0, \dots, \xi_n$ and all measurable B. Both sides of this equality are countably additive in A. If I could prove the equality separately for each piece $\{A \text{ and } \xi_n = j\}$, I could finish by summing out j. But $\{\xi_n = j\}$ is measurable on $\xi_0, \dots, \xi_n$, so $\{A \text{ and } \xi_n = j\}$ is the typical subset of $\{\xi_n = j\}$ measurable on $\xi_0, \dots, \xi_n$; therefore, I only have to prove the equality for subsets A of $\{\xi_n = j\}$ measurable on $\xi_0, \dots, \xi_n$. The integrand on the right is now constant, so the integral is

$$P_p\{A\} \cdot P_j\{B\}.$$

What I have left to prove is this identity:

(15) $$P_p\{A \text{ and } T^n \in B\} = P_p\{A\} \cdot P_j\{B\}$$

for all subsets A of $\{\xi_n = j\}$ which are measurable on $\xi_0, \dots, \xi_n$, and all

$B \in \sigma(I^\infty)$. Consider special A, of the form

$$\{\xi_0 = i_0, \ldots, \xi_n = i_n\} \quad \text{with } i_n = j;$$

and special B, of the form

$$\{\xi_0 = j_0, \ldots, \xi_m = j_m\}.$$

Here, $i_0, \ldots, i_n, j_0, \ldots, j_m$ are all in I, and m is variable. Then

$$\{T^n \in B\} = \{\xi_n = j_0, \ldots, \xi_{n+m} = j_m\}.$$

If $j_0 \neq j$, both sides of (15) vanish. If $j_0 = j$, each side of (15) can be computed separately from (11), remembering $i_n = j_0 = j$, and works out to

$$p(i_0) \cdot [\Pi_{v=0}^{n-1} P(i_v, i_{v+1})] \cdot [\Pi_{v=0}^{m-1} P(j_v, j_{v+1})].$$

So, I have verified (15) for special A and special B.

Any subset A of $\{\xi_n = j\}$ which is measurable on $\xi_0, \ldots, \xi_n$ is empty, or a finite or countably infinite union of different special A, which are automatically pairwise disjoint. For each B, both sides of (15) are countably additive in general A. Therefore, (15) holds for general A and special B. Fix one general A. Both sides of (15) are countably additive in general B, and agree at special B. This is still true if I call the null set and the whole space special. The special B generate the full σ-field $\sigma(I^\infty)$. The intersection of two special B is special: indeed, two different special B are either disjoint or nested. Use (10.16) to complete the proof.   ★

Let A be a subset of $\{\xi_n = j\}$ which is measurable on $\xi_0, \ldots, \xi_n$, and let f be a nonnegative, measurable function on I^∞. Then (15) can be rewritten as

$$(15^*) \qquad \int_A f(T^n) \, dP_p = P_p(A) \cdot \int f \, dP_j.$$

This is (15) when f is an indicator. As a function of f, each side of (15^*) is linear and continuous for nondecreasing passages to the limit.

For (16) and later use, call a transition from i to j *possible* iff $P(i, j) > 0$. All transitions in ω are *possible* iff $P[\xi_n(\omega), \xi_{n+1}(\omega)] > 0$ for all n.

(16) Proposition. *The set G of ω such that all transitions in ω are possible is measurable, and $P_p\{G\} = 1$.*

PROOF.   Clearly, $G = \bigcap_{n=0}^\infty G_n$, where

$$G_n = \{P(\xi_n, \xi_{n+1}) > 0\}.$$

Now

$$\begin{aligned}
P_i\{G_0\} &= P_i\{P(i, \xi_1) > 0\} \\
&= \Sigma_j \{P(i, j) : P(i, j) > 0\} \\
&= \Sigma_j P(i, j) \\
&= 1.
\end{aligned}$$

Check
$$G_n = \{T^n \in G_0\}.$$
So,
$$\begin{aligned}
P_p\{G_n\} &= \Sigma_i \, P_p\{\xi_n = i \text{ and } G_n\} \\
&= \Sigma_i \, P_p\{\xi_n = i\} P_i\{G_0\} \quad \text{by (15)} \\
&= 1.
\end{aligned}$$
★

If P and Q are substochastic matrices on I, so is PQ, where:

$$PQ(i, k) = \Sigma_{j \in I} \, P(i, j) Q(j, k).$$

If P and Q are stochastic, so is PQ. If P is a substochastic matrix on I, so is P^n, where $P^1 = P$ and $P^{n+1} = PP^n = P^nP$. If P is stochastic, so is P^n.

(17) Theorem (*Semigroup property*).   *For all $i \in I$ and $n = 1, 2, \dots$*

$$P_i\{\xi_n = j\} = P^n(i, j).$$

PROOF.   This is trivial for $n = 1$. Use induction:

$$\begin{aligned}
P_i\{\xi_{n+1} = k\} &= \Sigma_j \, P_i\{\xi_n = j \text{ and } \xi_{n+1} = k\} \\
&= \Sigma_j \, P_i\{\xi_n = j\} \cdot P(j, k) \quad \text{by (15)} \\
&= \Sigma_k \, P^n(i, k) P(k, j) \quad \text{by inductive assumption} \\
&= P^{n+1}(i, j).
\end{aligned}$$
★

WARNING.   This does not characterize Markov chains. See (Feller, 1959).

The strong Markov property (21) strengthens (14). To state it, make the following definitions: A random variable τ on $(I^\infty, \sigma(I^\infty))$ is a *Markov time*, or just *Markov*, iff τ takes only the values $0, 1, \dots, \infty$; for every $n = 0, 1, \dots$, the set $\{\tau \leq n\}$ is in the σ-field $\mathscr{F}_n$ generated by $\xi_0, \dots, \xi_n$. The *pre-τ sigma-field* $\mathscr{F}_\tau$ is the σ-field of all sets $A \in \sigma(I^\infty)$ such that

$$A \cap \{\tau \leq n\} \in \mathscr{F}_n \quad \text{for every } n = 0, 1, \dots.$$

NOTES.   (a) Suppose τ is a function on I^∞, taking the values $0, 1, \dots, \infty$. Do not assume τ is measurable. Then τ is Markov iff $\tau(\omega) = n$ and $\xi_m(\omega') = \xi_m(\omega)$ for $m = 0, \dots, n$ force $\tau(\omega') = n$. Indeed, if τ satisfies the condition, then $\{\tau = n\}$ is a union of sets

$$\{\xi_m = i_m \text{ for } m = 0, \dots, n\} \in \mathscr{F}_n.$$

Conversely, if τ is Markov, then $\{\tau = n\} \in \mathscr{F}_n$ is a union of such sets.
  (b) If τ isn't Markov, then $\mathscr{F}_\tau$ isn't a σ-field; in fact, $I^\infty \notin \mathscr{F}_\tau$.
  (c) $\mathscr{F}_\tau$ specifies the sample function up to and including time τ, when τ is

finite. More formally, the atoms of $\mathscr{F}_\tau$ are the singletons in $\{\tau = \infty\}$, and all sets

$$\{\xi_0 = i_0, \xi_1 = i_1, \ldots, \xi_n = i_n \text{ and } \tau = n < \infty\}.$$

(d) Suppose τ is Markov, and A is a subset of $\{\tau < \infty\}$. Do not assume A is measurable. Then $A \in \mathscr{F}_\tau$ iff $\omega \in A$ and $\tau(\omega) = n$ and $\xi_m(\omega) = \xi_m(\omega')$ for $m = 0, \ldots, n$ force $\omega' \in A$.

(18) Illustration. Let f be a nonnegative function on I, and let τ be the least n if any with $\Sigma_{m=0}^n f(\xi_m) \geqq 17$; let $\tau = \infty$ if none. Then τ is Markov.

(19) Illustrations. Let τ be a Markov time.

(a) $\Delta = \{\tau < \infty\} \in \mathscr{F}_\tau$.

(b) Any measurable subset of $I^\infty \backslash \Delta$, the complement of Δ, is in $\mathscr{F}_\tau$.

(c) The time τ is $\mathscr{F}_\tau$-measurable.

(d) The sum $\Sigma_{n=0}^\tau f(\xi_n)$ is $\mathscr{F}_\tau$-measurable, for any function f on I. This includes (c): put $f \equiv 1$.

(e) This event is in $\mathscr{F}_\tau$: the process $\{\xi_n\}$ visits both i and j on or before time τ, and the first i occurs before the first j.

Let $\zeta_n = \xi_{\tau+n}$, defined on $\Delta = \{\tau < \infty\}$. More explicitly,

$$\zeta_n(\omega) = \xi_{\tau(\omega)+n}(\omega) \quad \text{for } \omega \in \Delta \text{ and } n = 0, 1, \ldots.$$

Of course, ζ_n is measurable:

$$\{\zeta_n = k\} = \bigcup_{m=0}^\infty \{\tau = m \text{ and } \xi_{m+n} = k\}.$$

Let ζ be the whole *post-τ process*. Informally,

$$\zeta = (\zeta_0, \zeta_1, \ldots).$$

Formally, ζ is this mapping of Δ into I^∞:

$$\zeta(\omega) = T^{\tau(\omega)}(\omega) \quad \text{for } \omega \in \Delta,$$

where T is the shift, as defined for (14). Verify that

$$\zeta_n = \xi_n \circ \zeta,$$

so ζ is measurable.

(20) Illustration. $\zeta_0 = \xi_\tau$ is $\mathscr{F}_\tau$-measurable; that is,

$$\{\Delta \text{ and } \zeta_0 = i\} \in \mathscr{F}_\tau.$$

Theorem (21) uses the notation P_{ζ_0}. This is an abbreviation for a function Q of pairs (ω, B), with $\omega \in \Delta$ and $B \in \sigma(I^\infty)$:

$$Q(\omega, B) = P_{\zeta_0(\omega)}(B).$$

(21) Theorem. (*Strong Markov property*).   *Let τ be Markov and let ζ be the post-τ process. Given $\mathscr{F}_\tau$, a regular conditional P_p-distribution for ζ on Δ is P_{ζ_0}.*

PROOF.   As in (14), I can reduce this to proving

$$(22)\qquad\qquad P_p\{A \text{ and } \zeta \in B\} = P_p\{A\} \cdot P_i\{B\}$$

for all $A \in \mathscr{F}_\tau$ with $A \subset \{\Delta \text{ and } \zeta_0 = i\}$, and all $B \in \sigma(I^\infty)$. But

$$\{A \text{ and } \zeta \in B\} = \bigcup_{n=0}^\infty \{A \text{ and } \tau = n \text{ and } T^n \in B\}.$$

Now $\xi_n = \xi_\tau = \zeta_0 = i$ on $\{A \text{ and } \tau = n\}$, and this set is in $\mathscr{F}_n$, because $A \in \mathscr{F}_\tau$. With this and (15) as props:

$$\begin{aligned} P_p\{A \text{ and } \zeta \in B\} &= \Sigma_{n=0}^\infty P_p\{A \text{ and } \tau = n \text{ and } T^n \in B\} \\ &= \Sigma_{n=0}^\infty P_p\{A \text{ and } \tau = n\} \cdot P_i\{B\} \\ &= P_p\{A\} \cdot P_i\{B\}. \qquad\qquad \bigstar \end{aligned}$$

Let $A \in \mathscr{F}_\tau$ and $A \subset \{\tau < \infty \text{ and } \xi_\tau = j\}$. Let f be a nonnegative, measurable function on I^∞. Then (22) can be rewritten as

$$(22^*)\qquad\qquad \int_A f(\zeta)\, dP_p = P_p(A) \cdot \int f\, dP_j.$$

This is (22) when f is an indicator. As a function of f, each side of (22*) is linear and continuous for nondecreasing passages to the limit.

Proposition (23) is preliminary to (24), which illustrates the use of strong Markov (21). Neither result will be used again until Chapter 7. For (23), fix $j \in I$ with $P(j,j) < 1$. Let q_j be the probability on I which assigns mass 0 to j, and mass $P(j,k)/[1 - P(j,j)]$ to $k \neq j$. Let τ be the least n if any with $\xi_n \neq \xi_0$, and let $\tau = \infty$ if none. Let ζ be the post-τ process: $\zeta = T^\tau$. Say that U is *geometric with parameter* θ iff U is u with probability $(1 - \theta)\theta^u$ for $u = 0, 1, \ldots$.

(23) Proposition.   *With respect to P_j:*

$\tau - 1$ is geometric with parameter $P(j,j)$;
ζ is Markov with stationary transitions P and starting probability q_j;
τ and ζ are independent.

PROOF.   Let $n = 1, 2, \ldots$; let $i_0 \neq j$ and let $i_1, \ldots, i_m \in I$. Then

$$\{\xi_0 = j \text{ and } \tau = n \text{ and } \zeta_0 = i_0, \ldots, \zeta_m = i_m\}$$
$$= \{\xi_0 = \cdots = \xi_{n-1} = j, \xi_n = i_0, \ldots, \xi_{n+m} = i_m\},$$

an event of P_j-probability

$$P(j,j)^{n-1} P(j, i_0)P(i_0, i_1) \cdots P(i_{m-1}, i_m). \qquad\qquad \bigstar$$

For (24), keep $j \in I$ with $P(j, j) < 1$. Introduce the notion of a *j-sequence* of ξ; namely, a maximal interval of times n with $\xi_n = j$. Let $C_1, C_2, \ldots$ be the cardinalities of the first, second, ... *j*-sequences of ξ. Let $C_{n+1} = 0$ if there are n or fewer *j*-sequences. Let A_N be the event that there are N or more *j*-sequences in ξ.

(24) Proposition.   *Given A_N, the variables $C_1 - 1, C_2 - 1, \ldots, C_N - 1$ are conditionally P_p-independent and geometrically distributed, with common parameter $P(j, j)$.*

PROOF.   Fix positive integers N and $c_1, \ldots, c_N$. I claim

$$(25)\quad P_p\{C_1 = c_1, \ldots, C_N = c_N \mid A_N\} = P_j\{C_1 = c_1, \ldots, C_N = c_N \mid A_N\}.$$

Let

$$B = \{C_1 = c_1, \ldots, C_N = c_N \text{ and } A_N\}.$$

Let σ be the least n if any with $\xi_n = j$, and $\sigma = \infty$ if none. Then σ is Markov. Let η be the post-σ process. Now

$$B \subset A_1 = \{\sigma < \infty\}$$
$$\eta_0 = j \text{ on } \{\sigma < \infty\}$$
$$A_N = \{\sigma < \infty \text{ and } \eta \in A_N\}.$$

On A_N,

$$C_n = C_n \circ \eta \quad \text{for } n = 1, \ldots, N.$$

So

$$B = A_1 \cap \{\eta \in B\}.$$

By strong Markov (21),

$$(26)\qquad\qquad P_p\{B\} = P_p\{A_1\} \cdot P_j\{B\}.$$

Sum out $c_1, \ldots, c_N$:

$$(27)\qquad\qquad P_p\{A_N\} = P_p\{A_1\} \cdot P_j\{A_N\}.$$

Divide (26) by (27) to substantiate the claim (25). The case $N = 1$ is now immediate from (23).

Abbreviate $\theta = P(j, j)$ and $q = q_j$, as defined for (23). I claim

$$(28)\quad P_j\{C_1 = c_1, \ldots, C_{N+1} = c_{N+1} \mid A_{N+1}\}$$
$$= (1 - \theta)\theta^{c_1 - 1} P_q\{C_1 = c_2, \ldots, C_N = c_{N+1} \mid A_N\}.$$

This and (25) prove (24) inductively. To prove (28), let τ be the least n if any with $\xi_n \neq j$, and $\tau = \infty$ if none. Let ζ be the post-τ process. On $\{\xi_0 = j\}$,

$$C_1 = \tau \quad \text{and} \quad A_{N+1} = \{\zeta \in A_N\}.$$

On $\{\xi_0 = j$ and $A_{N+1}\}$,

$$C_{n+1} = C_n \circ \zeta \quad \text{for } n = 1, \ldots, N.$$

By (23),

(29) $\quad P_j\{C_1 = c_1, \ldots, C_{N+1} = c_{N+1}$ and $A_{N+1}\}$

$$= (1 - \theta)\theta^{c_1 - 1} P_q\{C_1 = c_2, \ldots, C_N = c_{N+1} \text{ and } A_N\}.$$

NOTE.   Check the indices.

Sum out $c_1, \ldots, c_{N+1}$:

(30) $\qquad\qquad\qquad P_j\{A_{N+1}\} = P_q\{A_N\}.$

Divide (29) by (30) to prove (28).   ★

One of the most useful applications of (21) is (31), a result that goes back to (Doeblin, 1938). To state it, introduce the notion of an *i-block*, namely, a finite or infinite *I*-sequence, which begins with i but contains no further i. The space of *i*-blocks is a measurable subset of $I^\infty \cup I^*$; give it the relative σ-field. Let $\tau_1, \tau_2, \ldots$ be the times n at which ξ_n visits i. The mth *i-block* B_m is the sample sequence ξ from τ_m to just before τ_{m+1}, shifted to the left so as to start at time 0. Formally, let τ_1 be the least n if any with $\xi_n = i$; if none, $\tau_1 = \infty$. Suppose $\tau_1, \ldots, \tau_m$ defined. If $\tau_m = \infty$, then $\tau_{m+1} = \infty$. If $\tau_m < \infty$, then τ_{m+1} is the least $n > \tau_m$ with $\xi_n = i$, if any; if none, $\tau_{m+1} = \infty$. On $\tau_m < \infty$, let B_m be the sequence of length $\tau_{m+1} - \tau_m$, whose nth term is $\xi_{\tau_m + n}$ for $0 \le n < \tau_{m+1} - \tau_m$. On $\tau_m = \infty$, let $B_m = \varnothing$, the empty sequence. Thus, $B_1, B_2, \ldots$ are random variables, with values $\varnothing$ or *i*-blocks. Let $\mu = P_i B_1^{-1}$, the P_i-distribution of B_1, a probability on the space of *i*-blocks.

(31) Theorem. *(Blocks). Given $B_1, \ldots, B_{n-1}$, where B_{n-1} is a finite i-block, a regular conditional P_p-distribution for B_n is μ.*

PROOF.   Clearly, τ_n is Markov, and

$$\{B_{n-1} \text{ is a finite } i\text{-block}\} = \{\tau_n < \infty\}.$$

Check that $B_1, \ldots, B_{n-1}$ are $\mathscr{F}_{\tau_n}$-measurable. Let ζ be the post-τ_n process:

$$\zeta = T^{\tau_n} \quad \text{on } \{\tau_n < \infty\}.$$

On $\{\tau_n < \infty\}$,

$$\zeta_0 = \xi_{\tau_n} = i$$

and

$$B_n = B_1 \circ \zeta.$$

Let C be a measurable subset of the space of *i*-blocks, and let $A \in \mathscr{F}_{\tau_n}$ with $A \subset \{\tau_n < \infty\}$.

NOTE. A and C are sets; τ_n, ζ, B_1, B_n are functions.

Then
$$\{\tau_n < \infty \text{ and } B_n \in C\} = \{\tau_n < \infty \text{ and } \zeta \in B_1^{-1}C\}.$$

With the help of (22):

$$P_p\{A \text{ and } B_n \in C\} = P_p\{A \text{ and } \zeta \in B_1^{-1}C\}$$
$$= P_p\{A\} \cdot P_i\{B_1^{-1}C\}$$
$$= P_p\{A\} \cdot \mu\{C\}. \qquad \bigstar$$

To identify μ in (31), anticipate a more general definition. Let $P\{i\}$ be this substochastic matrix on I: for $j \in I$ and $k \neq i$, let $P\{i\}(j, k) = P(j, k)$; while $P\{i\}(j, i) = 0$.

(32) Proposition. *With respect to P_i, the first i-block has distribution $P\{i\}_i$: it is Markov with stationary substochastic transitions $P\{i\}$, starting from i.*

PROOF.   Confine ω to the set where $\xi_0 = i$, so $\tau_1 = 0$. This set has P_i-probability 1. Let β_m be the mth term of the first i-block, so

$$\beta_m = \xi_m \quad \text{when } \tau_2 > m$$
$$\beta_m \text{ is undefined} \quad \text{when } \tau_2 \leqq m.$$

Now

$$P_i\{\beta_m = j_m \text{ for } m = 0, \dots, M \text{ and } \tau_2 > M\}$$

is 0 unless $j_0 = i$, while $j_1, \dots, j_m$ all differ from i; in which case, this probability is

$$\Pi_{m=0}^{M-1} P(j_m, j_{m+1})$$

by (11). That is, when $j_0 = i$,

$$P_i\{\beta_m = j_m \text{ for } m = 0, \dots, M \text{ and } \tau_2 > M\} = \Pi_{m=0}^{M-1} P\{i\}(j_m, j_{m+1}). \quad \bigstar$$

4.   CLASSIFICATION OF STATES

(33) Definition.

   $i \rightarrow j$ *iff* $P^n(i, j) > 0$ *for some* $n > 0$.
   $i \leftrightarrow j$ *iff* $i \rightarrow j$ *and* $j \rightarrow i$.
   i *is* essential *iff* $i \rightarrow j$ *implies* $j \rightarrow i$.

You should check the following properties of $\rightarrow$.

(34)   For any i, there is a j with $i \rightarrow j$:

because $\Sigma_j P(i, j) = 1$.

(35)   If $i \rightarrow j$ and $j \rightarrow k$, then $i \rightarrow k$:

because

(36) $$P^{n+m}(i, k) \geqq P^{n}(i, j) \cdot P^{m}(j, k).$$

(37) If i is essential, $i \to i$:

use (34), the definition, and (35).

(38) Lemma. $\leftrightarrow$ *is an equivalence relation when restricted to the essential states.*

PROOF. Use properties (35, 37).   ★

The $\leftrightarrow$ equivalence classes of essential states will be called *communicating classes* or sometimes just *classes*. The communicating class containing i is sometimes written $C(i)$. You should check

(39) $$\Sigma_j \{P(i, j) : j \in C(i)\} = 1:$$

indeed, $P(i, j) > 0$ implies $i \to j$, so $j \to i$ because i is essential; and $j \in C(i)$.

(40) Lemma. *If i is essential, and $i \to j$, then j is essential.*

PROOF. Suppose $j \to k$. Then $i \to k$ by (35), so $k \to i$ because i is essential. And $k \to j$ by (35) again.   ★

(41) Definition. *If $i \to i$, then period i is the* g.c.d. (*greatest common divisor*) *of $\{n : n > 0 \text{ and } P^n(i, i) > 0\}$.*

(42) Lemma. $i \leftrightarrow j$ *implies period i = period j.*

PROOF. Clearly,

(43) $$P^{a+m+b}(i, i) \geqq P^{a}(i, j) \cdot P^{m}(j, j) \cdot P^{b}(j, i).$$

Choose a and b so that $P^{a}(i, j) > 0$ and $P^{b}(j, i) > 0$. If $P^{m}(j, j) > 0$, then $P^{2m}(j, j) > 0$ by (36), so (43) implies

$$P^{a+m+b}(i, i) > 0 \quad \text{and} \quad P^{a+2m+b}(i, i) > 0.$$

Therefore, period i divides $a + m + b$ and $a + 2m + b$. So period i divides the difference m. That is, period i is no more than period j. Equally, period j is no more than period i.   ★

Consequently, the *period of a class* can safely be defined as the period of any of its members. As usual, $m \equiv n \ (d)$ means that $m - n$ is divisible by d. For (44) and (45), fix $i \in I$ and suppose

I forms one class of essential states, with period d.

(44) Lemma. *To each $j \in I$ there corresponds an $r_j = 0, 1, \ldots, d - 1$, such that: $P^n(i, j) > 0$ implies $n \equiv r_j \ (d)$.*

PROOF.   Choose s so that $P^s(j, i) > 0$. If $P^m(i, j) > 0$ and $P^n(i, j) > 0$, then $P^{m+s}(i, i) > 0$ and $P^{n+s}(i, i) > 0$ by (36). So period $i = d$ divides $m + s$ and $n + s$. Consequently, d divides the difference $m - n$, and $m \equiv n$ (d). You can define r_j as the remainder when n is divided by d, for any n with $P^n(i, j) > 0$.   ★

Let C_r be the set of j with $r_j \equiv r$ (d), for each integer r. Thus, $C_0 = C_d$.

Sometimes, $C_r(i)$ is written for C_r, to show the dependence on i. These sets are called the *cyclically moving subclasses* of I, and C_{r+1} is called the subclass *following* C_r.

(45) Theorem.   (a) $C_0, \ldots, C_{d-1}$ *are disjoint and their union is I.*
                                (b) $j \in C_r$ *and* $P(j, k) > 0$ *imply* $k \in C_{r+1}$.

PROOF.   *Assertion* (a).   Use (44).

Assertion (b).   If $P^n(i, j) > 0$ and $P(j, k) > 0$, then $P^{n+1}(i, k) > 0$ by (36). Since $n \equiv r$ (d), therefore $n + 1 \equiv r + 1$ (d) and $r_k \equiv r + 1$ (d), using (44) again.   ★

(46) Proposition.   *Let* $A_0, \ldots, A_{d-1}$ *be disjoint sets whose union is I. For integer r and s, let* $A_r = A_s$ *when* $r \equiv s$ (d). *Suppose* $j \in A_r$ *and* $P(j, k) > 0$ *imply* $k \in A_{r+1}$. *Fix* $i_0 \in A_0$. *Then* $A_n = C_n(i_0)$.

PROOF.   I say $C_n(i_0) \subset A_n$. Let $j \in C_n(i_0)$. If necessary, change n by a multiple of d, so $P^n(i_0, j) > 0$. This changes neither $C_n(i_0)$ nor A_n. Now there are $i_1, \ldots, i_{n-1}$ with

$$P(i_0, i_1) \cdots P(i_{n-1}, j) > 0,$$

so $j \in A_n$. That is, $C_n(i_0) \subset A_n$. Now (45) and the first condition on the sets $A_0, \ldots, A_{d-1}$ imply $C_n(i_0) = A_n$.   ★

Corollary.   *If* $j \in C_r(i)$, *then* $C_s(j) = C_{r+s}(i)$.

PROOF.   Use (46), with $A_s = C_{r+s}(i)$.   ★

(47) Lemma.   *Let I form one communicating class of period 1. Let p be a probability on I, and let* $j \in I$. *Then there is a positive integer* n^* *such that*

$$P_p\{\xi_n = j\} > 0 \text{ for all } n > n^*.$$

PROOF.   Fix i with $p(i) > 0$. Find a positive integer a with $P^a(i, j) > 0$. The set of n with $P^n(j, j) > 0$ is a semigroup by (36) and has g.c.d. 1, so it includes $\{b, b + 1, \ldots\}$ for some positive integer b, by (59). Then $n^* = a + b$ works, by (36).   ★

(48) Proposition. *States j and k are in the same C_r iff there is h in I and an $n > 0$ such that $P^n(j, h) > 0$ and $P^n(k, h) > 0$.*

PROOF. The *if* part is clear. For *only if*, suppose $j \in C_0(k)$. Then $P^{ad}(j, k) > 0$ for some positive integer a. But (59) implies $P^{nd}(k, k) > 0$ for all positive integers $n \geq n_0$. Thus, (36) makes

$$P^{(a+n_0)d}(j, k) > 0 \quad \text{and} \quad P^{(a+n_0)d}(k, k) > 0.$$

5. RECURRENCE

(49) Definition. *For substochastic P, define matrices eP, f^nP, and fP on I as follows: The entry $eP(i, j)$ is the P_i-mean number of visits to j, which may be infinite. Algebraically, $eP = \Sigma_{n=0}^{\infty} P^n$, where $P^0 = \Delta$, the identity matrix. The entry $f^nP(i, j)$ is the P_i-probability of a first visit to j in positive time at time n. Algebraically, $f^nP(i, j)$ is the sum of $\Pi_{m=0}^{n-1} P(i_m, i_{m+1})$ over all I-sequences $i_0, \ldots, i_n$ with $i_0 = i$, $i_n = j$, and $i_m \neq j$ for $0 < m < n$. The entry $fP(i, j)$ is the P_i-probability of ever visiting j in positive time. Thus $fP(i, i)$ is the P_i-probability of a return to i. Algebraically, $fP = \Sigma_{n=1}^{\infty} f^nP$.*

(50) Definition. *A state j is* recurrent *iff*

$$P_j\{\xi_n = j \text{ for infinitely many } n\} = 1.$$

A state j is transient *iff*

$$P_j\{\xi_n = j \text{ for finitely many } n\} = 1.$$

Equivalently, j is transient iff

$$P_j\{\xi_n = j \text{ for infinitely many } n\} = 0.$$

NOTE. Theorem (51) shows this classification to be exhaustive. Namely, j is recurrent or transient according as $\Sigma_n P^n(j, j)$ is infinite or finite.

(51) Theorem. (a) *$fP(j, j) = 1$ implies j is recurrent, and j is recurrent implies $eP(j, j) = \infty$.*

(b) *$fP(j, j) < 1$ implies $eP(j, j) < \infty$, and $eP(j, j) < \infty$ implies j is transient.*

(c) *$eP(j, j) = 1/[1 - fP(j, j)]$.*

(d) *$eP(i, j) = fP(i, j) \cdot eP(j, j)$ for $i \neq j$.*

PROOF. *Assertion (a).* Suppose $fP(j, j) = 1$. Let τ be the least $n > 0$ with $\xi_n = j$, and $\tau = \infty$ if none. Now $P_j\{\tau < \infty\} = fP(j, j) = 1$, so the first j-block is finite with P_j-probability 1. Consequently, by the block theorem (31), all j-blocks are finite with P_j-probability 1, that is, ξ visits infinitely many j's with P_j-probability 1. This also proves (c) when $fP(j,j) = 1$.

Assertion (b) and (c).   Clearly, $eP(j,j)$ is the P_j-mean number of j-blocks. But (31) implies

$$P_j\{B_{n+1} \text{ is infinite} \mid B_1, \ldots, B_n \text{ are finite}\} = 1 - fP(j,j).$$

Consequently, the number of j-blocks is distributed with respect to P_j like the number of tosses of a p-coin needed to produce a head, with the identification $p = 1 - fP(j,j)$. To complete the proof of (c) when $fP(j,j) < 1$, use this easy fact:

(52)   Toss a p-coin until a head is first obtained. The mean number of trials is $1/p$.

In particular, $fP(j,j) < 1$ implies $eP(j,j) < \infty$, so the number of visits to j is finite P_j—almost surely.

Assertion (d).   Use strong Markov on the time of first hitting j. More precisely, let τ be the least n with $\xi_n = j$, and $\tau = \infty$ if none. Then τ is Markov. Let $\zeta_n = \xi_{\tau+n}$ on $\{\tau < \infty\}$. So $\zeta_0 = j$ on $\{\tau < \infty\}$. Let $\delta_j(k)$ be 1 or 0, according as $j = k$ or $j \neq k$. Then

$$eP(i,j) = \int \Sigma_{n=0}^{\infty} \delta_j(\xi_n)\, dP_i$$

$$= \int_{\{\tau<\infty\}} \Sigma_{n=\tau}^{\infty} \delta_j(\xi_n)\, dP_i$$

$$= \int_{\{\tau<\infty\}} \Sigma_{n=0}^{\infty} \delta_j(\zeta_n)\, dP_i$$

$$= P_i\{\tau < \infty\} \cdot \int \Sigma_{n=0}^{\infty} \delta_j(\xi_n)\, dP_j \quad \text{by (22*)}$$

$$= fP(i,j) \cdot eP(j,j). \qquad \qquad \bigstar$$

For a generalization of (52), see (72).

(53) **Lemma.** $fP(i,k) \geq fP(i,j) \cdot fP(j,k)$.

PROOF.   Let K be the event that $\xi_n = k$ for some $n > 0$. Let τ be the least positive n if any with $\xi_n = j$, and $\tau = \infty$ if none. Let K^* be the event that $\tau < \infty$ and $\xi_n = k$ for some $n > \tau$. Then $K \supset K^*$, so $fP(i,k) \geq P_i(K^*)$. Check that τ is Markov; let ζ be the post-τ process: $\zeta = T^\tau$. Check that

$$K^* = \{\tau < \infty\} \cap \{\zeta \in K\}$$

and

$$\zeta_0 = \xi_\tau = j \quad \text{on } \{\tau < \infty\}.$$

By strong Markov (22),

$$P_i(K^*) = P_i\{\tau < \infty\} \cdot P_j(K) = fP(i,j) \cdot fP(j,k). \qquad \bigstar$$

(54) Corollary. *Fix two states i and j. If $fP(i,j) = fP(j,i) = 1$, then i and j are recurrent.*

PROOF.   Using (53),

$$fP(i,i) \geqq fP(i,j) \cdot fP(j,i) = 1.$$

Interchange i and j. Finally, use (51a). $\bigstar$

The next result implies that recurrence is a class property; that is, if one state in a class is recurrent, all are.

(55) Theorem. *$fP(j,j) = 1$ and $j \to k$ implies*

$$fP(j,k) = fP(k,j) = fP(k,k) = 1.$$

PROOF.   Suppose $k \neq j$. Let $B_1, B_2, \ldots$ be the j-blocks. Since $fP(j,j) = 1$, by the block theorem (31), the B_n are independent, identically distributed, finite blocks relative to P_j. Since $j \to k$, therefore B_1 contains a k with positive P_j-probability. The events $\{B_n$ contains a $k\}$ are P_j-independent and have common positive P_j-probability. Then (52) implies that with P_j-probability 1, there is an n such that B_n contains a k. Let τ be the least n if any with $\xi_n = k$, and $\tau = \infty$ if none. Plainly, τ is Markov. The first part of the argument shows $fP(j,k) = P_j\{\tau < \infty\} = 1$. The strong Markov property (22) implies $fP(k,j)$ is the P_j-probability that $\xi_{\tau+n} = j$ for some $n = 0, 1, \ldots$. So $fP(k,j) = 1$. Finally, use (53). $\bigstar$

NOTE.   If j is recurrent, then j is essential.

(56) Proposition. *For finite I, there is at least one essential state; and a state is recurrent iff it is essential.*

PROOF.   Suppose $i_0 \in I$ is not essential. Then $i_0 \to i_1 \nrightarrow i_0$; in particular, $i_1 \neq i_0$. If i_1 is not essential, $i_1 \to i_2 \nrightarrow i_1$, in particular, $i_2 \neq i_0$ and $i_2 \neq i_1$. And so on. This has to stop, so there is an essential state. Next, suppose J is a finite communicating class. Any infinite J-sequence contains infinitely many j, for some $j \in J$. Fix $i \in J$. There is one $j \in J$ such that

$$P_i\{\xi_n = j \text{ for infinitely many } n\} > 0.$$

Use (51) to make j recurrent. Use (55) to see all $j \in J$ are recurrent. $\bigstar$

If $J \subset I$ is a communicating class, and all $j \in J$ are recurrent, call J a *recurrent class*.

6. THE RENEWAL THEOREM

This section contains a proof of the renewal theorem based on (Feller, 1961). To state the theorem, let Y_1, Y_2, ... be independent, identically distributed, positive integer-valued random variables, on the triple $(\Omega, \mathscr{F}, \mathscr{P})$. Let μ be the expectation of Y_i, and $1/\mu = 0$ if $\mu = \infty$. Let $S_0 = 0$, and $S_n = Y_1 + \cdots + Y_n$, and let

$$U(m) = \mathscr{P}\{S_n = m \text{ for some } n = 0, 1, \ldots\}.$$

In particular, $U(0) = 1$. Of course, $\{S_n\}$ is a transient Markov chain with stationary transitions, say Q, and Q_j is the distribution of

$$\{j + S_n, n = 0, 1, \ldots\}.$$

(57) Theorem. (*Renewal theorem*). *If* g.c.d. $\{n: \mathscr{P}[Y_i = n] > 0\} = 1$, *then* $\lim_{m \to \infty} U(m) = 1/\mu$.

This result is immediate from (65) and (66). Lemma (65) follows from (58–64), and (66) is proved by a similar argument. To state (58–59), let F be a subset of the integers, containing at least one nonzero element. Let *group F* (respectively, *semigroup F*) be the smallest subgroup (respectively, subsemigroup) of the additive integers including F.

More constructively, semigroup F is the set of all integers n which can be represented as $f_1 + \cdots + f_m$ for some positive integer m and $f_1, \ldots, f_m \in F$. And group F is the set of all integers n which can be represented as $a - b$, with a, $b \in$ semigroup F. If $\lambda \in$ group F and q is a positive integer, then

$$\lambda q = \Sigma_{n=1}^{q} \lambda \in \text{group } F.$$

(58) Lemma. g.c.d. *F is the least positive element* λ *of group F.*

PROOF. Plainly, g.c.d. F divides λ, so g.c.d. $F \leqq \lambda$. To verify that λ divides any f in F, let $f = \lambda q + r$, where q is an integer and r is one of $0, \ldots, \lambda - 1$. Now $r = f - \lambda q \in$ group F, and $0 \leqq r < \lambda$, so $r = 0$. Consequently, $\lambda \leqq$ g.c.d. F.   ★

(59) Lemma. *Suppose each f in F is positive. Let* g.c.d. $F = 1$. *Then for some positive integer* m_0, *semigroup F contains all* $m \geqq m_0$.

PROOF. Use (58) to find a and b in semigroup F with $a - b = 1$. Plainly, semigroup $F \supset$ semigroup $\{a, b\}$. I say semigroup $\{a, b\}$ contains all $m \geqq b^2$. For if $m \geqq b^2$, then $m = qb + r$, where q is a nonnegative integer and r is one of $0, \ldots, b - 1$. Necessarily, $q \geqq b$, so $q - r > 0$. Then

$$m = qb + r(a - b) = ra + (q - r)b \in \text{semigroup } \{a, b\}.   ★$$

For (60–66), suppose

$$\text{g.c.d. } \{n: \mathscr{P}[Y_i = n] > 0\} = 1.$$

(60) Lemma. *There is a positive integer m_0 such that: for all $m \geqq m_0$, there is a positive integer $n = n(m)$ with $\mathscr{P}\{S_n = m\} > 0$.*

PROOF. Let G be the set of m such that $\mathscr{P}\{S_n = m\} > 0$ for some n. Then G is a semigroup, because

$$\mathscr{P}\{S_{n+n'} = m + m'\} \geqq \mathscr{P}\{S_n = m\} \cdot \mathscr{P}\{S_{n'} = m'\}.$$

And $G \supset \{m: \mathscr{P}[Y_i = m] > 0\}$. So g.c.d. $G = 1$. Now use (59).   ★

For (61–65), use the m_0 of (60), and let

$$L = \lim \sup_{n \to \infty} U(n).$$

(61) Lemma. *Let n' be a subsequence with $\lim_{n \to \infty} U(n') = L$. Then $\lim_{n \to \infty} U(n' - m) = L$ for any $m \geqq m_0$.*

Here n' is to be thought of as a function of n.

PROOF. Fix $m \geqq m_0$. Using (60), choose $N = N(m)$ so $\mathscr{P}\{S_N = m\} > 0$. Thus, $N \geqq m$. Using the diagonal argument (10. 56), find a subsequence n'' of n' such that

$$\lambda(t) = \lim_{n \to \infty} U(n'' - t)$$

exists for all t, and

$$\lambda(m) = \lim \inf_{n \to \infty} U(n'' - m).$$

Fix an integer $j > N$. Clearly,

$$\{S_r = j \text{ for some } r\}$$

is the disjoint union of

$$\{S_r = j \text{ for some } r \leqq N\}$$

and

$$\{S_r = j \text{ for no } r \leqq N, \text{ but } S_r = j \text{ for some } r > N\}.$$

The last set is $\bigcup_{t=N}^{j-1} A_t$, where A_t is the event:

$$S_N = t \text{ and } Y_{N+1} + \cdots + Y_{N+n} = j - t \text{ for some } n.$$

Consequently,

$$(62) \quad U(j) = \mathscr{P}\{S_r = j \text{ for some } r \leqq N\} + \Sigma_{t=N}^{j-1} \mathscr{P}\{S_N = t\} \cdot U(j - t).$$

Now $\max_r \{S_r : r \leqq N\}$ is a finite random variable. Put $j = n''$ in (62); this is safe for large n. Let $n \to \infty$ and use dominated convergence:

$$L = \Sigma_{t=N}^{\infty} \mathscr{P}\{S_N = t\} \cdot \lambda(t).$$

Dominated convergence is legitimate because $U \leq 1$; and n'' is a subsequence of n', so $L = \lim_{n \to \infty} U(n'')$. But $\lambda(t) \leq L$ for all t, and $\mathscr{P}\{S_N = m\} > 0$; so $\lambda(m) < L$ is impossible.   ★

(63) Lemma.   *There is a subsequence n^* such that $\lim_{n \to \infty} U(n^* - m) = L$ for every $m = 0, 1, \ldots$.*

PROOF.   Find a subsequence n' with $\lim_{n \to \infty} U(n') = L$ and $n' > m_0$ for all n. Let $n^* = n' - m_0$. Use (61).   ★

(64) Lemma. $\sum_{m=0}^{n} \mathscr{P}\{Y_i > m\} \cdot U(n - m) = 1$.

PROOF.   For $m = 0, \ldots, n$, let A_m be the event that $S_0, S_1, \ldots$ hits $n - m$, but does not hit $n - m + 1, \ldots, n$. Then

$$\mathscr{P}\{A_m\} = U(n - m) \cdot \mathscr{P}\{Y_i > m\},$$

by strong Markov (22) or a direct argument. The A_m are pairwise disjoint, and their union is Ω. So $\sum_{m=0}^{n} \mathscr{P}\{A_m\} = 1$.   ★

(65) Lemma. $L = 1/\mu$.

PROOF.   As usual, $\sum_{m=0}^{\infty} \mathscr{P}\{Y_i > m\} = \mu$. Suppose $\mu < \infty$. In (64), replace n by n^* of (63). Then let $n \to \infty$. Dominated convergence implies $\sum_{m=0}^{\infty} \mathscr{P}\{Y_i > m\} \cdot L = 1$, so $L = 1/\mu$. When $\mu = \infty$, use the same argument and Fatou to get

$$\sum_{m=0}^{\infty} \mathscr{P}\{Y_i > m\} \cdot L \leq 1,$$

forcing $L = 0 = 1/\mu$.   ★

(66) Lemma. $\liminf_{n \to \infty} U(n) = 1/\mu$.

PROOF.   This follows the pattern for (65). Now let L stand for $\liminf_{n \to \infty} U(n) = 1/\mu$. Lemma (61) still holds, with essentially the same proof: make

$$\lambda(m) = \limsup_{n \to \infty} U(n' - m);$$

and reverse the inequalities at the end. Lemmas (63, 65) still hold, with the same proof.   ★

This completes the proof of (57). I will restate matters for use in (69). Abbreviate $p_n = \mathscr{P}\{Y_i = n\}$. Drop the assumption that

$$\text{g.c.d. } \{n : p_n > 0\} = 1.$$

(67) Proposition.   *Let $d = \text{g.c.d.} \{n : p_n > 0\}$.*
   (a) $d = \text{g.c.d.} \{m : U(m) > 0\}$.
   (b) $\lim_{n \to \infty} U(nd) = d/\mu$.

PROOF.   Plainly, $\{m : U(m) > 0\} = \{0\} \cup$ semigroup $\{n : p_n > 0\}$. If F is a nonempty subset of the positive integers, g.c.d. $F =$ g.c.d. semigroup F. This does (a), and (b) is a weaker statement. Claim (c) reduces to (57) when you divide the Y_i by d.   ★

7. THE LIMITS OF P^n

The renewal theorem gives considerable insight into the limiting behavior of P^n. To state the results, let $mP(i, j)$ be the P_i-expectation of τ_j, where τ_j is the least $n > 0$ if any with $\xi_n = j$, and $\tau_j = \infty$ if none. For $n = 0, 1, \ldots$ let

$$\phi^n P(i, j) = P_i\{\xi_n = j \text{ and } \xi_m \neq j \text{ for } m < n\}.$$

Thus, $\phi^0 P(i, j)$ is 1 or 0, according as $i = j$ or $i \neq j$. And $\phi^n P(i, i) = 0$ for $n > 0$. Let

$$\phi P(i, j) = P_i\{\xi_n = j \text{ for some } n \geq 0\} = \Sigma_{n=0}^{\infty} \phi^n P(i, j).$$

There is no essential difference between ϕ and the f of (49). But this section goes more smoothly with ϕ.

(68) Theorem.   *If j is transient,* $\lim_{n \to \infty} P^n(i, j) = 0$.

PROOF.   Theorem (51) implies $eP(i, j) < \infty$. But $eP(i, j) = \Sigma_{n=0}^{\infty} P^n(i, j)$. ★

(69) Theorem.   *Suppose j is recurrent.*

(a) $\lim_{n \to \infty} \frac{1}{n} \Sigma_{m=1}^{n} P^m(i, j) = \phi P(i, j)/mP(j, j)$.

(b) *If $mP(j, j) = \infty$, then* $\lim_{n \to \infty} P^n(i, j) = 0$.
(c) *If $mP(j, j) < \infty$ and period $j = 1$, then*

$$\lim_{n \to \infty} P^n(i, j) = \phi P(i, j)/mP(j, j).$$

(d) *If $mP(j, j) < \infty$ and period $j = d$, then for $r = 0, 1, \ldots, d - 1$,*

$$\lim_{n \to \infty} P^{nd+r}(i, j) = d\Sigma_{m=0}^{\infty} \phi^{md+r} P(i, j)/mP(j, j).$$

PROOF.   *Claim (a)* follows from (b) and (d), or can be proved directly, as in (Doob, 1953, p. 175).

Claim (b) is similar to (d), and the proof is omitted.

Claim (c) is the leading special case of (d). Suppose (c) were known for $i = j$. Then (c) would hold for any i, by using dominated convergence on the identity

(70) $P^n(i, j) = \Sigma_{m=0}^{n} \phi^m P(i, j) P^{n-m}(j, j).$

ARGUMENT FOR (70).   This identity is trivial when $i = j$. Suppose $i \neq j$, so $\phi^0 P(i, j) = 0$. Clearly,

$$\{\xi_0 = i \text{ and } \xi_n = j\} = \bigcup_{m=1}^{n} A_m,$$

where

$$A_m = \{\xi_0 = i \text{ and } \xi_m = \xi_n = j, \text{ but } \xi_r \neq j \text{ for } r < m\}.$$

Markov (15) implies

$$P_i\{A_m\} = \phi^m P(i, j) \, P^{n-m}(j, j).$$

This completes the proof of (70).

The special case $i = j$ of (c) follows from the renewal theorem (57). Take the lengths of the successive j-blocks for the random variables $Y_1, Y_2, \ldots$. Use blocks (31) to verify that the Y's are independent and identically distributed. Check that $U(n) = P^n(j, j)$. From (67a):

$$\text{g.c.d. } \{n : P_j[Y_1 = n] > 0\} = 1.$$

This completes the argument for (c).

Claim (d) is similar to (c). In (70), if $n \equiv r(d)$, then $P^{n-m}(j, j) = 0$ unless $m \equiv r(d)$. Use (67) to make

$$\lim_{n \to \infty} P^{nd}(j, j) = d/mP(j, j). \qquad \bigstar$$

8.   POSITIVE RECURRENCE

Call j *positive recurrent* iff j is recurrent and $mP(j, j) < \infty$. Call j *null recurrent* iff j is recurrent and $mP(j, j) = \infty$. Is positive recurrence a class property? That is, suppose C is a class and $j \in C$ is positive recurrent. Does it follow that all $k \in C$ are positive recurrent? The affirmative answer is provided by (76), to which (71–73) are preliminary. Theorem (76) also makes the harder assertion: in a positive recurrent class, $mP(i, j) < \infty$ for all pairs i, j.

Lemma (71) is Wald's (1944) identity. To state it, let $Y_1, Y_2, \ldots$ be independent and identically distributed on $(\Omega, \mathscr{F}, \mathscr{P})$. Let

$$S_n = Y_1 + \cdots + Y_n,$$

so $S_0 = 0$. Let τ be a random variable on $(\Omega, \mathscr{F})$ whose values are non-negative integers or ∞. Suppose $\{\tau < n\}$ is independent of Y_n for all n; so $\{\tau \geq n\}$ is also independent of Y_n. Use E for expectation relative to $\mathscr{P}$.

(71) Lemma. $E(S_\tau) = E(Y_n) \cdot E(\tau)$, *provided* (a) *or* (b) *holds.*

(a) $Y_n \geq 0$ *and* $\mathscr{P}\{Y_n > 0\} > 0$.
(b) $E(|Y_n|) < \infty$ *and* $E \cdot (\tau) < \infty$.

PROOF.   Here is a formal computation: as usual, 1_A is 1 on A and 0 off A.

$$
\begin{aligned}
E(S_\tau) &= E(\Sigma_{n=1}^{\infty} Y_n 1_{\{\tau \geq n\}}) \\
&= \Sigma_{n=1}^{\infty} E(Y_n 1_{\{\tau \geq n\}}) \\
&= E(Y_n) \cdot \Sigma_{n=1}^{\infty} \mathscr{P}\{\tau \geq n\} \\
&= E(Y_n) E(\tau).
\end{aligned}
$$

If $Y_n \geq 0$, the interchange of E and Σ is justified by Fubini. If $E(|Y_n|) < \infty$ and $(E\tau) < \infty$, then $\Sigma_{n=1}^{\infty} E |Y_n 1_{\{\tau \geq n\}}| < \infty$, so Fubini still works.   ★

(72) Example.   A p-coin is tossed independently until n heads are obtained. The expected number of tosses is n/p.

PROOF.   Suppose $p > 0$. Let Y_m be 1 or 0 according as the mth toss is head or tail. Let τ be the least m with $S_m = Y_1 + \cdots + Y_m = n$. Then $S_\tau = n$, so $n = E(S_\tau) = E(Y_1) \cdot E(\tau) = p \cdot E(\tau)$, using (71).   ★

To state (73), let I_∞ be the set of $\omega \in I^\infty$ such that $\omega(n) = k$ for infinitely many n, for every $k \in I$.

(73) Lemma.   If I is a recurrent class, $P_i\{I_\infty\} = 1$.

PROOF.   Let A_m be the event: the mth i-block contains a j. From the block theorem (31), with respect to P_i,

(74)   the A_m are independent and have common probability.

Because $i \leftrightarrow j$,

(75) $P_i\{A_m\} > 0$.

Consequently, $P_i\{\limsup A_m\} = 1$.   ★

This repeats part of (55).

(76) Theorem.   Let I be a recurrent class. Either $mP(i,j) < \infty$ for all i and j in I, or $mP(j,j) = \infty$ for all j in I.

For a generalization, see (2.98).

PROOF.   Suppose $mP(i,i) < \infty$ for an i in I. Fix $j \neq i$. What must be proved is that $mP(i,j)$, $mP(j,i)$, and $mP(j,j)$ are all finite. To start the proof, confine ω to $I_\infty \cap \{\xi_0 = i\}$. Let τ be the least n such that the nth i-block contains a j. Using the A_m of (73), and the notation $C\backslash D$ for the set of points in C but not in D,

$$
\{\tau = n\} = (I_\infty\backslash A_1) \cap \cdots \cap (I_\infty\backslash A_{n-1}) \cap A_n.
$$

Relation (74) implies that τ is P_i-distributed like the waiting time for the first head in tossing a p-coin, where $0 < p = P_i(A_m)$ by (75). Now (72) implies $\int \tau \, dP_i < \infty$. Let $Y_1, Y_2, \ldots$ be the lengths of the successive i-blocks.

By the block theorem (31), the Y_m's are P_i-independent and identically distributed; and $\{\tau < n\}$ is P_i-independent of Y_n. By definition, $\int Y_1\, dP_i = mP(i, i)$. Now Wald's identity (71) forces

$$\int \Sigma_{m=1}^{\tau} Y_m\, dP_i < \infty.$$

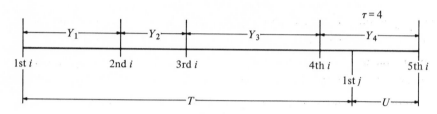

Figure 3

As in Figure 3, let $T(\omega)$ be the least n with $\omega(n) = j$. Let $T(\omega) + U(\omega)$ be the least $n > T(\omega)$ with $\omega(n) = i$. Then

$$\Sigma_{m=1}^{\tau} Y_m = T + U; \quad \text{so} \quad \int T\, dP_i < \infty \quad \text{and} \quad \int U\, dP_i < \infty.$$

By definition, $\int T\, dP_i = mP(i, j)$. Use the strong Markov property to see $\int U\, dP_i = mP(j, i)$. This proves $mP(i, j)$ and $mP(j, i)$ are finite. To settle $mP(j, j)$, check $Y_1 \leqq T + U$, so $mP(i, i) \leqq mP(i, j) + mP(j, i)$. Interchange i and j to get $mP(j, j) \leqq mP(i, j) + mP(j, i) < \infty$. ★

(77) Remark. The argument shows: if $mP(i, j) < \infty$ and $mP(j, i) < \infty$ for some i and j, then i and j are positive recurrent.

If $J \subset I$ is a communicating class, and all $j \in J$ are positive (respectively, null) recurrent, call J a *positive* (respectively, *null*) *recurrent class.*

(78) Proposition. *Suppose I is finite and $j \in I$ is recurrent. Then j is positive recurrent.*

PROOF. Reduce I to the communicating class containing j, and use (79) below. ★

(79) Proposition. *Let τ_j be the least n if any with $\xi_n = j$, and $\tau_j = \infty$ if none. Suppose I is finite, j is a given state in I, and $i \to j$ for all $i \in I$. Then there are constants A and r with $0 < A < \infty$ and $0 < r < 1$, such that*

$$P_i\{\tau_j > n\} \leqq Ar^n \quad \text{for all } i \in I \text{ and } n = 0, 1, \dots.$$

PROOF. Let $\bar{P}$ agree with P except in the jth row, where $\bar{P}(j, j) = 1$. Then

$$\bar{P}_i\{\xi_0 = i_0, \dots, \xi_m = i_m\} = P_i\{\xi_0 = i_0, \dots, \xi_m = i_m\}$$

provided $i_0, i_1, \ldots, i_{m-1}$ are all different from j; however, $i_m = j$ is allowed. Sum over all such sequences with $i_m = j$ and $m \leq n$:

$$P_i\{\tau_j \leq n\} = \bar{P}_i\{\tau_j \leq n\} = \bar{P}^n(i, j).$$

So $i \to j$ for $\bar{P}$, and I only have to get the inequality for $\bar{P}$.

You should see that $\bar{P}^n(i, j)$ is nondecreasing with n, and is positive for $n \geq n_i$, for some positive integer n_i. Let $N = \max_i n_i$, so

$$0 < \varepsilon = \min_i \bar{P}^N(i, j),$$

using the finitude of I twice. Thus

$$1 - \varepsilon \geq \bar{P}_i\{\tau_j > N\} \quad \text{for all } i.$$

Recall the shift T, introduced for the Markov property (14). Check

$$\{\tau_j > (n + 1)N\} = \{\tau_j > nN\} \cap T^{-nN} \{\tau_j > N\}.$$

Make sure that $\{\tau_j > nN\}$ is measurable on $\xi_0, \ldots, \xi_{nN}$. Therefore,

$$\begin{aligned}
\bar{P}_i\{\tau_j > (n + 1)N\} &= \Sigma_k \bar{P}_i\{\xi_{nN} = k \text{ and } \tau_j > (n + 1)N\} \\
&= \Sigma_k \bar{P}_i\{\xi_{nN} = k \text{ and } \tau_j > nN\} \cdot \bar{P}_k\{\tau_j > N\} \quad \text{by (15)} \\
&\leq (1 - \varepsilon) \Sigma_k \bar{P}_i\{\xi_{nN} = k \text{ and } \tau_j > nN\} \\
&= (1 - \varepsilon)\bar{P}_i\{\tau_j > nN\} \\
&\leq (1 - \varepsilon)^{n+1} \quad \text{by induction.}
\end{aligned}$$

Suppose $nN \leq m < (n + 1)N$. Then

$$\begin{aligned}
\bar{P}_i\{\tau_j > m\} &\leq \bar{P}_i\{\tau_j > nN\} \\
&\leq (1 - \varepsilon)^n \\
&= \frac{1}{1 - \varepsilon} [(1 - \varepsilon)^{1/N}]^{(n+1)N} \\
&\leq \frac{1}{1 - \varepsilon} [(1 - \varepsilon)^{1/N}]^m.
\end{aligned}$$

                                                         ★

9. INVARIANT PROBABILITIES

(80) **Definition.** *A measure μ on I is* invariant *iff*

$$\mu(j) = \Sigma_{i \in I} \mu(i)P(i, j) \quad \text{for all } j,$$

and subinvariant *iff $\mu(j) \geq \Sigma_{i \in I} \mu(i)P(i, j)$. The convention $\infty \cdot 0 = 0$ applies.*

Abbreviate $\pi(j) = 1/mP(j, j)$. The main result on invariant probabilities is:

(81) **Theorem.** *If I is a positive recurrent class, then π is an invariant probability, and any invariant signed measure with finite mass is a scalar multiple of π.*

NOTES.   Suppose I forms one recurrent class.

(a) $\pi(j) > 0$ for all j.

(b) As will be seen in (2.24), any subinvariant measure is automatically finite and invariant, and a nonnegative scalar multiple of π.

Measures are nonnegative, unless specified otherwise. The proof of (81) consists of lemmas (82–87). In all of them, assume I is a positive recurrent class.

(82) Lemma. π *is a subprobability.*

PROOF.   Because P^m is stochastic,

$$\Sigma_{j \in I} \left[\frac{1}{n} \Sigma_{m=1}^{n} P^m(i, j) \right] = 1.$$

Send n to ∞. By (69a),

$$\frac{1}{n} \Sigma_{m=1}^{n} P^m(i, j) \to \phi P(i, j) / m P(j, j).$$

But $\phi P(i, j) = 1$ because I is a recurrent class, and $1/m P(j, j) = \pi(j)$ by definition. So Fatou makes

$$\Sigma_{j \in I} \pi(j) \leqq 1. \qquad \qquad \bigstar$$

(83) Lemma. π *is subinvariant.*

PROOF.   Because

$$\Sigma_{j \in I} P^m(i, j) P(j, k) = P^{m+1}(i, k),$$

therefore

$$\Sigma_{j \in I} \left[\frac{1}{n} \Sigma_{m=1}^{n} P^m(i, j) \right] P(j, k) = \frac{1}{n} \Sigma_{m=2}^{n+1} P^m(i, k).$$

Send n to ∞. Use (69a) and Fatou, as in (82):

$$\Sigma_{j \in I} \pi(j) P(j, k) \leqq \pi(k). \qquad \qquad \bigstar$$

(84) Lemma. π *is invariant.*

PROOF.   Lemma (83) makes $\pi(k) \geqq \Sigma_{j \in I} \pi(j) P(j, k)$. If inequality occurs anywhere, sum out k and get

$$\Sigma_{k \in I} \pi(k) > \Sigma_{j \in I} \pi(j).$$

This contradicts (82). $\bigstar$

For (85), let μ be an invariant signed measure with finite mass.

(85) Lemma. $\mu(j) = [\Sigma_{i \in I} \mu(i)] \pi(j).$

PROOF. By iteration,

(86) $\mu(j) = \Sigma_{i \in I} \mu(i) P^m(i, j),$

so

$$\mu(j) = \Sigma_{i \in I} \mu(i) \left[\frac{1}{n} \Sigma_{m=1}^{n} P^m(i, j) \right].$$

Send n to ∞. Use (69a) as in (82), and dominated convergence:

$$\mu(j) = \Sigma_{i \in I} \mu(i) \pi(j). \qquad \qquad \star$$

(87) Lemma. π *is a probability.*

PROOF. Using (82) and (84), put π for μ in (85):

$$\pi(j) = [\Sigma_{i \in I} \pi(i)] \, \pi(j).$$

But $\pi(j) = 1/mP(j,j) > 0$, because I is positive recurrent. So

$$\Sigma_{i \in I} \pi(i) = 1. \qquad \qquad \star$$

PROOF OF (81). Use (84), (87), and (85). $\qquad \qquad \star$

Now drop the assumption that I is a positive recurrent class. Let μ be a signed measure on I with finite mass. The next theorem describes all invariant μ. To state it, let C be the set of all positive recurrent classes $J \subset I$. For $J \in C$, define a probability π_J on I by: $\pi_J(j) = 1/mP(j,j)$ for $j \in J$, and $\pi_J(j) = 0$ for $j \notin J$.

(88) Theorem. μ *is invariant iff* $\mu = \Sigma_{J \in C} \mu(J) \, \pi_J.$

The proof of (88) is deferred.

(89) Lemma. *Let μ be an invariant signed measure on I, with finite mass. Then μ assigns measure 0 to the transient and null recurrent states.*

PROOF. Send m to ∞ in (86). Then use dominated convergence, and (68) or (69b). $\qquad \qquad \star$

For (90–91), define a matrix P_J on $J \in C$ by $P_J(i, j) = P(i, j)$, for i and j in J. As (39) implies, P_J is stochastic. For (90), fix $J \in C$. Let v_J be a signed measure on J with finite mass, invariant with respect to P_J. Define a measure v on I by: $v(i) = v_J(i)$ for $i \in J$, and $v(i) = 0$ for $i \notin J$.

(90) Lemma. v *is invariant.*

PROOF. Suppose $j \in J$. Then

$$\Sigma_{i \in I} v(i) P(i, j) = \Sigma_{i \in J} v_J(i) P_J(i, j) = v_J(j) = v(j).$$

Suppose $j \notin J$. If $\nu(i) > 0$, then $i \in J$, so $P(i,j) = 0$ by (39). That is, $\nu(i)P(i,j) = 0$ for all i. And

$$\Sigma_{i \in I} \, \nu(i)P(i,j) = 0 = \nu(j). \qquad \bigstar$$

For (91), let μ be an invariant signed measure on I, with finite mass. Fix $J \in C$.

(91) Lemma. μ retracted to J is invariant with respect to P_J.

PROOF.   Suppose $j \in J$. Then $P(i,j) = 0$ when $i \in K \in C \backslash \{J\}$ by (39). And $\mu(i) = 0$ when $i \notin \bigcup \{K : K \in C\}$, by (89). So $\mu(i)P(i,j) = 0$ unless $i \in J$. That is,

$$\begin{aligned} \mu(j) &= \Sigma_{i \in I} \, \mu(i)P(i,j) \\ &= \Sigma_{i \in J} \, \mu(i)P(i,j) \\ &= \Sigma_{i \in J} \, \mu(i)P_J(i,j). \qquad \bigstar \end{aligned}$$

PROOF OF (88).   Let μ be an invariant signed measure on I, with finite mass. As (89) implies, μ concentrates on $\bigcup \{J : J \in C\}$. When $J \in C$, let

$$\begin{aligned} \mu_J(j) &= \mu(j) \quad \text{for } j \in J \\ &= 0 \qquad \text{for } j \notin J. \end{aligned}$$

Then

$$\mu = \Sigma_{J \in C} \, \mu_J.$$

As (91) implies, the retract of μ to J is P_J-invariant. So, (81) on P_J implies

$$\mu_J = \mu(J) \, \pi_J.$$

Therefore

$$\mu = \Sigma_{J \in C} \, \mu(J) \, \pi_J.$$

Conversely, the retract of π_J to J is P_J-invariant by (81). So π_J is P-invariant by (90). If $\Sigma \, |d_J| < \infty$, then

$$\Sigma_{J \in C} \, d_J \, \pi_J$$

is also P-invariant. $\qquad \bigstar$

10.   THE BERNOULLI WALK

In this section, I is the set of integers and $0 < p < 1$. Define the stochastic matrix $[p]$ on I by: $[p](i, i+1) = p$, and $[p](i, i-1) = 1 - p$. This notation is strictly temporary. Plainly, I is a communicating class of period 2.

(92) Theorem. I is recurrent for $[p]$ iff $p = \frac{1}{2}$.

PROOF. *Only if.* You should check that

$[p]_0\{\xi_0 = 0,\ \xi_1 = i_1,\ \dots,\ \xi_{2n-1} = i_{2n-1},\ \xi_{2n} = 0\}$
$$= (4p(1-p))^n\ [\tfrac{1}{2}]_0\{\xi_0 = 0,\ \xi_1 = i_1,\ \dots,\ \xi_{2n-1} = i_{2n-1},\ \xi_{2n} = 0\}.$$

Sum over all these sequences, with $i_m \neq 0$ for $0 < m < 2n$;

$$f^{2n}[p](0, 0) = (4p(1-p))^n f^{2n}[\tfrac{1}{2}](0, 0).$$

The definition of f^m and f is in (49). If $p \neq \tfrac{1}{2}$, then $4p(1-p) < 1$, so

$$f[p](0, 0) = \Sigma_{n=1}^{\infty} f^{2n}[p](0, 0)$$
$$< \Sigma_{n=1}^{\infty} f^{2n}[\tfrac{1}{2}](0, 0)$$
$$\leqq 1.$$

Use (51) to see that $p \neq \tfrac{1}{2}$ implies I is transient. This argument was suggested by T. F. Hou.

If. I say that $x = f[p](0, 1)$ satisfies

(93) $x = p + (1-p)x^2.$

To begin with, $x = f[p](-1, 0)$: indeed, the $[p]_0$-distribution of $\xi_0 - 1$, $\xi_1 - 1, \dots$ is $[p]_{-1}$; and $\xi_0 - 1, \xi_1 - 1, \dots$ hits 0 iff $\xi_0, \xi_1, \dots$ hits 1. Next,

$$f[p](-1, 1) = f[p](-1, 0) \cdot f[p](0, 1) = x^2,$$

by strong Markov (22). Use Markov (15) in line 3:

$$x = [p]_0\{\xi_n = 1 \text{ for some } n\}$$
$$= [p]_0\{\xi_1 = 1\} + [p]_0\{\xi_1 = -1 \text{ and } \xi_n = 1 \text{ for some } n\}$$
$$= p + (1-p)\,[p]_{-1}\{\xi_n = 1 \text{ for some } n\}$$
$$= p + (1-p)f[p](-1, 1)$$
$$= p + (1-p)x^2.$$

This proves (93).

For the rest of the proof, suppose $p = \tfrac{1}{2}$. Then (93) has only one solution, $x = 1$. Moreover,

$$f[\tfrac{1}{2}](1, 0) = f[\tfrac{1}{2}](-1, 0) = f[\tfrac{1}{2}](0, 1);$$

the second equality is old; the first one works because the $[\tfrac{1}{2}]_1$-distribution of $-\xi_0, -\xi_1, \dots$ is $[\tfrac{1}{2}]_{-1}$, and $-\xi_0, -\xi_1, \dots$ hits 0 iff $\xi_0, \xi_1, \dots$ hits 0. Now use (54).   ★

(94) The class I is null recurrent for $[\tfrac{1}{2}]$.

Here is an argument for (94) that I learned from Harry Reuter. By previous reasoning, $P^n(j, j)$ does not depend on j. So $\lim_{n\to\infty} P^{2n}(j, j)$ does not depend on j. If I were positive recurrent, the invariant probability would have to assign equal mass to all integers by (69d) and (81). This is untenable.

Suppose $\frac{1}{2} < p < 1$. The two solutions of (93) are 1 and $p/(1 - p) > 1$. Thus, $f[p](0, 1) = 1$. Previous arguments promote this to

(95) $$f[p](i, j) = 1 \quad \text{for } i < j.$$

Now $y = f[p](0, -1) < 1$, for otherwise I would be recurrent by (54). Interchange right and left, so p and $1 - p$, in (93):

$$y = 1 - p + py^2.$$

The two solutions are $y = 1$ and $y = (1 - p)/p$, so $f[p](0, -1) = (1 - p)/p$. Previous arguments promote this to

(96) $$f[p](i, j) = \left(\frac{1 - p}{p}\right)^{i-j} \quad \text{for } i > j.$$

Moreover, $f[p](0, 0) = pf[p](1, 0) + (1 - p)f[p](-1, 0) = 2(1 - p).$

Previous arguments promote this to

(97) $$f[p](i, i) = 2(1 - p).$$

Use (51) to get:

(98) $$e[p](i, j) = \frac{1}{2p - 1} \qquad \text{for } i \leqq j$$

$$= \left(\frac{1 - p}{p}\right)^{i-j} \frac{1}{2p - 1} \quad \text{for } i > j.$$

11. FORBIDDEN TRANSITIONS

The material in this section will be used in Section 12, and referred to in Chapter 3 and 4. It is taken from (Chung, 1960, Section 1.9).

(99) Definition.   *For any subset H of I, define a substochastic matrix PH on I: for $i \in I$ and $j \notin H$, let $PH(i, j) = P(i, j)$: but for $j \in H$, let $PH(i, j) = 0$.*

Let τ be the least $n > 0$ if any with $\xi_n \in H$, and $\tau = \infty$ if none. With respect to P_i, the fragment $\{\xi_n : 0 \leqq n < \tau\}$ is Markov with stationary transitions PH. Thus, $ePH(i, k)$ is the P_i-mean number of $n \geqq 0$, but less than the first positive m with $\xi_m \in H$, such that $\xi_n = k$. Moreover, $fPH(i, k)$ is the P_i-probability that there is an $n > 0$, but less than the first positive m with $\xi_m \in H$, such that $\xi_n = k$. The operators e and f were defined in (49).

In the proof of (100), I will use some theorems proved for stochastic P on substochastic P. To legitimate this, adjoin a new state ∂ to I. Define

$$P_\partial(i,j) = P(i,j) \qquad\qquad \text{for } i \text{ and } j \text{ in } I$$
$$P_\partial(i, \partial) = 1 - \Sigma_{i\in I} P(i,j) \quad \text{for } i \text{ in } I$$
$$P_\partial(\partial, i) = 0 \qquad\qquad\qquad \text{for } i \text{ in } I$$
$$P_\partial(\partial, \partial) = 1.$$

Suppose $\{X_n\}$ is a Markov chain with substochastic transitions P. Let $Y_n = X_n$ when X_n is defined, and $Y_n = \partial$ when X_n is undefined. Then $\{Y_n\}$ is Markov with stochastic transitions P_∂. Use the old theorems on P_∂.

(100) Lemma. *If $k \to h$ for some $h \in H$, then $ePH(i, k) < \infty$. More precisely*:
$$ePH(k, k) = 1/[1 - fPH(k, k)];$$
and for $i \neq k$,
$$ePH(i, k) = fPH(i, k)/[1 - fPH(k, k)].$$

PROOF.   First, suppose $k \in H$. Then $fPH(k, k) = 0$ and $ePH(k, k) = 1$, proving the first display. Let $i \neq k$. Then $fPH(i, k) = ePH(i, k) = 0$, proving the second display.

Now, suppose $k \notin H$. A k-block contains no $h \in H$ with probability $fPH(k, k) < 1$. Use (51c) to verify that

$$ePH(k, k) = 1/[1 - fPH(k, k)].$$

Let $i \neq k$. Use (51d) to get

$$ePH(i, k) = fPH(i, k) \cdot ePH(k, k). \qquad\qquad \bigstar$$

Give I the discrete topology, and let $\bar{I} = I \cup \{\varphi\}$ be its one-point compactification; for example, let $\bar{I} = \{i_1, i_2, \ldots, i_\infty\}$, where $i_\infty = \varphi$; metrize $\bar{I}$ with $\rho(i_n, i_m) = \left| \dfrac{1}{n} - \dfrac{1}{m} \right|$ and $\dfrac{1}{\infty} = 0$. A sequence $k_n \in I$ converges to φ iff $k_n = j$ for only finitely many n, for each $j \in I$.

(101) Lemma. $fP(i, j) = \lim_{k\to\varphi} fP\{k\}(i, j)$.

PROOF.   Let D_n be a sequence of finite sets swelling to I. As n increases, the event E_n that $\{\xi_m\}$ hits j before hitting $I \setminus D_n$ increases to the event E that $\{\xi_m\}$ hits j. So, $P_i(E_n) \to fP(i, j)$. If $k \notin D_n$, then E_n is included in the event that $\{\xi_m\}$ hits j before hitting k. So,

$$fP(i, j) \geqq fP\{k\}(i, j) \geqq P_i(E_n). \qquad\qquad \bigstar$$

(102) Lemma. $mP(i, i) = \Sigma_k eP\{i\}(i, k).$

PROOF. Let τ be the least $n > 0$ if any with $\xi_n = i$, and $\tau = \infty$ if none. Let $\zeta(n, k)$ be the indicator of the event: $n < \tau$ and $\xi_n = k$. Thus,

$$\tau = \Sigma_{k \in I} \Sigma_{n=0}^{\infty} \zeta(n, k),$$

and

$$mP(i, i) = \int \tau \, dP_i = \Sigma_{k \in I} \Sigma_{n=0}^{\infty} \int \zeta(n, k) \, dP_i = \Sigma_{k \in I} eP\{i\}(i, k). \qquad \bigstar$$

(103) Lemma. *If I is recurrent and $i \neq k$,*

$$fP\{i\}(k, k) + fP\{k\}(k, i) = 1.$$

PROOF. With respect to P_k, almost all paths hit either i or k first, in positive time. $\qquad \bigstar$

12. THE HARRIS WALK

The next example was studied by Harris (1952), using Brownian motion. To describe the example, let $0 < a_j < 1$ and $b_j = 1 - a_j$ for $j = 1, 2, \ldots$. Let I be the set of nonnegative integers. Define the stochastic matrix P on I by: $P(0, 1) = 1$, while $P(j, j + 1) = a_j$ and $P(j, j - 1) = b_j$ for $1 \leq j < \infty$. Plainly, I is an essential class of period 2. When is it recurrent? To state the answer, let $r_0 = 1$; let $r_n = (b_1 \cdots b_n)/(a_1 \cdots a_n)$ for $n = 1, 2, \ldots$; let $R(0) = 0$; let $R(n) = r_0 + \cdots + r_{n-1}$ for $n = 1, 2, \ldots$; and let $R(\infty) = \Sigma_{n=0}^{\infty} r_n$.

(104) Theorem. *I is recurrent or transient according as $R(\infty) = \infty$ or $R(\infty) < \infty$. If I is recurrent, it is null or positive according as $\Sigma_{n=1}^{\infty} 1/(a_n r_n)$ is infinite or finite.*

The proof of this theorem is presented as a series of lemmas. Of these, (105–111) deal with the criterion for recurrence, and (112) deals with distinguishing null from positive recurrence. It is convenient to introduce a stochastic matrix Q on I, which agrees with P except in the 0th row, when $Q(0, 0) = 1$.

(105) Lemma. *For each $i \in I$, the process $R(\xi_0), R(\xi_1), \ldots$ is a martingale relative to Q_i.*

PROOF. On $\{\xi_n = j\}$, the conditional Q_i-expectation of $R(\xi_{n+1})$ given $\xi_0, \ldots, \xi_n$ is $\Sigma_k Q(j, k)R(k)$, by Markov (15). When $j = 0$, this sum is clearly $0 = R(j)$. When $j > 0$, this sum is

$$a_j R(j + 1) + b_j R(j - 1) = a_j[R(j) + r_j] + b_j[R(j) - r_{j-1}]$$
$$= R(j) + a_j r_j - b_j r_{j-1}$$
$$= R(j). \qquad \bigstar$$

Let $0 \le i < j < k$ in (106–111).

(106) Lemma. *With Q_j-probability 1, there is an n such that ξ_n is i or k.*

PROOF.   Let π be the product $a_{i+1} \cdots a_{k-1}$, and let $d = k - i - 1$. Given $\xi_0, \ldots, \xi_n$, on $i < \xi_n < k$, the conditional Q_j-probability that

$$i < \xi_n, \ldots, \xi_{n+d} < k$$

is no more than $1 - \pi$. Indeed, π underestimates the conditional probability that $\xi_{n+1}, \xi_{n+2}, \ldots$ proceed steadily to the right until reaching k. So, the Q_j-probability that $i < \xi_0, \ldots, \xi_{md} < k$ is no more than $(1 - \pi)^m$.   ★
Restate (106) as

(107) $fQ\{i\}(j, k) + fQ\{k\}(j, i) = 1.$

Of course, $fQ\{i\}(j, k)$ is the Q_j-probability that ξ hits k before i.

(108) Lemma. $fQ\{k\}(j, i) = [R(k) - R(j)]/[R(k) - R(i)].$

PROOF.   Let $x = fQ\{k\}(j, i)$, so $1 - x = fQ\{i\}(j, k)$ by (107). Let τ be the least n with $\xi_n = i$ or k, and $\tau = \infty$ if none. Stop $\{R(\xi_n)\}$ at τ, using (106) and (10.28). Thus

$$R(j) = \int R(\xi_\tau) \, dQ_j$$
$$= xR(i) + (1 - x)R(k).$$

Solve for x.   ★

(109) Lemma. $fQ\{k\}(j, i) = fP\{k\}(j, i).$

PROOF.   Let $i_0, i_1, \ldots, i_n$ be any I-sequence which does not contain 0 except possibly for i_n. Then

$$P_j\{\xi_0 = i_0, \xi_1 = i_1, \ldots, \xi_n = i_n\}$$
$$= Q_j\{\xi_0 = i_0, \xi_1 = i_1, \ldots, \xi_n = i_n\}.$$

Sum over all such sequences, with $i_0 = j$ and $i_n = i$ and $i_m \ne k$ for $0 < m < n$; even n is allowed to vary.   ★

(110) Lemma. (a) $fP\{k\}(j, i) = [R(k) - R(j)]/[R(k) - R(i)].$

(b) $fP(j, i) = [R(\infty) - R(j)]/[R(\infty) - R(i)]$ *if $R(\infty) < \infty$.*

(c) $fP(j, i) = 1$ *if $R(\infty) = \infty$.*

PROOF.   Use (108, 109) to get (a). Let $k \to \infty$ and use (101) to get (b) and (c).   ★

(111) Lemma. $fP(i, j) = 1.$

PROOF.   As in (106).   ★

ARGUMENT FOR RECURRENCE CRITERION IN (104). Suppose $R(\infty) = \infty$. Then $fP(i, j) = fP(j, i) = 1$ by (110c, 111). And I is recurrent by (54).

Suppose $R(\infty) < \infty$. Clearly, $R(i) < R(j)$; so $fP(j, i) < 1$ by (110b). Now I is transient by (55). ★

Suppose I is recurrent for (112).

(112) **Lemma.** $mP(0, 0) = 1 + \Sigma_{k=1}^{\infty} 1/(a_k r_k)$.

This is sharper than the null recurrence criterion of (104).

PROOF.   Begin by computing some hitting probabilities. Let $0 < j < k$. Then

$$fP\{0\}(j, k) = 1 - fP\{k\}(j, 0)   \text{(103) and recurrence}$$

$$= 1 - \frac{R(k) - R(j)}{R(k)}   \text{(110}a\text{)}.$$

So

(113) $$fP\{0\}(j, k) = \frac{R(j)}{R(k)}.$$

Clearly, $fP\{0\}(0, k) = f\{0\}(1, k)$; so (113) makes

(114) $$fP\{0\}(0, k) = \frac{1}{R(k)}.$$

By (16) and recurrence, $fP\{0\}(k, k) = a_k + b_k fP\{0\}(k - 1, k)$; so (113) makes

$$1 - fP\{0\}(k, k) = b_k \left[1 - \frac{R(k - 1)}{R(k)} \right];$$

and by algebra,

(115) $$1 - fP\{0\}(k, k) = \frac{a_k r_k}{R(k)}.$$

Now compute $mP(0, 0)$, as follows:

$$mP(0, 0) = \Sigma_{k=0}^{\infty} eP\{0\}(0, k)   \text{(102)}$$

$$= eP\{0\}(0, 0) + \Sigma_{k=1}^{\infty} eP\{0\}(0, k)$$

$$= 1 + \Sigma_{k=1}^{\infty} \frac{fP\{0\}(0, k)}{1 - fP\{0\}(k, k)}   \text{(100)}$$

$$= 1 + \Sigma_{k=1}^{\infty} \frac{1}{a_k r_k}   \text{(114, 115). ★}$$

In a similar way, $mP(i, j)$ can be computed. It is easy to drop the condition $0 < a_j < 1$, and using the idea of (94), it is easy to handle the case where I is all the integers. For further information on random walks, consult (Chung and Fuchs, 1951), (Feller, 1966), or (Spitzer, 1964).

13. THE TAIL σ-FIELD AND A THEOREM OF OREY

Let $\{X_n : n = 0, 1, \ldots\}$ be a stochastic process. Let $\Sigma^{(n)}$ be the σ-field spanned by $X_n, X_{n+1}, \ldots$. The *tail σ-field* of X is $\Sigma^{(\infty)} = \bigcap_n \Sigma^{(n)}$. Let $\mathscr{F}^{(\infty)}$ be the tail σ-field of the coordinate process $\{\xi_n\}$ on I^∞. The *invariant σ-field* $\mathscr{I}$ of I^∞ is the σ-field of measurable sets B such that $(i_0, i_1, \ldots) \in B$ iff $(i_1, i_2, \ldots) \in B$. If each X_n is I-valued, then the *invariant σ-field* of X is $X^{-1}\mathscr{I}$. Of course, $X = (X_0, X_1, \ldots)$ is a measurable object with values in I^∞. The *exchangeable σ-field* $\mathscr{E}$ of I^∞ is the σ-field of measurable sets B such that $(i_0, i_1, \ldots) \in B$ iff $(i_{\pi 0}, i_{\pi 1}, \ldots) \in B$ for all finite permutations π of $(0, 1, \ldots)$: namely, those 1-1 mappings π of $(0, 1, \ldots)$ onto itself with $\pi(n) \neq n$ for only finitely many n. If each X_n is I-valued, the *exchangeable σ-field* of X is $X^{-1}\mathscr{E}$. The object of this section is to describe $\mathscr{F}^{(\infty)}$, $\mathscr{I}$, and $\mathscr{E}$ up to null sets for recurrent chains, and to prove theorem (128) of Orey. The discussion here and in Section 14 is based on (Blackwell and Freedman, 1964).

Clearly, $\mathscr{I} \subset \mathscr{F}^{(\infty)} \subset \mathscr{E}$. The inclusions are strict if I has two or more elements. For simplicity, assume $I \supset \{-1, 1\}$ during the next three illustrations.

(116) Illustration. Let A be the set where $\xi_n = 1$ for infinitely many n. Then $A \in \mathscr{I}$.

(117) Illustration. Let B be the set where $\xi_{2n} = 1$ for infinitely many n. Then $B \in \mathscr{F}^{(\infty)}$, but $B \notin \mathscr{I}$. Indeed, define $\omega^* \in I^\infty$ as follows:

$$\omega^*(n) = 1 \quad \text{for even } n.$$
$$= -1 \quad \text{for odd } n.$$

Then $\omega^* \in B$ but $T\omega^* = \omega^*(1 + \cdot) \notin B$.

(118) Illustration. Let C be the set where $\xi_n = -1$ or 1 for all n and $\xi_0 + \cdots + \xi_n = 0$ for infinitely many n. Then $C \in \mathscr{E}$, but $C \notin \mathscr{F}^{(\infty)}$. Indeed, define ω^* as in (117). Let $\omega^+(n) = \omega^*(n)$ except at $n = 1$. Let $\omega^+(1) = 1$. Then $\omega^* \in C$ but $\omega^+ \notin C$.

Throughout this section, unless specified otherwise, I is a countable set and p is a probability on I; and P is a stochastic matrix on I, for which

ASSUMPTION. All states are recurrent.

To state the results, let $\{I_m : m \in M\}$ be the partition of I into its cyclically moving subclasses; recall from (48) that i and j are in the same I_m iff there is an $n \geq 0$ and a state k in I with $P^n(i, k) > 0$ and $P^n(j, k) > 0$. For $m \in M$, let $t(m)$ be the index of the subclass following I_m; so $i \in I_m$ and $P(i, j) > 0$ imply $j \in I_{t(m)}$, by (45). Thus, t is a 1-1 mapping of M onto itself. Let T be the shift, so $T\omega = \omega(1 + \cdot)$. Check $T^{-1}\mathscr{F}^{(\infty)} \subset \mathscr{F}^{(\infty)}$. Let $\{I_c : c \in C\}$ be the partition of I into its communicating classes; recall from (33) that i and j are in the same I_c iff there are nonnegative n and m with $P^n(i, j) > 0$ and $P^m(j, i) > 0$. Finally, say $i \sim j$ iff there is a state k and finite sequences ρ and σ of states, such that $i\rho$ is a permutation of $j\sigma$ and all the transitions in $i\rho k$ and in $j\sigma k$ are possible. Of course, the transition from a to b is *possible* iff $P(a, b) > 0$.

(119) Theorem. *Each $\mathscr{F}^{(\infty)}$-set differs by a P_p-null set from some union of sets $\{\xi_0 \in I_m\}$. More precisely, let $A \in \mathscr{F}^{(\infty)}$. Let $M(A)$ be the set of $m \in M$ such that $P_i(A) > 0$ for some $i \in I_m$. Then A differs from $\bigcup \{I_m : m \in M(A)\}$ by a P_p-null set from an $\mathscr{F}^{(\infty)}$-set. Conversely, each set $\{\xi_0 \in I_m\}$ differs by a P_p-null set from an $\mathscr{F}^{(\infty)}$-set. Finally, $M(T^{-1}A) = t^{-1}M(A)$.*

NOTE. $T^{-1}\{\xi_0 \in I_m\} = \{\xi_1 \in I_m\} = \{\xi_0 \in I_{t^{-1}(m)}\}$: the last equality is a.e.

WARNING. $P_p(A) = 0$ and $P_p(T^{-1}A) = 1$ is a possibility, because $p(I_m) = 0$ and $p(I_{t^{-1}(m)}) = 1$ is a possibility.

(120) Theorem. *Each $\mathscr{I}$-set differs by a P_p-null set from some union of sets $\{\xi_0 \in I_c\}$. Conversely, each set $\{\xi_0 \in I_c\}$ differs by a P_p-null set from an $\mathscr{I}$-set.*

(121) Theorem. *The relation $\sim$ is an equivalence relation: let $\{I_e : e \in E\}$ be the partition it induces on I. Each $\mathscr{E}$-set differs by a P_p-null set from some union of sets $\{\xi_0 \in I_e\}$: and each set $\{\xi_0 \in I_e\}$ differs by a P_p-null set from an $\mathscr{E}$-set.*

NOTE.   The partition $\{I_e : e \in E\}$ is finer than the partition $\{I_m : m \in M\}$, which in turn is finer than $\{I_c : c \in C_c\}$.

Turn now to the proofs. The first result is the 0-1 law of Hewitt and Savage (1955). For future use, I will state the result quite generally. Let $(V, \mathscr{A})$ be a measurable space. Let V^∞ be the space of V-sequences, endowed with the product σ-field $\mathscr{A}^\infty$. A *finite permutation* π on $Z = (0, 1, \ldots)$ is a 1-1 mapping of Z onto Z, with $\pi(n) = n$ for all but finitely many n. Each π induces a 1-1 bimeasurable mapping π^* of V^∞ onto V^∞:

$$\pi^*(v_0, v_1, \ldots) = (v_{\pi 0}, v_{\pi 1}, \ldots).$$

The σ-field $\mathscr{E}$ of *exchangeable sets* is the σ-field of $A \in \mathscr{A}^\infty$ with $\pi^*A = A$ for all finite permutations π of Z. Let $(\Omega, \mathscr{F}, \mathscr{P})$ be a probability triple. Let $X_0, X_1, \ldots$ be measurable mappings from $(\Omega, \mathscr{F})$ to $(V, \mathscr{A})$. Then $X = (X_0, X_1, \ldots)$ is a measurable mapping from $(\Omega, \mathscr{F})$ to $(V^\infty, \mathscr{A}^\infty)$. An *exchangeable X-set* is a set $X^{-1}A$ with $A \in \mathscr{E}$.

(122) Theorem.   *Suppose* X_0, X_1, $\cdots$ *are* $\mathscr{P}$-*independent and identically distributed. Then any exchangeable X-set has* $\mathscr{P}$-*probability* 0 *or* 1.

PROOF.   Let $Q = \mathscr{P}X^{-1}$. It is enough to prove (122) when X_n is the coordinate process on $(V^\infty, \mathscr{A}^\infty, Q)$. Let $A \in \mathscr{E}$ and let B be measurable and depend only on finitely many coordinates. The rest of the proof shows $Q(A \cap B) = Q(A)Q(B)$. The equality then holds for all measurable B, especially $B = A$. Therefore, $Q(A) = Q(A)^2 = 0$ or 1.

Fix $\varepsilon > 0$. Find a measurable set A_ε, depending on only finitely many coordinates, with $Q(A_\varepsilon \triangle A) < \varepsilon$. Here, $C \triangle D = (C - D) \cup (D - C)$. Now Q and A are invariant under π^*, so

$$\varepsilon > Q(\pi^*A_\varepsilon \triangle \pi^*A) = Q(\pi^*A_\varepsilon \triangle A).$$

Consequently,

$$|Q(B \cap A) - Q(B \cap \pi^*A_\varepsilon)| < \varepsilon.$$

I will construct a π with

$$Q(B \cap \pi^*A_\varepsilon) = Q(B)Q(\pi^*A_\varepsilon) = Q(B)Q(A_\varepsilon).$$

Indeed, suppose B depends only on coordinates $n \leq b$, and A_ε depends only on coordinates $n \leq a$. Let $c > \max\{a, b\}$. Let

$$\pi(n) = n + c \quad \text{for } 0 \leq n < c$$
$$= n - c \quad \text{for } c \leq n < 2c$$
$$= n \quad\quad\text{for } n \geq 2c.$$

Then π^*A_ε depends only on coordinates n with $c \leq n < 2c$, and is independent of B. For this π,

$$|Q(B \cap \pi^*A_\varepsilon) - Q(B)Q(A)| < Q(B)\,|Q(A_\varepsilon) - Q(A)| < \varepsilon.$$

As usual,

$$Q(B \cap A) - Q(B)Q(A) = Q(B \cap A) - Q(B \cap \pi^*A_\varepsilon)$$
$$+ Q(B \cap \pi^*A_\varepsilon) - Q(B)Q(A).$$

Thus,

$$|Q(B \cap A) - Q(B)Q(A)| < 2\varepsilon. \qquad\qquad \bigstar$$

(123) Lemma.   *If* $A \in \mathscr{F}^{(\infty)}$, *then* $P_i(A) = 0$ *or* 1.

PROOF.   Let Ω be the set of $\omega \in I^\infty$ such that $\omega(n) = i$ for infinitely many n, and $\omega(0) = i$. Then $P_i(\Omega) = 1$, by (50). So

$$P_i(A) = P_i(A \cap \Omega).$$

I say $A \cap \Omega$ is in the exchangeable σ-field of the process of i-blocks retracted to Ω. Indeed, let $\beta_0, \beta_1, \ldots$ be the successive i-blocks. Now $\beta = (\beta_0, \beta_1, \ldots)$

is a 1-1 bimeasurable mapping of Ω onto the space of sequences of finite i-blocks. Let $B = \beta(A \cap \Omega)$. Let w be a sequence of finite i-blocks, and let π be a finite permutation of Z. I have to show that

$$w \in B \quad \text{iff} \quad \pi^* w \in B.$$

But $w = \beta\omega$ and $\pi^* w = \beta\omega^*$ for unique ω and ω^* in Ω. By thinking, $\omega^* = \rho^*\omega$ for some finite permutation ρ of Z with $\rho(0) = 0$.

NOTE. ρ depends on π and ω; for each ω, as π ranges over all the finite permutations, ρ typically ranges only over a small subset of the finite permutations.

Finally, $\omega \in A \cap \Omega$ iff $\rho^*\omega \in A \cap \Omega$, because $A \in \mathscr{F}^{(\infty)}$. Blocks (31) and Hewitt-Savage (122) now force

$$P_i(A \cap \Omega) = 0 \text{ or } 1. \qquad \bigstar$$

(124) Lemma.   *Let i and j be in the same cyclically moving subclass. Then $P_i = P_j$ on $\mathscr{F}^{(\infty)}$.*

PROOF.   Suppose $P^n(i, k) > 0$ and $P^n(j, k) > 0$. Let $A \in \mathscr{F}^{(\infty)}$ and suppose $P_i\{A\} > 0$. Using (123),

$$1 = P_i\{A\} = P^n(i, k) \cdot P_i\{A \mid \xi_n = k\} + [1 - P^n(i, k)] \cdot P_i\{A \mid \xi_n \neq k\}.$$

So, $0 < P_i\{A \mid \xi_n = k\} = P_j\{A \mid \xi_n = k\}$ by Markov (15); and $P_j\{A\} > 0$, forcing $P_j\{A\} = 1$ by (123). $\qquad \bigstar$

This proof of (124) was suggested by Volker Strassen.

PROOF OF (119).   Let $A \in \mathscr{F}^{(\infty)}$. Then M, the set of cyclically moving subclasses I_m, is the disjoint union of M_0 and M_1, where $m \in M_0$ iff $P_i\{A\} = 0$ for all $i \in I_m$, and $m \in M_1$ iff $P_i\{A\} = 1$ for all $i \in I_m$. This uses (123–124). Thus, $M(A) = M_1$. If $m \in M_0$, then $P_p\{A \text{ and } \xi_0 \in I_m\} = 0$. If $m \in M_1$, then $P_p\{A \text{ and } \xi_0 \in I_m\} = P_p\{\xi_0 \in I_m\}$. Thus, A differs by a P_p-null set from the union over $m \in M_1$ of $\{\xi_0 \in I_m\}$.

For the converse assertion, suppose $i \in I_m$ has period d. Then

$$\{\xi_{nd} \in I_m \text{ for infinitely many } n\}$$

is a tail set which differs from $\{\xi_0 \in I_m\}$ by a P_p-null set; for instance, use (16) and (45b).

For the final assertion, use Markov (15):

$$P_i\{T^{-1}A\} = \Sigma_j P_i\{\xi_1 = j \text{ and } T^{-1}A\}$$
$$= \Sigma_j P(i, j) \cdot P_j\{A\}.$$

Suppose $i \in I_m$. Then j can be confined to $I_{t(m)}$, for otherwise $P(i, j) = 0$

by (45). Now $P_j\{A\}$ is constant at 0 or 1. So $P_i\{T^{-1}A\}$ is 0 or 1, according as $i \in I_m$ with $t(m) \in M_0$ or $t(m) \in M_1$.   ★

PROOF OF (120).   Suppose $I_m \subset I_c$, and I_c has period d. Then (45) shows

$$I_c = \bigcup_{n=0}^{d-1} I_{t^{-n}(m)}.$$

Let A be an invariant set. Then $A \in \mathscr{F}^{(\infty)}$. Use (119) to check $M(A) = M(T^{-1}A) = t^{-1}M(A)$. Consequently,

$$\bigcup \{I_m : m \in M(A)\} = \bigcup \{I_c : c \in C(A)\},$$

where $c \in C(A)$ iff $I_c \supset I_m$ for some $m \in M(A)$.

Conversely, $\{\xi_0 \in I_c\}$ differs by a P_p-null set from the invariant set

$$\{\xi_n \in I_c \text{ for infinitely many } n\};$$

For instance, use (16) and (39).   ★

(125) Lemma.   *If $A \in \mathscr{E}$, then $P_i\{A\} = 0$ or 1.*

PROOF.   As in (123).   ★

(126) Lemma. $i \not\sim j$ *implies* $P_i = P_j$ *on* $\mathscr{E}$.

PROOF.   Find a state k and finite state sequences ρ and σ, with $i\rho$ a permutation of $j\sigma$ and all transitions in $i\rho k$ and $j\sigma k$ possible. Let N be length of ρ, also of σ. Let $A \in \mathscr{E}$, with $P_i\{A\} = 1$. The problem is to show $P_j\{A\} > 0$; this reduction depends on (125).

For any finite state sequence φ, let A_φ be the set of infinite state sequences ψ with $\varphi\psi \in A$. Here, $\varphi\psi$ means the infinite state sequence φ followed by ψ. If φ is a permutation of φ^*, then $A_\varphi = A_{\varphi^*}$ by exchangeability. Because $P_i\{A\} = 1$ and all transitions in $i\rho k$ are possible,

$$\begin{aligned}
1 &= P_i\{A \mid \xi_0 \cdots \xi_{N+1} = i\rho k\} \\
&= P_k\{A_{i\rho k}\} \\
&= P_k\{A_{j\sigma k}\} \\
&= P_j\{A \mid \xi_0 \cdots \xi_{N+1} = j\sigma k\};
\end{aligned}$$

this uses Markov (15). Because all transitions in $j\sigma k$ are possible, $P_j\{A\} > 0$.   ★

(127) Lemma. $i \not\sim j$ *implies* $P_i \neq P_j$ *on* $\mathscr{E}$.

PROOF.   Fix any state k in the same communicating class as i. Let A_0 be the set of all infinite state sequences with infinitely many k's, starting from i, and all transitions possible. So $P_i\{A_0\} = 1$ by (16, 73). Let A_1 be the smallest exchangeable set including A_0, so $P_i\{A_1\} = 1$. Now A_1 consists of all infinite state sequences such that k occurs infinitely often, and for all remote

k's, the preceding part of the sequence is obtained by permuting some finite state sequence $i\rho$, all the transitions in $i\rho k$ being possible. I say $P_j\{A_1\} = 0$. For $A_1 \subset \bigcup_{n=1}^{\infty} B_n$, where B_n is the set of infinite state sequences with k in the nth place, the preceding part of the sequence being a permutation of some finite state sequence $i\rho$, all the transitions in $i\rho k$ being possible. But, $i \nrightarrow j$ implies $P_j\{B_n\} = 0$.   ★

PROOF OF (121).   As for (119), using (125–127).   ★

Similar results apply to random walks. Let G be a countable Abelian group and let π be a probability on G. Let V be the set of $i \in G$ with $\pi(i) > 0$. Let $\{G_m : m \in M\}$ partition G into the cosets of the group spanned by $V - V$. Let $\{G_c : c \in C\}$ partition G into the cosets of the group spanned by V. Let $\{Z_n : 0 \le n < \infty\}$ be independent random variables with values in G, the Z_n with $n \ge 1$ having common distribution π. Let $X_n = \Sigma_{\nu=0}^{n} Z_\nu$. Now, $\{X_n\}$ is a Markov chain, but is in general transient. The tail σ-field of $\{X_n\}$ is equivalent to the atomic σ-field generated by the sets $\{X_0 \in G_m\}$ for $m \in M$. The invariant σ-field of $\{X_n\}$ is equivalent to the atomic σ-field generated by the sets $\{X_0 \in G_c\}$ for $c \in C$. Proofs are omitted, being virtually identical to those for (119) and (120). An analog of (128) can also be obtained, using similar ideas. This leads to another proof of the renewal theorem.

The next theorem is due to (Orey, 1962).

(128) Theorem.   *If i and j are in the same cyclically moving subclass,*

$$\lim_{n \to \infty} \Sigma_{k \in I} |P^n(i, k) - P^n(j, k)| = 0.$$

This auxiliary result will be helpful:

(129) Lemma.   *Let $\{X_n : n = 0, 1, \ldots\}$ be a stochastic process on $(\Omega, \mathscr{F}, \mathscr{P})$. The tail σ-field of X is trivial under $\mathscr{P}$ if*

(130) $\lim_{n \to \infty} \sup_{A \in \Sigma^{(n)}} |\mathscr{P}\{A \cap B\} - \mathscr{P}\{A\}\mathscr{P}\{B\}| = 0$

for each $B \in \Sigma^{(\infty)}$. If the tail σ-field of X is trivial under $\mathscr{P}$, then (130) holds for each $B \in \mathscr{F}$.

PROOF.   If (130) holds, use it on $A = B \in \Sigma^{(\infty)}$ to get $\mathscr{P}\{B\} = \mathscr{P}\{B\}^2 = 0$ or 1. Conversely, if $\Sigma^{(\infty)}$ is trivial, fix $B \in \mathscr{F}$, and let 1_B be 1 on B and 0 off B. For $A \in \Sigma^{(n)}$,

$$|\mathscr{P}\{A \cap B\} - \mathscr{P}\{A\}\mathscr{P}\{B\}| = \left| \int_A [1_B - \mathscr{P}\{B\}] \, d\mathscr{P} \right|$$

$$= \left| \int_A [\mathscr{P}\{B \,|\, \Sigma^{(n)}\} - \mathscr{P}\{B\}] \, d\mathscr{P} \right|$$

$$\le \int_A |\mathscr{P}\{B \,|\, \Sigma^{(n)}\} - \mathscr{P}\{B\}| \, d\mathscr{P}.$$

Use backward martingales (10.34).   ★

PROOF OF (128).   Let $Q = \frac{1}{2}P_i + \frac{1}{2}P_j$. As (119) implies, $\mathscr{F}^{(\infty)}$ is trivial for Q. Use (129) twice, with $A = \{\xi_n \in S\}$ and $B = \{\xi_0 = i\}$ or $B = \{\xi_0 = j\}$, to get

$$\lim_{n\to\infty} \sup_{S\subset I} |P_i\{\xi_n \in S\} - P_j\{\xi_n \in S\}| = 0.$$

Finally, $\Sigma_{k\in I} |q(k) - r(k)| = 2 \sup_{S\subset I} |q(S) - r(S)|$ for any probabilities q and r on all subsets S of I.   ★

14. EXAMPLES

(131) Example.   There is a transition matrix P for which I is a recurrent class of states, having period 1, such that two independent particles, starting in different states and moving according to P, may never meet.

DISCUSSION.   The states are the nonnegative integers, and $P(i, i - 1) = 1$ for $i \geqq 1$,   while $P(0, n) = a_{n+1} > 0$ for $n = 0, 1, \ldots$ will be defined below. Consider two independent Markov processes $\{X_n : 0 \leqq n < \infty\}$ and $\{Y_n : 0 \leqq n < \infty\}$ with P for matrix of stationary transition probabilities and initial states 0 and 1, respectively. If the two processes ever meet, they stay together until reaching 0. Let S_ν be the time of the νth return of $\{X_n\}$ to 0; while T_ν is the time of the νth visit of $\{Y_n\}$ to 0. Then, $S_k = \Sigma_{\nu=1}^k U_\nu$ and $T_k = 1 + \Sigma_{\nu=1}^{k-1} V_\nu$, where $\{U_\nu : 1 \leqq \nu < \infty; V_\nu : 1 \leqq \nu < \infty\}$ are independent, with this common distribution: the value n is taken with probability a_n for $n = 1, 2, \ldots$. The $\{a_n\}$ will be chosen to make the event $\{S_k = T_m$ for some $k \geqq 1$ and $m \geqq 1\}$ have probability less than 1. Consider the random walk on the planar lattice where a point moves to each of its four neighbors with probability $1/5$ and stays fixed with probability $1/5$. Let a_n be the probability of a first return to the origin at time n. If two independent such walks start simultaneously from the origin, the probability that for some n the first returns to the origin at time n, while the second returns at time $n - 1$, is precisely the probability of

$$\{S_k = T_m \text{ for some } k \geqq 1 \text{ and } m \geqq 1\}.$$

This probability cannot be 1; for suppose it were. By symmetry, with probability 1 there is some m such that the first walk returns to the origin at time $m - 1$ and the second returns at m. By strong Markov, the two walks would return to the origin at the same time with probability 1, which is known to be false (Chung and Fuchs, 1951, Theorem 6).   ★

NOTE.   Let P be a stochastic matrix on I. Let $\{X_n\}$ and $\{Y_n\}$ be independent P-chains. Then $\{(X_n, Y_n)\}$ is also a Markov chain, with state space $I \times I$ and transitions T, where

$$T^n[(i,j), (i',j')] = P^n(i, i') P^n(j, j').$$

Suppose I is a positive recurrent class of period 1 for P. As (68–69) imply, $I \times I$ is a positive recurrent class of period 1 for T. In particular, $\{X_n\}$ and $\{Y_n\}$ meet with probability 1.

(132) Example. There is a transition matrix P and starting probability p for which I is a transient essential class, the invariant σ-field is trivial, while the tail σ-field is full and nonatomic.

DISCUSSION. The states are pairs (m, n) of positive integers with $n \leqq 2 \cdot 3^m$. The matrix P is defined by:

$$P[(m, 1), (m, 3^m)] = P[(m, 1), (m, 2 \cdot 3^m)] = \tfrac{1}{2};$$

while

$$P[(m, n), (m, n - 1)] = 1 \quad \text{for } n \geqq 3$$

and

$$P[(m, 2), (m + 1, 1)] = 1.$$

The starting probability concentrates on $(1, 1)$.

For $m \geqq 1$, let U_m be 1 or 2 according as $(m, 1)$ is followed by $(m, 3^m)$ or $(m, 2 \cdot 3^m)$ in $\{\xi_n : n = 0, 1, \ldots\}$. Consider $P_{(1,1)}$. The time T_m at which $(m + 1, 1)$ is reached equals $\Sigma_{\nu=1}^m U_\nu 3^\nu$ a.e., and T_m determines $U_1, \ldots, U_m$ a.e. The tail σ-field of $\{\xi_n\}$ is therefore equivalent to the σ-field determined by $\{U_n : 1 \leqq n < \infty\}$, which is equivalent to the full σ-field. Since $U_1, U_2, \ldots$ are independent and identically distributed, and nonconstant, the σ-field they generate is nonatomic. Let A be an invariant set. Then

$$1_A = 1_A(\xi_{T_m}, \xi_{T_m+1}, \ldots)$$

is measurable on $U_{m+1}, U_{m+2}, \ldots$, and A has probability 0 or 1 by the Kolmogorov 0-1 law.   ★

(133) Example. There is a transition matrix P and starting probability p on $I = \{1, 2, 3\}$ for which I is a communicating class of period 1, but the exchangeable σ-field is nontrivial.

DISCUSSION. P is the matrix

$$\begin{pmatrix} 0 & 1 & 0 \\ 0 & 0 & 1 \\ \tfrac{1}{2} & \tfrac{1}{2} & 0 \end{pmatrix}.$$

Let ρ and σ be finite sequences of states with all transitions possible in $1\rho3$ and $3\sigma3$. In a finite sequence of states with all transitions possible, any 3 is either at the beginning of the sequence or is preceded by a 2. Consequently, in 1ρ there is one more 2 than 3. In 3σ, however, there are as many 2's as 3's. Thus 1ρ cannot be permuted into 3σ, and in view of (121), the σ-field of exchangeable events is not trivial for P_p, provided $p(1) > 0$ and $p(3) > 0$.   ★

2

RATIO LIMIT THEOREMS

1. INTRODUCTION

Throughout this chapter, unless noted otherwise, I is a recurrent class relative to the stochastic matrix P. Interest centers on the null recurrent case and on measures with infinite mass. Fix a reference state $s \in I$. Remember that $\{\xi_n\}$ is Markov with stationary transitions P and starting state s relative to the probability P_s. Remember that the first s-block runs from the first s to just before the second s. Remember the definition (1.80) of invariance and subinvariance. Define the measure μ on I by the relation: $\mu(i)$ is the P_s-mean number of i's in the first s-block; that is,

$$\mu(i) = e(P\{s\})(s, i) = eP\{s\}(s, i).$$

As (1.100) implies, $\mu(i) < \infty$.

(1) Theorem (*Derman*, 1954).   *There is one and only one invariant measure whose value at s is 1, namely μ. Any subinvariant measure is invariant.*

(2) Theorem (*Doeblin*, 1938).   *Let $i, j, k, l \in I$. Then*

$$\lim_{n \to \infty} [\Sigma_{m=0}^n P^m(i, j)]/[\Sigma_{m=0}^n P^m(k, l)] = \mu(j)/\mu(l).$$

Section 2 contains some preliminary material on reversing time, and makes no assumptions about I. Theorems (1) and (2) are then proved together in Section 3. Section 4 contains some variations on (1) and (2); on first reading the book, do only (21–24) in Section 4. Section 5 is about restricting the range of a chain $\{\xi_n\}$: fix a set J of states, and delete the times n with $\xi_n \notin J$. This idea leads to another proof of (1), and reappears in *ACM*. On first reading the book, skip Section 5.

I want to thank Allan Izenman for checking the final draft of this chapter.

(3) Theorem (*Kingman and Orey*, 1964).   *Let N be a positive integer, and let $\varepsilon > 0$. Suppose I has period 1, and $P^N(i, i) > \varepsilon$ for all i. Then for each $m = 0, 1, \ldots$ and $i, j, k, l \in I$:*

$$\lim_{n \to \infty} P^{n+m}(i, j)/P^n(k, l) = \mu(j)/\mu(l).$$

This result is proved in Section 6. Section 7 contains an example like Dyson's in (Chung, 1960, p. 55), which shows why you should assume $P^N(i, i) > \varepsilon$. This section can be skipped without logical loss.

(4) Theorem (*Harris*, 1952) *and* (*Lévy*, 1951).   *Let $V_n(i)$ be the number of $m \leqq n$ with $\xi_m = i$. With P_s-probability 1,*

$$\lim_{n \to \infty} V_n(j)/V_n(k) = \mu(j)/\mu(k).$$

This result is proved as (83) in Section 8. Section 9 contains related material which is used in a marginal way in Chapter 3. Part (a) of (92) is in (Chung, 1960, p. 79).

2.   REVERSAL OF TIME

In this section, let P be a stochastic matrix on the countable set I; but do not make any assumptions about the recurrence properties of I. Let ν be a measure on I. Say ν is *strictly positive* iff $\nu(i) > 0$ for all $i \in I$. Say ν is *locally finite* iff $\nu(i) < \infty$ for all $i \in I$. Say ν is *finite* iff $\nu(I) < \infty$. Do not assume ν is finite without authorization. Throughout this section, suppose

$$\nu \text{ is subinvariant.}$$

If Q is substochastic, write $i \to j$ *relative to* Q iff $Q^n(i, j) > 0$ for some $n > 0$. Write $i \to j$ iff $i \to j$ for P.

(5) Lemma.   (a) $\nu(i) > 0$ *and* $i \to j$ *imply* $\nu(j) > 0$.
            (b) $\nu(j) < \infty$ *and* $i \to j$ *imply* $\nu(i) < \infty$.

PROOF. $\nu(j) \geqq \Sigma_k \nu(k)P^n(k, j) \geqq \nu(i)P^n(i, j)$.   ★

(6) Lemma.   *If I is a communicating class, either ν is identically 0, or ν is strictly positive and locally finite, or $\nu(i) = \infty$ for all i.*

PROOF.   Use (5).   ★

Let $_\nu I = \{i : 0 < \nu(i) < \infty\}$. For any matrix M on I, define the matrix, $_\nu M$ on $_\nu I$ by:

$$_\nu M(i, j) = \nu(j)M(j, i)/\nu(i).$$

Call $_\nu M$ the *reversal of M by ν*.

(7) Lemma.   (a) $_vP$ is stochastic.

   (b) *Suppose v is locally finite, and not identically zero. Then $_vP$ is stochastic iff v is invariant.*

PROOF.   Abbreviate $J = {}_vI$ and fix $i \in J$. If $v(j) = 0$, then $v(j)P(j, i) = 0$. If $v(j) = \infty$, then $P(j, i) = 0$ by (5b), so $v(j)P(j, i) = 0$ by convention. Consequently,

(8)
$$\Sigma_{j \in J} \, {}_vP(i, j) = \Sigma_{j \in J} \, v(j)P(j, i)/v(i)$$
$$= \Sigma_{j \in I} \, v(j)P(j, i)/v(i).$$

This proves (a), and the *if* part of (b), because J is nonempty. For the *only if* part of (b), suppose v is locally finite and $_vP$ is stochastic. Then (8) makes

$$v(i) = \Sigma_{j \in J} \, v(j)P(j, i) \quad \text{for } v(i) > 0.$$

Subinvariance by itself makes

$$v(i) = \Sigma_{j \in I} \, v(j)P(j, i) \quad \text{for } v(i) = 0. \qquad \bigstar$$

Let $e^nP = \Sigma_{m=0}^{n} P^m$, where P^0 is the identity matrix. So, $e^nP(i, j)$ is the P_i-mean number of visits to j up to time n. Remember eP from (1.49). So $eP = \lim_n e^nP$. Check

(9) $(_vP)^n = {}_v(P^n)$ and $e^n(_vP) = {}_v(e^nP)$ and $e(_vP) = {}_v(eP)$.

Suppose $i, j \in {}_vI$. Use (9) and (1.51, 1.69):

(10) $i \to j$ for P and $_vP$ simultaneously; i is transient, null recurrent or positive recurrent for P and $_vP$ simultaneously.

(11) Lemma.   *Let Q be a substochastic matrix on I, which is not stochastic. Suppose that $i \to j$ relative to Q for any pair i, j. Then $eQ < \infty$.*

PROOF.   Let $\partial \notin I$. Extend Q to a matrix $\bar{Q}$ on $\bar{I} = I \cup \{\partial\}$ as follows:

$$\bar{Q}(i, j) = Q(i, j) \qquad\qquad \text{for } i \text{ and } j \text{ in } I$$
$$\bar{Q}(i, \partial) = 1 - \Sigma_{j \in I} Q(i, j) \quad \text{for } i \text{ in } I$$
$$\bar{Q}(\partial, i) = 0 \qquad\qquad\qquad \text{for } j \text{ in } I$$
$$\bar{Q}(\partial, \partial) = 1.$$

Then $\bar{Q}$ is stochastic. By assumption, there is a $k \in I$ with $\bar{Q}(k, \partial) > 0$. Let $i \in I$. By assumption, $i \to k$ for Q. Then $\bar{Q}$ makes $i \to k$ and $k \to \partial$, so $i \to \partial$; but $\partial \not\to i$: now (1.55) shows i to be transient. Let $j \in I$. Use (1.51) to get $e\bar{Q}(i, j) < \infty$. Check $Q^n(i, j) = \bar{Q}^n(i, j)$. So $eQ(i, j) < \infty$. $\qquad \bigstar$

3. PROOFS OF DERMAN AND DOEBLIN

Suppose again that I is a recurrent class of states relative to P.

(12) Lemma. *Any subvariant measure is invariant.*

PROOF. Let ν be subinvariant. Suppose ν is strictly positive and locally finite: (6) takes care of the other cases. Then (9) and (1.51) make

$$e(_\nu P)(i, j) = \infty.$$

Lemma (11) makes $_\nu P$ stochastic, and (7b) makes ν invariant.   ★

If ν is a measure, f is a function, and M is a matrix on I, let

$$\nu M f = \Sigma_{i,j} \, \nu(i) M(i, j) f(j).$$

Let the probability δ_i assign mass 1 to i and vanish elsewhere. Let the function δ_i be 1 at i and vanish elsewhere. So $pe^n P \delta_i$ is the P_p-mean number of visits to i up to time n.

(13) Lemma. *Let p and q be probabilities on I. Then*

$$\lim_{n\to\infty} (pe^n P \delta_i)/(qe^n P \delta_i) = 1.$$

PROOF. It is enough to do the case $q = \delta_i$:

$$\frac{pe^n P \delta_i}{qe^n P \delta_i} = \frac{pe^n P \delta_i}{\delta_i e^n P \delta_i} \cdot \frac{\delta_i e^n P \delta_i}{qe^n P \delta_i}.$$

Let τ be the least n if any with $\xi_n = i$, and $\tau = \infty$ if none. I say

(14) $$pe^n P \delta_i = \Sigma_{t=0}^n P_p\{\tau = t\} \cdot (e^{n-t} P)(i, i).$$

Indeed, $\delta_i(\xi_m) = 0$ for $m < \tau$, so

$$pe^n P \delta_i = \int \Sigma_{m=0}^n \delta_i(\xi_m) \, dP_p$$

$$= \Sigma_{t=0}^n \int_{\{\tau=t\}} \Sigma_{m=t}^n \delta_i(\xi_m) \, dP_p.$$

Check that $\{\tau = t\}$ is measurable on $\{\xi_0, \ldots, \xi_t\}$; and $\xi_t = i$ on $\{\tau = t\}$. So Markov (1.15*) implies

$$\int_{\{\tau=t\}} \Sigma_{m=t}^n \delta_i(\xi_m) \, dP_p = P_p\{\tau = t\} \cdot \int \Sigma_{m=0}^{n-t} \delta_i(\xi_m) \, dP_i$$

$$= P_p\{\tau = t\} \cdot (e^{n-t} P)(i, i).$$

This proves (14).

Divide both sides of (14) by

$$\delta_i e^n P \delta_i = (e^n P)(i, i).$$

For each $t = 0, 1, \ldots$, the ratio $(e^{n-t}P)(i, i)/(e^n P)(i, i)$ is no more than 1, and converges to 1 as $n \to \infty$: because

$$0 \leqq (e^n P)(i, i) - (e^{n-t}P)(i, i) \leqq t$$

and

$$e^n P(i, i) \to \infty \quad \text{as } n \to \infty$$

by (1.51). Next, (1.55) makes

$$\Sigma_{0 \leqq t < \infty} P_p\{\tau = t\} = 1.$$

Dominated convergence now proves (13) in the case $q = \delta_i$.   ★

For measures v and functions f on I, let $v(f) = \Sigma_{i \in I} f(i)v(i)$, and let $f \times v$ be the measure on I which assigns mass $f(i)v(i)$ to i.

(15) Lemma. *Let f and g be functions on I. Let v be a strictly positive and locally finite invariant measure on I. Suppose $v(|f|) < \infty$ and $v(|g|) < \infty$ and $v(g) \neq 0$. Then*

$$\lim_{n \to \infty} (\delta_i e^n Pf)/(\delta_i e^n Pg) = v(f)/v(g).$$

PROOF.   Let M be a substochastic matrix on I, and let h be a nonnegative function on I, with $v(h) < \infty$. By algebra,

$$\delta_i Mh = (h \times v)(_v M)\delta_i/v(i).$$

Using (9),

(16) $$\delta_i e^n Ph = (h \times v)e^n(_v P)\delta_i/v(i) < \infty.$$

Suppose h' is another nonnegative function on I, with $v(h') < \infty$. As (16) implies, $\delta_i e^n Ph' < \infty$. Consequently,

(17) $$\delta_i e^n P(h - h') = \delta_i e^n Ph - \delta_i e^n Ph'.$$

Case 1: f and g are nonnegative.   Use (13), with $(f \times v)/v(f)$ for p and $(g \times v)/v(g)$ for q and $_v P$ for P; this is legitimate because (10) makes I a recurrent class for $_v P$. The yield is

$$\frac{(f \times v)e^n(_v P)\delta_i}{(g \times v)e^n(_v P)\delta_i} \to \frac{v(f)}{v(g)}.$$

Now use (16) with f or g for h.

Case 2: g is nonnegative.   Split f into its positive and negative parts. Use (17) and case 1.

Case 3: f is nonnegative.   Take reciprocals and use case 2.

The general case. Split f into its positive and negative parts. Use (17) and case 3. ★

Remember that $\mu(i) = eP\{s\}(s, i)$ is the P_s-mean number of i's in the first s-block.

(18) Lemma. **(a)** *Let $k \neq s$. The P_s-mean number of pairs j followed by k in the first s-block is $\mu(j)P(j, k)$.*

(b) *The P_s-probability that the first s-block ends with j is $\mu(j)P(j, s)$.*

PROOF. *Claim (a).* Let τ be the least $n > 0$ if any with $\xi_n = s$ and $\tau = \infty$ if none. Let α_n be the indicator function of the event

$$A_n = \{\xi_n = j \text{ and } n < \tau\}.$$

Let β_n be the indicator function of the event

$$B_n = \{A_n \text{ and } \xi_{n+1} = k\}.$$

Confirm $\tau > n + 1$ on B_n, because $k \neq s$. On $\{\xi_0 = s\}$;

$\sum_{n=0}^{\infty} \alpha_n$ is the number of j's in the first s-block;

$\sum_{n=0}^{\infty} \beta_n$ is the number of pairs j followed by k in the first s-block.

Check that $\{\tau > n\}$ is in the σ-field generated by $\xi_0, \ldots, \xi_n$. Now Markov (1.15) and monotone convergence imply

$$\int (\sum_{n=0}^{\infty} \beta_n)\, dP_s = \sum_{n=0}^{\infty} P_s\{A_n \text{ and } \xi_{n+1} = k\}$$

$$= \sum_{n=0}^{\infty} P_s(A_n) \cdot P(j, k)$$

$$= \left[\int (\sum_{n=0}^{\infty} \alpha_n)\, dP_s\right] \cdot P(j, k)$$

$$= \mu(j)P(j, k).$$

Claim (b) is similar. ★

(19) Lemma. $\mu(s) = 1$ *and μ is invariant.*

PROOF. By definition, the first s-block begins at the first s, and ends just before the second s. There is exactly one s in the first s-block, so $\mu(s) = 1$. If $k = s$, then $\mu(j)P(j, k)$ is the P_s-probability that the first s-block ends with j by (18b). The sum on j is therefore 1, that is, $\mu(s)$. If $k \neq s$, then $\mu(j)P(j, k)$ is the P_s-mean number of pairs j followed by k in the first s-block by (18a). The sum on j is therefore the P_s-mean number of k's in the first s-block, that is, $\mu(k)$. ★

(20) Lemma. *If ν is a strictly positive and locally finite invariant measure then $\nu(j)/\nu(k) = \mu(j)/\mu(k)$.*

PROOF. Use (15), with $f = \delta_j$ and $g = \delta_k$. Then use (19) to put μ for ν in (15).   ★

PROOF OF (1). Use (12, 19, 20).   ★

PROOF OF (2). Abbreviate $\theta(i, j) = \delta_i e^n P \delta_j$. Then

$$\frac{\theta(i, j)}{\theta(k, l)} = \frac{\theta(i, j)}{\theta(k, j)} \cdot \frac{\theta(k, j)}{\theta(k, l)}.$$

Use (13) with δ_i for p and δ_k for q and j for i to make the first factor converge to 1. Use (15) with k for i and δ_j for f and δ_l for g to make the second factor converge to $\mu(j)/\mu(l)$.   ★

4. VARIATIONS

(21) Theorem. $eP\{i\}(i, k) = eP\{i\}(i, j) \cdot eP\{j\}(j, k)$.

FIRST PROOF. Define a measure ν on I by:

$$\nu(k) = eP\{i\}(i, k)/eP\{i\}(i, j).$$

Plainly, $\nu(j) = 1$. By (19), the measure ν is invariant, so (1) implies $\nu(k) = eP\{j\}(j, k)$.   ★

SECOND PROOF. The nth j-block is the sample sequence from the nth j until just before the $n + 1$st j, shifted to the left so as to start from time 0. Here, $n = 1, 2, \ldots$. Suppose i, j, k are all different. Let Z_n be the number of k's in the nth j-block. Let τ be the number of j's in the first i-block. As in Figure 1, let A be the number of k's on or after the first i, but before the first j after the first i. Let B be the number of k's on or after the second i, but before the first j after the second i. On $\{\xi_0 = i\}$, the number of k's in the first i-block is

$$\Sigma_{n=1}^\tau Z_n + A - B.$$

So

$$eP\{i\}(i, k) = \int (\Sigma_{n=1}^\tau Z_n + A - B) \, dP_i$$

$$= \int \Sigma_{n=1}^\tau Z_n \, dP_i + \int A \, dP_i - \int B \, dP_i.$$

By strong Markov (1.22*):

$$\int A \, dP_i = \int B \, dP_i;$$

$$\int Z_1 \, dP_i = eP\{j\}(j, k).$$

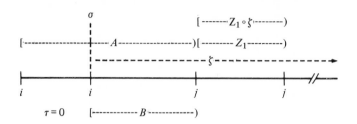

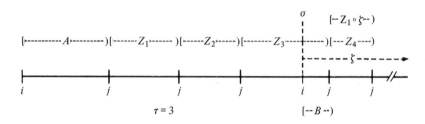

Visits to i and j are shown, but not to k. The variables A, B, Z_1, $Z_2 \ldots$ are the number of k's in the intervals shown.

Figure 1.

By blocks (1.31),

(22)   the variables Z_1, Z_2, ... are P_i-independent and identically distributed.

I claim

(23) $\{\tau < n\}$ is independent of Z_n.

Granting (23), Wald (1.71) implies

$$\int \Sigma_{n=1}^{\tau} Z_n \, dP_i = \int \tau \, dP_i \cdot \int Z_1 \, dP$$

$$= eP\{i\}(i, j) \cdot eP\{j\}(j, k).$$

Relation (23) follows from

$$P_i\{\tau = m \text{ and } Z_n = z\} = P_i\{\tau = m\} \cdot P_i\{Z_n = z\}$$

for $m < n$. Let σ be the time of the second i. Then σ is Markov, and $\xi_\sigma = i$ on $\{\sigma < \infty\}$. Clearly, τ is measurable on the pre-σ sigma field. Let ζ be the post-σ process. Then $Z_n = Z_{n-m} \circ \zeta$ on $\{\xi_0 = i \text{ and } \tau = m\}$, as in Figure 1.

Use strong Markov:

$$P_i\{\tau = m \text{ and } Z_n = z\} = P_i\{\tau = m \text{ and } \zeta \in [Z_{n-m} = z]\}$$
$$= P_i\{\tau = m\} \cdot P_i\{Z_{n-m} = z\} \quad \text{by (1.22)}$$
$$= P_i\{\tau = m\} \cdot P_i\{Z_n = z\} \quad \text{by (22).}$$

This proves (23).   ★

For (24), let I be a positive recurrent class. That is, the P_i-mean waiting time $mP(i, i)$ for a return to i is finite. And $\pi(i) = 1/mP(i, i)$ is the unique invariant probability.

(24) Theorem.   *Suppose I is positive recurrent:*

(a) $eP\{i\}(i, j) = \pi(j)/\pi(i) = mP(i, i)/mP(j, j)$.

(b) *Any subinvariant measure is a nonnegative scalar multiple of π, and is automatically of finite total mass and invariant.*

PROOF.   *Claim (a).*   As in (21), using (1.84) for the first argument.
Claim (b).   Use (1) and (a).   ★

At a first reading of the book, skip to Section 6.
Remember $e^n P = P^0 + \cdots + P^n$, so $(e^n P)(i, j)$ is the P_i-mean number of j's through time n, and

$$pMf = \Sigma_{i,j}\, p(i)M(i, j)f(j).$$

(25) Theorem.   *For any probability p on I,*

$$\lim_{n\to\infty} (pe^n P\delta_i)/(pe^n P\delta_j) = eP\{j\}(j, i).$$

PROOF.   Suppose $i \neq j$. Let $\tau(n)$ be the number of j's up to and including n. Let Z_m be the number of i's in the mth j-block. As in Figure 2, let A be the number of i's before the first j, and let $B(n)$ be the number of i's after n but before the first j after n. Then the number of i's up to and including time n is

$$\Sigma_{m=1}^{\tau(n)} Z_m + A - B(n).$$

So

(26) $$pe^n P\delta_i = \int \Sigma_{m=1}^{\tau(n)} Z_m \, dP_p + \int A \, dP_p - \int B(n) \, dP_p.$$

By blocks (1.31),

(27) $Z_1, Z_2, \ldots$ are P_p-independent and identically distributed.

I claim

(28) $\{\tau(n) < m\}$ is P_p-independent of Z_m.

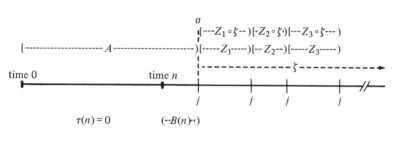

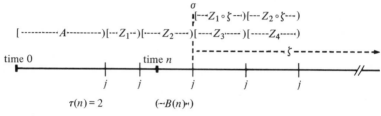

Visits to j are shown, but not to i. Variables A, B, (n), $Z_1, Z_2 \ldots$ are the number of i's in the intervals shown.

Figure 2.

This follows from

(29) $P_p\{\tau(n) = t \text{ and } Z_m = z\} = P_p\{\tau(n) = t\} \cdot P_p\{Z_m = z\}$

for $t < m$. To prove (29), let σ be the time of the $t + 1$st j. Then σ is Markov, and $\xi_\sigma = j$ on $\{\sigma < \infty\}$. Moreover, $\{\tau(n) = t\}$ is in the pre-σ sigma field. Let ζ be the post-σ process. Then $Z_m = Z_{m-t} \circ \zeta$ on $\{\tau(n) = t\}$, as in Figure 2. By strong Markov (1.22):

$$P_p\{\tau(n) = t \text{ and } Z_m = z\} = P_p\{\tau(n) = t \text{ and } \zeta \in [Z_{m-t} = z]\}$$
$$= P_p\{\tau(n) = t\} \cdot P_j\{Z_{m-t} = z\}.$$

and

$$P_j\{Z_{m-t} = z\} = P_p\{Z_m = z\}.$$

This proves (29), and with it (28). By strong Markov (1.22*),

$$\int Z_1 \, dP_p = eP\{j\}(j, i).$$

Use Wald (1.71):

(30a)
$$\int \Sigma_{m=1}^{\tau(n)} Z_m \, dP_p = \int \tau(n) \, dP_p \cdot \int Z_1 \, dP_p$$
$$= pe^n P\delta_j \cdot eP\{j\}(j, i).$$

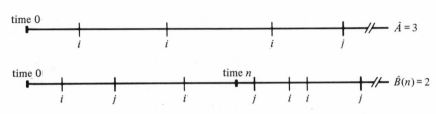

Figure 3.

As in Figure 3, let $\hat{A}$ be the number of i's on or after the first i but before the first j after the first i, and let $\hat{B}(n)$ be the number of i's on or after the first i after n, but before the first j after the first i after n. By the strong Markov property (1.22*),

$$\int \hat{A} \, dP_p = \int \hat{B}(n) \, dP_p = eP\{j\}(i, i),$$

which is finite by (1.100). Plainly, $A \leqq \hat{A}$ and $B(n) \leqq \hat{B}(n)$. So

(30b) $$\int A \, dP_p \leqq eP\{j\}(i, i) \quad \text{and} \quad \int B(n) \, dP_p \leqq eP\{j\}(i, i).$$

Using (26, 30),

$$\left| \frac{pe^n P\delta_i}{pe^n P\delta_j} - eP\{j\}(j, i) \right| \leqq \frac{eP\{j\}(i, i)}{pe^n P\delta_j}.$$

But $pe^n P\delta_j \to \infty$ as $n \to \infty$, by (1.51).   ★

Remember $\mu(i) = eP\{s\}(s, i)$; and $\mu(g) = \Sigma \, g(i)\mu(i)$; and $g \times \mu$ assigns mass $g(i)\mu(i)$ to i,

(31) Theorem. *Let f be a function on I, with $\mu(|f|) < \infty$ and $\mu(f) \neq 0$. Then*

$$\lim_{n \to \infty} (\delta_i e^n Pf)/(\delta_j e^n Pf) = 1.$$

PROOF.   *The case* $f \geqq 0$. By   (16),

$$\delta_i e^n Pf = (f \times \mu) e^n (_\mu P)\delta_i / \mu(i) < \infty.$$

Use (25) with $(f \times \mu)/\mu(f)$ for p and $_\mu P$ for P. This is legitimate because

(10) makes I a recurrent class for $_\mu P$. You learn

$$(\delta_i e^n Pf)/(\delta_j e^n Pf) \to \mu(j) e_\mu P\{j\}(j, i)/\mu(i).$$

But, I say,

$$e_\mu P\{j\}(j, i) = \mu(i)/\mu(j).$$

For $\mu(\cdot)/\mu(j)$ is 1 at j by inspection, and is $_\mu P$-invariant by computation. So (1) works on $_\mu P$.

The general case. Let f^+ and f^- be the positive and negative parts of f, so $f = f^+ - f^-$. Let

$$a_n = \delta_i e^n Pf^+ \quad \text{and} \quad b_n = \delta_i e^n Pf^-$$

$$c_n = \delta_j e^n Pf^+ \quad \text{and} \quad d_n = \delta_j e^n Pf^-.$$

Then

$$\frac{\delta_i e^n Pf}{\delta_j e^n Pf} = \frac{a_n - b_n}{c_n - d_n} = \frac{\dfrac{a_n}{c_n} - \dfrac{b_n}{d_n}\dfrac{d_n}{c_n}}{1 - \dfrac{d_n}{c_n}}.$$

Use the special case to get $a_n/c_n \to 1$ and $b_n/d_n \to 1$. Use (15) to get $d_n/c_n \to \mu(f^-)/\mu(f^+) \neq 1$.   ★

Let f and g be functions on I, with $\mu(|f|) < \infty$, and $\mu(|g|) < \infty$, and $\mu(g) \neq 0$. Let p and q be probabilities on I. Combining (13) and (25) gives

(32) $\lim_{n \to \infty} (pe^n P\delta_i)/(qe^n P\delta_j) = \mu(i)/\mu(j).$

Combining (15) and (31) gives

(33) $\lim_{n \to \infty} (\delta_i e^n Pf)/(\delta_j e^n Pg) = \mu(f)/\mu(g).$

Of course, (33) can be obtained from (32) by reversal. It is tempting to combine (32) and (33). However, according to (Krengel 1966), there are probabilities p and q bounded above setwise by a multiple of μ, and a set $A \subset I$ with $\mu(A) < \infty$, such that $(pe^n P1_A)/(qe^n P1_A)$ fails to converge. The same paper contains this analog of (31):

(34) Theorem. *If g is nonnegative, positive somewhere, and bounded, then* $\lim_{n \to \infty} (\delta_i e^n Pg)/(\delta_j e^n Pg) = 1.$

PROOF.   Suppose $i \neq j$. Let τ be the least t if any with $\xi_t = j$ and $\tau = \infty$ if none. I claim

(35) $\delta_i e^n Pg \geqq \sum_{t=1}^{n} P_i\{\tau = t\} \cdot \delta_j e^{n-t} Pg.$

Indeed,

$$\delta_i e^n P g = \int \Sigma_{m=0}^n g(\xi_m) \, dP_i$$

$$\geq \int_{\{\tau \leq n\}} \Sigma_{m=0}^n g(\xi_m) \, dP_i$$

$$= \Sigma_{t=1}^n \int_{\{\tau=t\}} \Sigma_{m=0}^n g(\xi_m) \, dP_i$$

$$\geq \Sigma_{t=1}^n \int_{\{\tau=t\}} \Sigma_{m=t}^n g(\xi_m) \, dP_i$$

$$= \Sigma_{t=1}^n P_i\{\tau = t\} \cdot \int \Sigma_{m=0}^{n-t} g(\xi_m) \, dP_i$$

by Markov (1.15*). This proves (35).

Abbreviate $\|g\| = \sup_k g(k)$. Divide both sides of (35) by $\delta_j e^n P g$. For each t, the ratio $(\delta_j e^{n-t} P g)/(\delta_j e^n P g)$ is at most 1, and tends to 1 as n increases, because

$$0 \leq \delta_j e^n P g - \delta_j e^{n-t} P g \leq t \cdot \|g\|$$

and $\delta_j e^n P g \to \infty$ as $n \to \infty$. By dominated convergence,

$$\liminf_{n \to \infty} (\delta_i e^n P g)/(\delta_j e^n P g) \geq 1.$$

Interchange i and j. ★

5. RESTRICTING THE RANGE

The computations in this section were suggested by related work of Farrell; the basic idea goes back to Kakutani (1943). Let I_n be the set of functions ω from $\{0, \ldots, n\}$ to I. For $\omega \in I_n$ and $0 \leq m \leq n$, let $\xi_m(\omega) = \omega(m)$. Let Ω consist of I^∞, all the I_n, and the empty sequence $\varnothing$. So ξ_m is partially defined on Ω:

$$\text{domain } \xi_m = I^\infty \cup [\mathsf{U}_{n \geq m} I_n].$$

Give Ω the smallest σ-field $\mathscr{F}$ over which all ξ_m are measurable. Let $\mathscr{P}$ be a finite or infinite measure on $\mathscr{F}$, such that

$$\mathscr{P}\{\xi_n = i\} < \infty \quad \text{for all } n \text{ and } i.$$

Say $\mathscr{P}$ is *subinvariant* iff for all n, and all I-sequences $i_0, \ldots, i_n$,

(36) $\quad \mathscr{P}\{\xi_m = i_m \text{ for } 0 \leq m \leq n\} \geq \mathscr{P}\{\xi_{m+1} = i_m \text{ for } 0 \leq m \leq n\}.$

Say $\mathscr{P}$ is *invariant* iff equality always holds in (36). Let Q be a substochastic matrix on I. Say $\mathscr{P}$ is *Markov with stationary transitions* Q iff for all n, and all I-sequences $i_0, \ldots, i_n, i_{n+1}$,

$$(37) \quad \mathscr{P}\{\xi_m = i_m \text{ for } 0 \leqq m \leqq n + 1\}$$
$$= \mathscr{P}\{\xi_m = i_m \text{ for } 0 \leqq m \leqq n\} \cdot Q(i_n, i_{n+1}).$$

Let $(\mathscr{X}, \mathscr{A}, \mu)$ be an abstract measure space. Let $Y_0, Y_1, \ldots$ be partially defined functions from $\mathscr{X}$ to I. Call $Y = (Y_0, Y_1, \ldots)$ a *partially defined I-process* iff domain $Y_{n+1} \subseteq$ domain Y_n, and $\{Y_n = i\} \in \mathscr{A}$ has finite μ-measure for all n and i. Then Y is a measurable mapping from $\mathscr{X}$ to Ω:

$$[Y(x)](n) = Y_n(x) \quad \text{if } x \in \text{domain } Y_n.$$

The *distribution* of Y is μY^{-1}. Say Y is subinvariant, invariant, or Markov with stationary transitions Q iff its distribution has the corresponding properties.

Let J be a subset of I. Define Markov times τ_m on Ω as follows: $\tau_0 = 0$; and τ_{m+1} is the least $n > \tau_m$ if any with $\xi_n \in J$, while $\tau_{m+1} = \infty$ if none. Let $\Omega/J = \{\xi_0 \in J\}$. Let $\mathscr{F}/J$ be $\mathscr{F}$ relativized to Ω/J. Let $\mathscr{P}/J$ be the retract of $\mathscr{P}$ to $(\Omega/J, \mathscr{F}/J)$. Let ξ/J be this partially defined I-process on $(\Omega/J, \mathscr{F}/J)$:

$$(\xi/J)_n = \xi_{\tau_n} \quad \text{provided} \quad \tau_n < \infty.$$

(38) Theorem. *Suppose $\mathscr{P}$ is subinvariant. Then ξ/J is subinvariant.*

PROOF. Let $i_0 \in J$, and let $i_1, \ldots, i_N \in I$. Abbreviate τ for τ_1. I claim

$$(39) \quad \mathscr{P}\{\xi_n = i_n \text{ for } 0 \leqq n \leqq N\}$$
$$\geqq \mathscr{P}\{\xi_0 \in J \text{ and } \tau < \infty \text{ and } \xi_{\tau+n} = i_n \text{ for } 0 \leqq n \leqq N\}.$$

Indeed, let:

$A_m = \{\xi_0 \in J \text{ and } \xi_t \notin J \text{ for } 1 \leqq t \leqq m - 1 \text{ and } \xi_{m+n} = i_n \text{ for } 0 \leqq n \leqq N\};$

$B_m = \{\xi_t \notin J \text{ for } 1 \leqq t \leqq m - 1 \text{ and } \xi_{m+n} = i_n \text{ for } 0 \leqq n \leqq N\};$

$C_m = \{\xi_t \notin J \text{ for } 0 \leqq t \leqq m - 1 \text{ and } \xi_{m+n} = i_n \text{ for } 0 \leqq n \leqq N\}.$

By inspection, $C_m \subseteq B_m$. By subinvariance,

$$(40) \qquad\qquad \mathscr{P}(B_m) \leqq \mathscr{P}(C_{m-1}).$$

So, $\mathscr{P}(C_m) \leqq \mathscr{P}(B_m) \leqq \mathscr{P}(C_{m-1})$. Consequently,

$$(41) \qquad \mathscr{P}(C_m) \text{ nondecreases to a nonnegative limit as } m \text{ increases.}$$

The A_m are pairwise disjoint, their union is the set on the right of (39), and $A_m = B_m$ and $B_m \backslash C_m$. So the right side of (39) is

$$\Sigma_{m=1}^\infty \mathscr{P}(A_m) = \Sigma_{m=1}^\infty [\mathscr{P}(B_m) - \mathscr{P}(C_m)]$$
$$\leqq \Sigma_{m=1}^\infty [\mathscr{P}(C_{m-1}) - \mathscr{P}(C_m)] \quad \text{by (40)}$$
$$= \mathscr{P}(C_0) - \lim_m \mathscr{P}(C_m) \qquad \text{by (41) and telescoping}$$
$$\leqq \mathscr{P}(C_0).$$

But C_0 is the set on the left of (39); this completes (39).

Let ζ be the post-τ process; that is, ζ maps $\{\tau < \infty\}$ into Ω by

$$[\zeta(\omega)](n) = \xi_{\tau(\omega)+n}(\omega) \quad \text{for } \omega \in \text{domain } \xi_{\tau(\omega)+n}.$$

Let $j \in J$ and $A \in \mathscr{F}$ with $A \subset \{\xi_0 = j\}$. Then (39) makes

$$\mathscr{P}\{A\} \geqq \mathscr{P}\{\xi_0 \in J \text{ and } \tau < \infty \text{ and } \zeta \in A\}.$$

Let $j_0, \ldots, j_n \in J$. Specialize $j = j_0$ and

$$A = \{\xi_0 \in J \text{ and } (\xi/J)_m = i_m \text{ for } 0 \leqq m \leqq n\}.$$

Then

$$\{\xi_0 \in J \text{ and } \tau < \infty \text{ and } \zeta \in A\} = \{\xi_0 \in J \text{ and } (\xi/J)_{m+1} = i_m \text{ for } 0 \leqq m \leqq n\}.$$

This gives (36) with $\mathscr{P}/J$ for $\mathscr{P}$ and ξ/J for ξ.   ★

(42) Theorem. *If $\mathscr{P}$ is Markov with stationary substochastic transitions, then ξ/J is Markov with stationary substochastic transitions relative to $\mathscr{P}/J$.*

If $\mathscr{P}$ has transitions Q, write Q/J for the transitions of ξ/J. Even if Q is stochastic, Q/J need not be.

PROOF.   Let Q be the transitions of $\mathscr{P}$. Remember that the probability Q_i on Ω makes ξ Markov with stationary transitions Q and starting state i. Let

$$R(i, j) = Q_i\{\xi_{\tau_1} = j\}$$

for i and j in J. Let $i_0, \ldots, i_n, i_{n+1} \in J$. I say

$$\mathscr{P}\{\xi_0 \in J \text{ and } (\xi/J)_m = i_m \text{ for } 0 \leqq m \leqq n + 1\}$$
$$= \mathscr{P}\{\xi_0 \in J \text{ and } (\xi/J)_m = i_m \text{ for } 0 \leqq m \leqq n\} \cdot R(i_n, i_{n+1}).$$

Indeed,

$$\mathscr{P}\{A\} = \Sigma_j \, \mathscr{P}\{\xi_0 = j\} \cdot Q_j(A);$$

so it is enough to do it with Q_j in place of $\mathscr{P}$. After this reduction, use strong Markov (1.22) on the Markov time τ_n. If $m \leqq n$, then $(\xi/J)_m$ is measurable on the pre-τ_n sigma-field.   ★

(43) Lemma. *Let $I = \{1, 2\}$. Let Q be a stochastic matrix on I, for which I is a communicating class. There is one and only one subinvariant measure, up to scalar multiplication, and it is necessarily invariant.*

PROOF.   A locally finite, subinvariant measure has finite total mass, so is invariant. Let $u = [u(1), u(2)]$ be an invariant measure. Since $Q(2, 1) > 0$, the equation $uQ = u$ implies

$$u(2) = Q(1, 2) \, Q(2, 1)^{-1} \, u(1).$$ ★

With these preliminaries out of the way, it is possible to give an

ALTERNATIVE PROOF OF (1). Remember that P is a stochastic matrix on I, for which I is a recurrent class. Let u be a locally finite subinvariant measure on I. I have to show u is invariant and unique. Define a measure $\mathscr{P}$ on Ω by:

$$\mathscr{P} = \Sigma_{i \in I} u(i) P_i.$$

Check that $\mathscr{P}$ is subinvariant and Markov with stationary transitions P. Let $J \subset I$. Relative to $\mathscr{P}/J$, the process ξ/J is subinvariant and Markov with stationary transitions P/J by (38, 42); its starting measure is the retract of u to J. I claim the retract of u to J is subinvariant for P/J. Indeed, if $j \in J$, then

$$\begin{aligned}
u(j) &= \mathscr{P}\{\xi_0 = j\} \\
&= \mathscr{P}\{\xi_0 \in J \text{ and } (\xi/J)_0 = j\} \\
&\geqq \mathscr{P}\{\xi_0 \in J \text{ and } (\xi/J)_1 = j\} \\
&= \Sigma_{i \in J} \mathscr{P}\{\xi_0 \in J \text{ and } (\xi/J)_0 = i \text{ and } (\xi/J)_1 = j\} \\
&= \Sigma_{i \in J} \mathscr{P}\{\xi_0 \in J \text{ and } (\xi/J)_0 = i\} \cdot (P/J)(i,j) \\
&= \Sigma_{i \in J} \mathscr{P}\{\xi_0 = i\} \cdot (P/J)(i,j) \\
&= \Sigma_{i \in J} u(i)(P/J)(i,j).
\end{aligned}$$

Each $j \in J$ is recurrent, so ξ/J is defined almost everywhere, and

(44) P/J is stochastic.

Because I is a communicating class for P, therefore J is a communicating class for P/J. If $J = \{i, j\}$, the ratio $u(i)/u(j)$ can be computed from P/J using (43), so in principle from P. So there is only one locally finite subinvariant measure, up to scalar multiplication.

Why is it invariant? Suppose

$$u(j) > \Sigma_{i \in I} u(i) P(i,j);$$

so

(45) $\mathscr{P}\{\xi_0 = j\} > \mathscr{P}\{\xi_1 = j\}.$

Let $J = \{j\}$. I say

$$\begin{aligned}
u(j) &= \mathscr{P}\{\xi_0 = j\} \\
&= \mathscr{P}\{\xi_0 \in J \text{ and } (\xi/J)_0 = j\} \\
&> \mathscr{P}\{\xi_0 \in J \text{ and } (\xi/J)_1 = j\} \\
&= \mathscr{P}\{\xi_0 \in J \text{ and } (\xi/J)_0 = j \text{ and } (\xi/J)_1 = j\} \\
&= \mathscr{P}\{\xi_0 \in J \text{ and } (\xi/J)_0 = j\} \cdot (P/J)(j,j) \\
&= \mathscr{P}\{\xi_0 = j\} \cdot (P/J)(j,j) \\
&= u(j).
\end{aligned}$$

To get line 3, argue as for (39): put $N = 0$ and $i_0 = j$, so $B_1 = \{\xi_1 = j\}$ and $C_0 = \{\xi_0 = j\}$, and $\mathcal{P}(B_1) < \mathcal{P}(C_0)$ by (45). To get line 4, remember $(\xi/J)_0 \in J = \{j\}$. To get line 5, use (42). For the last line, (44) makes P/J stochastic on $J = \{j\}$. The contradiction $u(j) > u(j)$ forces u to be invariant.   ★

EXAMPLE.   Let $I = \{1, 2, 3\}$. Define the stochastic matrix P on I by

$$P(1, 2) = P(2, 3) = P(3, 1) = 1;$$

all other entries in P vanish. Let p be the invariant probability:

$$p(1) = p(2) = p(3) = 1/3.$$

Let $\tau_0, \tau_1, \ldots$ be the times n with $\xi_n = 1$ or 2. Then $\{\xi_{\tau_n} : n = 0, 1, \ldots\}$ is not P_p-invariant:

$$P_p\{\xi_{\tau_0} = 1\} = \tfrac{2}{3} \quad \text{and} \quad P_p\{\xi_{\tau_0} = 2\} = \tfrac{1}{3}.$$

A digression on ergodic theory

The technique used in proving (39) gives a quick proof of a theorem of Kac (1947). To state it, let $Y_0, Y_1, \ldots$ be an invariant process of 0's and 1's on a probability triple $(\Omega, \mathcal{F}, \mathcal{P})$. Let τ be the least positive n if any with $Y_n = 1$, and $\tau = \infty$ if none.

(46) Theorem. $\int_{\{Y_0=1\}} \tau \, d\mathcal{P} = \mathcal{P}\{Y_n = 1 \text{ for some } n \geq 0\}$.

PROOF.   The left side of (46) is $\Sigma_{n=1}^{\infty} \mathcal{P}\{Y_0 = 1 \text{ and } \tau \geq n\}$. But

$$\mathcal{P}\{Y_0 = 1 \text{ and } \tau \geq 1\} = \mathcal{P}\{Y_0 = 1\}.$$

For $n \geq 2$,

$$\begin{aligned}
\mathcal{P}\{Y_0 = 1 &\text{ and } \tau \geq n\} \\
&= \mathcal{P}\{Y_0 = 1 \text{ and } Y_1 = \cdots = Y_{n-1} = 0\} \\
&= \mathcal{P}\{Y_1 = \cdots = Y_{n-1} = 0\} - \mathcal{P}\{Y_0 = \cdots = Y_{n-1} = 0\} \\
&= \mathcal{P}\{Y_0 = \cdots = Y_{n-2} = 0\} - \mathcal{P}\{Y_0 = \cdots = Y_{n-1} = 0\}.
\end{aligned}$$

So, $\Sigma_{n=1}^{\infty} \mathcal{P}\{Y_0 = 1 \text{ and } \tau \geq n\}$ telescopes to

$$\mathcal{P}\{Y_0 = 1\} + \mathcal{P}\{Y_0 = 0\} - \lim_{n \to \infty} \mathcal{P}\{Y_0 = \cdots = Y_n = 0\}$$

which is the same as

$$1 - \mathcal{P}\{Y_n = 0 \text{ for all } n \geq 0\} = \mathcal{P}\{Y_n = 1 \text{ for some } n \geq 0\}. \qquad ★$$

6. PROOF OF KINGMAN–OREY

I learned the proof of (3) from Don Ornstein: here are the preliminaries. Let

$$f(p, a) = p^a(1 - p)^{1-a}$$

for p and a in $[0, 1]$, except for $p = a = 0$ and $p = a = 1$.

(47) Lemma. *Fix a with $0 < a < 1$. As p runs through the closed interval $[0, 1]$, the function $p \rightarrow f(p, a)$ has a strict maximum at $p = a$.*

PROOF.   Calculus.   ★

Let

$$m(p, a) = f(p, a)/f(a, a).$$

(48) Lemma. *Let $0 < p < a < 1$. Let Y_1, Y_2, ... be independent and identically distributed random variables on the probability triple $(\Omega, \mathscr{F}, \mathscr{P})$, each taking the value 1 with probability p, and the value 0 with probability $1 - p$.*

(a) $\mathscr{P}\{Y_1 + \cdots + Y_n \geq na\} \leq m(p, a)^n.$

(b) $m(p, a) < 1.$

PROOF.   *Claim (a).* Abbreviate $Y = Y_1$ and $S = Y_1 + \cdots + Y_n$. Write E for expectation relative to $\mathscr{P}$. Let $1 < x < \infty$. Then $S \geq na$ iff $x^S \geq x^{na}$. By Chebychev, this event has probability at most

$$[x^{-a}E(x^Y)]^n.$$

Compute

$$E(x^Y) = 1 - p + px.$$

By calculus, the minimum of $x \rightarrow x^{-a}(1 - p + px)$ occurs at

$$x = \frac{a(1 - p)}{p(1 - a)} > 1$$

and is $m(p, a)$. Use this x.
    Claim (b).   Use (47).   ★

(49) Lemma. *Let $0 \leq f \leq \infty$ be a subadditive function on $\{1, 2, \ldots\}$, which is finite on $\{A, A + 1, \ldots\}$ for some A. Let*

$$\alpha = \inf_{n \geq 1} f(n)/n.$$

Then

$$\lim_{n \to \infty} f(n)/n = \alpha.$$

Subadditive means $f(a + b) \leq f(a) + f(b)$.

PROOF.   Fix $\delta > 0$. I will argue

$$\limsup_{n \to \infty} f(n)/n \leq \alpha + \delta.$$

To begin, choose a positive integer a with

$$f(a)/a \leq \alpha + \delta.$$

Since $f(ka) \leq kf(a)$, I can choose $a \geq A$. Let $\beta = \max\{f(n): a \leq n < 2a\}$. Let $n \geq 2a$; then

$$n = ma + b = (m - 1)a + (a + b)$$

for some positive integer m and an integer b with $0 \leq b < a$. Of course, m and b depend on n. So

$$f(n) \leq (m - 1)f(a) + f(a + b);$$
$$\frac{f(n)}{n} \leq \frac{(m - 1)a}{n} \cdot \frac{f(a)}{a} + \frac{f(a + b)}{n}$$
$$\leq \frac{(m - 1)a}{n} \cdot (\alpha + \delta) + \frac{\beta}{n}.$$

But $(m - 1)a/n \to 1$ and $\beta/n \to 0$ as $n \to \infty$.   ★

(50) Lemma.   *If i has period 1, then $[P^n(i, i)]^{1/n}$ converges as $n \to \infty$.*

PROOF.   Use (49) on $f(n) = -\log P^n(i, i)$.   ★

(51) Lemma.   *Suppose I is a communicating class of period 1. There is an L with $0 \leq L \leq 1$, such that $\lim_{n \to \infty} [P^n(i, j)]^{1/n} = L$ for all i and j in I.*

PROOF.   Let $k > 0$. Then

(52) $$k^{1/n} \to 1 \quad \text{as } n \to \infty.$$

Let $f(n)^{1/n} \to L$ as $n \to \infty$, and let c be an integer. Then

(53) $$f(n + c)^{1/n} \to L \quad \text{as } n \to \infty.$$

As (50) implies, $\lim_{n \to \infty} [P^n(i, i)]^{1/n} = L(i)$ exists for all i. Fix $i \neq j$. Choose a and b with

$$P^a(j, i) > 0 \quad \text{and} \quad P^b(i, j) > 0.$$

Then

$$P^n(j, j) \geq P^a(j, i) \cdot P^{n-a-b}(i, i) \cdot P^b(i, j).$$

Take nth roots and use (52–53) to get $L(j) \geq L(i)$. Interchange i and j to get $L(i) = L(j) = L$ say. Abbreviate $g(n) = [P^n(i, j)]^{1/n}$. I say $g(n) \to L$. Indeed,

$$P^n(i, j) \cdot P^a(j, i) \leq P^{n+a}(i, i)$$
$$P^n(i, j) \geq P^b(i, j) \cdot P^{n-b}(j, j).$$

Take nth roots and use (52–53):

$$\limsup g(n) \leq L(i) = L$$
$$\liminf g(n) \geq L(j) = L.$$ ★

(54) Lemma. *If I is a recurrent class of period* 1, *then $L = 1$.*

Proof. If $L < 1$, then $\Sigma_n P^n(i, i) < \infty$. Use (1.51). ★

Note. In some transient classes, $L = 1$.

For the balance of this section, suppose I is a recurrent class of period 1. The next two lemmas prove the case $N = 1$ of theorem (3). Remember that δ_j is 1 at j and 0 elsewhere. Remember

$$pMf = \Sigma_{i,j}\, p(i)M(i,j)f(j).$$

(55) Lemma. *Suppose $P(i, i) > \varepsilon > 0$ for all $i \in I$. Let p be a probability on I. Then* $\lim_{n\to\infty} (pP^{n+1}\delta_j)/(pP^n\delta_j) = 1$.

Proof. Suppose $\varepsilon < 1$. Let P^* be this stochastic matrix on I:

$$P^*(i, j) = P(i, j)/(1 - \varepsilon) \qquad \text{for } i \neq j$$
$$= [P(i, i) - \varepsilon]/(1 - \varepsilon) \quad \text{for } i = j.$$

Then $P = (1 - \varepsilon)P^* + \varepsilon\Delta$, where Δ is the identity matrix. Make the usual convention that $P^{*0} = \Delta$. Let

$$s_m = \binom{n+1}{m}(1 - \varepsilon)^m \varepsilon^{n+1-m}\, pP^{*m}\delta_j$$

$$t_m = \binom{n}{m}(1 - \varepsilon)^m \varepsilon^{n-m}\, pP^{*m}\delta_j.$$

So

$$pP^{n+1}\delta_j = \Sigma_{m=0}^{n+1} s_m \quad \text{and} \quad pP^n\delta_j = \Sigma_{m=0}^n t_m.$$

Of course, s_m and t_m depend on n.

Fix a positive ε' much smaller than ε. Let

$$M = \left\{m : m = 0, \ldots, n + 1 \text{ and } 1 - \varepsilon - \varepsilon' < \frac{m}{n+1} < 1 - \varepsilon + \varepsilon'\right\}$$

$$a_n = \Sigma_m \{s_m : m \in M\}$$
$$b_n = \Sigma_m \{s_m : m = 0, \ldots, n + 1 \text{ but } m \notin M\}$$
$$c_n = \Sigma_m \{t_m : m \in M\}$$
$$d_n = \Sigma_m \{t_m : m = 0, \ldots, n \text{ but } m \notin M\}.$$

Check that I is still a communicating class of period 1 for P^*. Use (1.47) to create a positive integer n^* such that:

$$n > n^* \quad \text{implies} \quad pP^{*n}\delta_j > 0.$$

Keep n so large that

(56a) $(n + 1)(1 - \varepsilon - \varepsilon') > n^*$

(56b) $(n + 1)(1 - \varepsilon + \varepsilon') < n$

(56c) $n - (n + 1)(1 - \varepsilon - \varepsilon') > n(\varepsilon + \tfrac{1}{2}\varepsilon').$

In particular, $m \in M$ makes $m < n$ and $pP^{*m}\delta_j > 0$. So

(57) $pP^{n+1}\delta_j = a_n + b_n \quad \text{and} \quad pP^n\delta_j = c_n + d_n.$

By algebra

$$\frac{s_m}{t_m} = \frac{\varepsilon}{1 - \dfrac{m}{n+1}},$$

so

$$\frac{\varepsilon}{\varepsilon + \varepsilon'} < \frac{s_m}{t_m} < \frac{\varepsilon}{\varepsilon - \varepsilon'} \quad \text{for } m \in M.$$

By Cauchy's inequality,

(58) $$\frac{\varepsilon}{\varepsilon + \varepsilon'} < \frac{a_n}{c_n} < \frac{\varepsilon}{\varepsilon - \varepsilon'}.$$

Let $Y_1, Y_2, \ldots$ be independent and identically distributed random variables on $(\Omega, \mathscr{F}, \mathscr{P})$, taking the value 1 with probability $1 - \varepsilon$ and the value 0 with probability ε. Let $\bar{y} = 1 - y$. Because P^* is stochastic,

$$0 \leq pP^{*m}\delta_j \leq 1.$$

So b_n and d_n are nonnegative. Furthermore,

$$b_n \leq \Sigma_m \left\{ \binom{n+1}{m}(1 - \varepsilon)^m \varepsilon^{n+1-m} : 0 \leq m \leq n + 1 \text{ but } m \notin M \right\}$$

$$= \mathscr{P}\{Y_1 + \cdots + Y_{n+1} \geq (n + 1)(1 - \varepsilon + \varepsilon')\}$$
$$+ \mathscr{P}\{Y_1 + \cdots + Y_{n+1} \leq (n + 1)(1 - \varepsilon - \varepsilon')\}$$

$$= \mathscr{P}\{Y_1 + \cdots + Y_{n+1} \geq (n + 1)(1 - \varepsilon + \varepsilon')\}$$
$$+ \mathscr{P}\{\bar{Y}_1 + \cdots + \bar{Y}_{n+1} \geq (n + 1)(\varepsilon + \varepsilon')\}.$$

Similarly,

$$d_n \leq \mathscr{P}\{Y_1 + \cdots + Y_n \geq (n+1)(1 - \varepsilon + \varepsilon')\}$$
$$+ \mathscr{P}\{\overline{Y}_1 + \cdots + \overline{Y}_n \geq n - (n+1)(1 - \varepsilon - \varepsilon')\}$$
$$\leq \mathscr{P}\{Y_1 + \cdots + Y_n \geq n(1 - \varepsilon + \varepsilon')\}$$
$$+ \mathscr{P}\{\overline{Y}_1 + \cdots + \overline{Y}_n \geq n(\varepsilon + \tfrac{1}{2}\varepsilon')\}$$

by (56). So (48) produces an $r = r(\varepsilon, \varepsilon') < 1$ such that

(59) $b_n \leq r^n$ and $d_n \leq r^n$.

But (51, 54, 57) makes $\liminf (a_n + b_n)^{1/n} \geq 1$, so $r^n = o(a_n + b_n)$; and $b_n \leq r^n$ by (59), so $b_n = o(a_n + b_n)$, forcing $b_n = o(a_n)$. Similarly, $d_n = o(c_n)$. Therefore,

$$\liminf \frac{a_n + b_n}{c_n + d_n} = \liminf \frac{a_n}{c_n} \quad \text{and} \quad \limsup \frac{a_n + b_n}{c_n + d_n} = \limsup \frac{a_n}{c_n}.$$

By (57, 58), the $\liminf$ and $\limsup$ of $(pP^{n+1}\delta_j)/(pP^n\delta_j)$ are both trapped between $\varepsilon/(\varepsilon + \varepsilon')$ and $\varepsilon/(\varepsilon - \varepsilon')$; these bounds are close to 1.   ★

(60) Lemma. *Suppose $P(i, i) > \varepsilon > 0$ for all $i \in I$. For any probability p on I,*

$$\lim_{n \to \infty} (pP^n\delta_j)/(pP^n\delta_i) = eP\{i\}(i, j).$$

PROOF.   For any subsequence, use the diagonal argument (10.56) to find a sub-subsequence n' such that $(pP^{n'}\delta_j)/(pP^{n'}\delta_i)$ converges as $n \to \infty$, say to $\mu(j)$, for all $j \in I$. Here $0 \leq \mu(j) \leq \infty$, and $\mu(i) = 1$. My problem is to show $\mu(j) = eP\{i\}(i, j)$. As (1.47) shows, $pP^n\delta_i > 0$ for all large n. Make the convention $\infty \cdot 0 = 0$, as required by (1.80), and estimate as follows:

$$\Sigma_j \mu(j)P(j, k) = \Sigma_j \lim_{n \to \infty} \left[\frac{pP^{n'}\delta_j}{pP^{n'}\delta_i}\right] \cdot P(j, k) \quad \text{(definition)}$$

$$= \Sigma_j \lim_{n \to \infty} \left[\frac{pP^{n'}\delta_j \cdot P(j, k)}{pP^{n'}\delta_i}\right] \quad \text{(convention)}$$

$$\leq \lim_{n \to \infty} \Sigma_j \left[\frac{pP^{n'}\delta_j \cdot P(j, k)}{pP^{n'}\delta_i}\right] \quad \text{(Fatou)}$$

$$= \lim_{n \to \infty} \left[\frac{pP^{n'+1}\delta_k}{pP^{n'}\delta_i}\right] \quad \text{(algebra)}$$

$$= \lim_{n \to \infty} \left[\frac{pP^{n'}\delta_k}{pP^{n'}\delta_i}\right] \quad \text{(55)}$$

$$= \mu(k) \quad \text{(definition)}.$$

So, μ is subinvariant. Use (1).   ★

I will now work on the case $N > 1$ in theorem (3). Suppose N is an integer greater than 1. Suppose I is a recurrent class of period 1 relative to P, and

$$P^N(i, i) > \varepsilon > 0 \quad \text{for all } i \in I.$$

Let p be any probability on I.

(61) Lemma. $\lim_{n\to\infty} (pP^{n+N}\delta_j)/(pP^n\delta_j) = 1$.

PROOF. Check that I is a recurrent class of period 1 for P^N. Let r be one of $0, \ldots, N-1$. Use (55) with P^N for P and pP^r for p, to see that

$$\lim_{m\to\infty} (pP^{N(m+1)+r}\delta_j)/(pP^{Nm+r}\delta_j) = 1. \qquad \bigstar$$

(62) Lemma. $\lim_{n\to\infty} (pP^n\delta_j)/(pP^n\delta_i) = eP\{i\}(i,j)$.

PROOF. As in (60), choose a subsequence n' such that $(pP^{n'}\delta_j)/(pP^{n'}\delta_i)$ converges, say to $\mu(j)$, for all j. Using (61), argue that μ is subinvariant with respect to P^N. Now $eP\{i\}(i, \cdot)$ is invariant with respect to P, so with respect to P^N. The uniqueness part of (1), applied to P^N, identifies $\mu(j)$ with $eP\{i\}(i,j)$. $\qquad \bigstar$

(63) Lemma. $\lim_{n\to\infty} P^n(k,j)/P^n(j,j) = 1$.

PROOF. Reverse (62), as in (15). $\qquad \bigstar$

(64) Lemma. $\lim_{n\to\infty} P^{n+1}(j,j)/P^n(j,j) = 1$.

PROOF. Let r be one of $1, \ldots, N-1$. By algebra,

$$P^{n+r}(j,j)/P^n(j,j) = \Sigma_k P^r(j,k) \cdot P^n(k,j)/P^n(j,j).$$

By (63) and Fatou,

(65) $$\liminf_{n\to\infty} P^{n+r}(j,j)/P^n(j,j) \geqq 1.$$

Use (61) to replace n in the denominator by $n + N$ without changing the lim inf:

$$\liminf_{m\to\infty} P^{n+r}(j,j)/P^{n+N}(j,j) \geqq 1.$$
Put $m = n + r$:
$$\liminf_{m\to\infty} P^m(j,j)/P^{m+N-r}(j,j) \geqq 1.$$
Invert:

(66) $$\limsup_{m\to\infty} P^{m+N-r}(j,j)/P^m(j,j) \leqq 1.$$

for $r = 1, \ldots, N-1$. Put $r = 1$ in (65) and $r = N-1$ in (66). $\qquad \bigstar$

Theorem (3) follows by algebra from (62–64).

7. AN EXAMPLE OF DYSON

In this section, I will follow Dyson's lead in constructing a countable set I, a state $s \in I$, and a stochastic matrix P on I, such that:

(67a) I is a recurrent class of period 1 for P;

(67b) $P(i, i) > 0$ for all i in I;

(67c) $\liminf_{n \to \infty} P^{n+1}(s, s)/P^n(s, s) \leq 2P(s, s)$;

(67d) $\limsup_{n \to \infty} P^{n+1}(s, s)/P^n(s, s) = \infty$.

NOTE. $P^{n+1}(s, s)/P^n(s, s) \geq P(s, s)$ for any P.

The construction for (67) has three parameters. The first parameter p is a positive function on the positive integers, with $\sum_{n=1}^{\infty} p(n) = 1$. There are no other strings on p: you choose it now. The second parameter f is a function from the positive integers to the positive integers, with $f(1) = 1$. For $n = 2$, $3, \ldots$, I will require

(68) $f(n) \geq f(n - 1) + 2.$

I get to choose f inductively; it will increase quickly. The third parameter q is a function from $\{2, 3, \ldots\}$ to $(0, 1)$. You can pick it, subject to

(69) $q(N)^{f(N)} > 1 - \dfrac{p(n)}{n}$ for $n = 1, \ldots, N$

The state space I consists of all pairs $[n, m]$ of positive integers with $1 \leq m \leq f(n)$. The special state s is $[1, 1]$. Here comes P. Let

$$P(s, j) = 0 \qquad \text{unless } j = [n, 1] \text{ for some } n;$$

$$P(s, [n, 1]) = p(n) \qquad \text{for } n = 1, 2, \ldots.$$

For $n > 1$ and $1 \leq m < f(n)$, let:

$$P([n, m], j) = 0 \qquad \text{unless } j = [n, m] \text{ or } [n, m + 1];$$

$$P([n, m], [n, m]) = 1 - q(n);$$

$$P([n, m], [n, m + 1]) = q(n).$$

Let $P([n, f(n)], j) = 0 \qquad \text{unless } j = [n, f(n)] \text{ or } s;$

$$P([n, f(n)], [n, f(n)]) = 1 - q(n);$$

$$P([n, f(n)], s) = q(n).$$

You check (67a–b), using (1.16). Abbreviate $\theta(n) = P^n(s, s)$. For my f and $n \geqq 2$, it will develop that

(70a) $$\theta[f(n)] < 2p(n)/n$$

(70b) $$\left(1 - \frac{p(n)}{n}\right)p(n) < \theta[f(n) + 1] < \left(1 + \frac{2}{n}\right)p(n)$$

(70c) $$2\left(1 - \frac{p(n)}{n}\right)p(n)P(s, s) < \theta[f(n) + 2] < 2p(n)\left[P(s, s) + \frac{1}{n}\right].$$

You can get (67c–d) from (70).

Remember that I^∞ is the space of I-sequences. And $\xi_t(\omega) = \omega(t)$ for $\omega \in I^\infty$ and $t = 0, 1, \ldots$. Let $\alpha(t, \omega)$ be the first component of $\omega(t)$, and let $\beta(t, \omega)$ be the second component, so

$$\xi_t(\omega) = [\alpha(t, \omega), \beta(t, \omega)].$$

Let $\tau_n(\omega)$ be the least t if any with $\alpha(t, \omega) \geqq n$, and let $\tau_n(\omega) = \infty$ if none. The probability P_s on I^∞ makes ξ Markov with stationary transitions P and starting state s. For $n = 2, 3, \ldots$, I will require

(71) $$P_s\{\tau_n \leqq f(n)\} > 1 - \frac{p(n)}{n}.$$

Let $\partial \notin I$. Let $\zeta_t^n = \xi_t$ for $t < \tau_n$, and $\zeta_t^n = \partial$ for $t \geqq \tau_n$. Let $N \geqq 2$. Suppose $f(n)$ and $q(n)$ have been chosen for $n < N$, so that (68, 69, 71) hold. I know this doesn't determine P, but the P_s-distribution of ζ^N is already fixed: ζ^N is Markov with stationary transitions and a finite state space; every state leads to the absorbing state ∂. I can use (1.79) to choose $f(N)$ so large that (68, 71) hold for N. Now you choose $q(N)$ so (69) holds for N. The construction is finished.

I must now argue (70). Fix $n \geqq 2$. Suppose $\tau_n(\omega) < \infty$. Abbreviate

$$\alpha(\omega) = \alpha[\tau_n(\omega), \omega].$$

Now $\alpha(\omega) \geqq n$. By (68),

(72a) $\tau_n(\omega) < \infty$ implies $f[\alpha(\omega)] \geqq f(n)$

(72b) $\tau_n(\omega) < \infty$ and $\alpha(\omega) > n$ imply $f[\alpha(\omega)] \geqq f(n) + 2$.

Let G be the set of ω such that

(73a) $\omega(0) = s$

(73b) $\tau_n(\omega) \leqq f(n)$

(73c) $\omega(t) = [\alpha(\omega), t - \tau_n(\omega) + 1]$ for
         $\tau_n(\omega) \leqq t \leqq \tau_n(\omega) + f[\alpha(\omega)] - 1$

(73d) $\omega(t) = s$ for $t = \tau_n(\omega) + f[\alpha(\omega)]$.

Relations (73a–b) force

(73e) $1 \leqq \tau_n(\omega) \leqq f(n).$

Check that τ_n is Markov. Remember $\alpha \geqq n$. By strong Markov (1.21) and (69),

$$P_s\{G \mid \mathscr{F}_{\tau_n}\} = q[\alpha(\omega)]^{f[\alpha(\omega)]} > 1 - \frac{p(n)}{n}$$

almost surely on $\{\tau_n \leqq f(n)\}$. By (71),

(74) $P_s\{G\} > \left[1 - \frac{p(n)}{n}\right]^2 > 1 - \frac{2p(n)}{n}.$

ARGUMENT FOR (70a).   Suppose $\omega \in G$. By (72a) and (73e),

$$\tau_n(\omega) + f[\alpha(\omega)] - 1 \geqq f(n).$$

And (73c) prevents $\omega[f(n)] = s$. That is,

$$\{\xi_{f(n)} = s\} \subset I^\infty \backslash G.$$

Use (74):

$$\theta[f(n)] \leqq 1 - P_s(G) < 2p(n)/n.$$

ARGUMENT FOR (70b).   Plainly,

$$\{\xi_m = (n, m) \text{ for } m = 1, \ldots, f(n) \text{ and } \xi_{f(n)+1} = s\} \subset \{\xi_{f(n)+1} = s\}.$$

So (69) does the first inequality in (70b). For the second, suppose $\omega \in G$. Suppose $\tau_n(\omega) = 1$ and $\xi_1(\omega) \neq [n, 1]$. Then (73c) makes $\alpha(\omega) > n$, and (72b) makes

(75) $\tau_n(\omega) + f[\alpha(\omega)] - 1 \geqq f(n) + 1;$

now (73c) prevents $\omega[f(n) + 1] = s$. Suppose $\tau_n(\omega) \geqq 2$. Then (72a) establishes (75), and (73c) still prevents $\omega[f(n) + 1] = s$. That is,

$$G \cap \{\xi_{f(n)+1} = s\} \subset \{\xi_1 = [n, 1]\}.$$

Use (74):

$$\theta[f(n) + 1] \leqq P_s\{\xi_1 = [n, 1]\} + P_s\{I^\infty \backslash G\}$$

$$< p(n)\left(1 + \frac{2}{n}\right).$$

ARGUMENT FOR (70c).   Let A be the event that

$$\xi_m = [n, m] \text{ for } m = 1, \ldots, f(n) \text{ and } \xi_{f(n)+1} = \xi_{f(n)+2} = s$$

and let B be the event that

$\xi_1 = s$ and $\xi_m = [n, m - 1]$ for $m = 2, \ldots, f(n) + 1$ and $\xi_{f(n)+2} = s$.

Use (69):

$$P_s(A) = P_s(B) = q(n)^{f(n)} p(n) P(s, s) > \left(1 - \frac{p(n)}{n}\right) p(n) P(s, s).$$

But A and B are disjoint, and $A \cup B \subset \{\xi_{f(n)+2} = s\}$. This proves the first inequality in (70c). For the second, suppose $\omega \in G$. Suppose $\tau_n(\omega) = 1$ and $\xi_1(\omega) \neq [n, 1]$, or $\tau_n(\omega) = 2$ and $\xi_2(\omega) \neq [n, 1]$. Then (73c) makes $\alpha(\omega) > n$, and (72b) makes $f[\alpha(\omega)] \geq f(n) + 2$. So

(76) $\tau_n(\omega) + f[\alpha(\omega)] - 1 \geq f(n) + 2.$

Now (73c) prevents $\omega[f(n) + 2] = s$. Suppose $\tau_n(\omega) \geq 3$. Then (72a) establishes (76), and (73c) still prevents $\omega[f(n) + 2] = s$. If $\omega \in G$ and $\xi_1(\omega) = [n, 1]$, then $\omega \in A$; this uses (73d). So

$$G \cap \{\xi_{f(n)+2} = s\} \subset A \cup \{\xi_2 = [n, 1]\}.$$

And

$$\theta[f(n) + 2] \leq P_s(A) + P_s\{\xi_2 = [n, 1]\} + 1 - P_s(G).$$

Check

(77) $P_s\{\xi_1 \neq s \text{ and } \xi_2 = [n, 1]\} = 0.$

Use (74) and (77):

$$P_s(A) < p(n) P(s, s)$$
$$P_s\{\xi_2 = [n, 1]\} = P_s\{\xi_1 = s \text{ and } \xi_2 = [n, 1]\} = p(n) P(s, s)$$
$$1 - P_s(G) < 2p(n)/n.$$

Add to get the second inequality in (70c).   ★

I think you can modify P to keep (67a, c, d), but strengthen (67b) to $P(i, j) > 0$ for all i, j.

8. ALMOST EVERYWHERE RATIO LIMIT THEOREMS

Let I be a recurrent class relative to the stochastic matrix P. Let μ be an invariant measure. I remind you that

$$\mu(h) = \Sigma_{i \in I} \mu(i) h(i).$$

The next result is due to Harris and Robbins (1953).

(78) Theorem. *Suppose f and g are functions on I, with $\mu(|f|) < \infty$ and $\mu(|g|) < \infty$. Suppose at least one of $\mu(f)$ and $\mu(g)$ is nonzero. Fix $s \in I$. With P_s-probability 1,*

$$\lim_{n \to \infty} \frac{\sum_{m=0}^{n} f(\xi_m)}{\sum_{m=0}^{n} g(\xi_m)} = \frac{\mu(f)}{\mu(g)}.$$

PROOF. Suppose $\mu(s) = 1$ and $\mu(g) \neq 0$. Using (1.73), confine ω to the set where $\xi_0 = s$ and $\xi_n = s$ for infinitely many n, which has P_s-probability 1. Let $0 = \tau_1 < \tau_2 < \cdots$ be the times n with $\xi_n = s$. Let $l(n)$ be the largest m with $\tau_m \leqq n$, so $l(n) \to \infty$ with n. Let h be a function on I with $\mu(|h|) < \infty$: the interesting h's are f, $|f|$, g, $|g|$. For $m = 1, 2, \ldots$, let

$$Y_m(h) = \Sigma_n \{h(\xi_n) : \tau_m \leqq n < \tau_{m+1}\}$$
$$V_m(h) = Y_1(h) + \cdots + Y_m(h).$$

I claim that with P_s-probability 1, as $n \to \infty$:

(79)
$$\frac{V_{l(n)}(f)}{V_{l(n)}(g)} \to \frac{\mu(f)}{\mu(g)},$$

and

(80)
$$\frac{Y_{l(n)}(h)}{V_{l(n)}(g)} \to 0.$$

Introduce E_s for expectation relative to P_s. Let ζ_j be the number of j's in the first s-block: namely, the number of $n < \tau_2$ with $\xi_n = j$. As (1) implies, $\mu(j) = E_s\{\zeta_j\}$. Clearly,

$$Y_1(h) = \Sigma_{j \in I} h(j)\zeta_j.$$

So

$$E_s\{Y_1(h)\} = \Sigma_{j \in I} h(j)\mu(j) = \mu(h).$$

By blocks (1.31), the variables $Y_m(h)$ are independent and identically distributed for $m = 1, 2, \ldots$. The strong law now implies that with P_s-probability 1,

(81)
$$\lim_{m \to \infty} V_m(h)/m = \mu(h).$$

Put $h = f$ or g and divide: with P_s-probability 1,

$$\lim_{m \to \infty} V_m(f)/V_m(g) = \mu(f)/\mu(g).$$

Put $m = l(n)$ to get (79). Next,

$$\frac{Y_m(h)}{m} = \frac{V_m(h)}{m} - \frac{m-1}{m} \cdot \frac{V_{m-1}(h)}{m-1}.$$

So (81) implies that with P_s-probability 1,

$$\lim_{m \to \infty} Y_m(h)/m = 0.$$

Put g for h in (81): with P_s-probability 1,

$$\lim_{m \to \infty} Y_m(h)/V_m(g) = 0.$$

Put $m = l(n)$ to get (80).

Abbreviate

$$S_n(h) = \Sigma_{m=0}^{n} h(\xi_m).$$

Check

(82) $|S_n(h) - V_{l(n)}(h)| \leqq Y_{l(n)}(|h|).$

Clearly,

$$\frac{S_n(f)}{S_n(g)} = \frac{S_n(f) - V_{l(n)}(f) + V_{l(n)}(f)}{S_n(g) - V_{l(n)}(g) + V_{l(n)}(g)}.$$

But $S_n(h) - V_{l(n)}(h) = o[V_{l(n)}(g)]$ almost surely for $h = f$ or g by (80, 82). Using (79),

$$\lim_{n \to \infty} \frac{S_n(f)}{S_n(g)} = \lim_{n \to \infty} \frac{V_{l(n)}(f)}{V_{l(n)}(g)} = \frac{\mu(f)}{\mu(g)}$$

with P_s-probability 1.   ★

(83) Corollary. *Let $V_n(i)$ be the number of $m \leqq n$ with $\xi_m = i$. Then with P_s-probability* 1,

$$\lim_{n \to \infty} V_n(j)/V_n(k) = \mu(j)/\mu(k).$$

PROOF. Put $f = \delta_j$ and $g = \delta_k$ in (78).   ★

(84) Corollary. *Suppose I is positive recurrent, with invariant probability π. Suppose $\pi(|f|) < \infty$. Then with P_s-probability* 1,

$$\lim_{n \to \infty} \frac{1}{n} \Sigma_{m=0}^{n} f(\xi_m) = \pi(f).$$

PROOF. Put $g \equiv 1$ in (78).   ★

9. THE SUM OF A FUNCTION OVER DIFFERENT j-BLOCKS

The distribution of $Y_n(h)$, as defined in Section 8, depends not only on h, but also on the reference state s. Certain facts about this dependence will be useful in Chapter 3, and are conveniently established as (92). Result (92a) is in (Chung, 1960, p. 79). Here are some standard preliminaries.

(85) Lemma. *Let u_i be real numbers, and $1 \leq p < \infty$. Then*

$$|\Sigma_{i=1}^n u_i|^p \leq n^{p-1}\Sigma_{i=1}^n |u_i|^p.$$

PROOF.   Use Jensen's inequality (10.9) on the convex function $x \to |x|^p$ to get

$$\left|\frac{1}{n}\Sigma_{i=1}^n u_i\right|^p \leq \frac{1}{n}\Sigma_{i=1}^n |u_i|^p. \qquad \bigstar$$

(86) Lemma. *Let u_i be real numbers, and $0 < p < 1$. Then*

$$|\Sigma_{i=1}^n u_i|^p \leq \Sigma_{i=1}^n |u_i|^p.$$

PROOF.   The general case follows from the case $n = 2$, by induction. The inequality $|u + v| \leq |u| + |v|$ further reduces the problem to nonnegative u and v. Divide by u to make the final reduction: I only need

$$(1 + x)^p \leq 1 + x^p \quad \text{for } x \geq 0.$$

Both sides agree at $x = 0$. The derivative of the left side is strictly less than that of the right side at positive x. $\bigstar$

From now on, $0 < p < \infty$. Let π be a probability on $(-\infty, \infty)$. Say $\pi \in L^p$ iff $\int_{-\infty}^{\infty} |x|^p \pi(dx) < \infty$.

(87) Lemma. *Let π_1 and π_2 be probabilities on $(-\infty, \infty)$. Let $0 \leq \theta \leq 1$ and let $\pi = \theta\pi_1 + (1 - \theta)\pi_2$.*

   (a) *If π_1 and π_2 are in L^p, then $\pi \in L^p$.*

   (b) *If $\theta > 0$ and $\pi \in L^p$, then $\pi_1 \in L^p$.*

PROOF.   Easy. $\bigstar$

Let $(\Omega, \mathscr{F}, \mathscr{P})$ be a probability triple. Write E for $\mathscr{P}$-expectation. Say a random variable $X \in L^p$ iff $E\{|X|^p\} < \infty$. For (88), let U and V be random variables on $(\Omega, \mathscr{F}, \mathscr{P})$.

(88) Lemma.   (a) *If $U \in L^p$ and $V \in L^p$, then $U + V \in L^p$.*

   (b) *If U and V are independent, and $U + V \in L^p$, then $U \in L^p$ and $V \in L^p$.*

PROOF   *Claim (a)* is clear from (85) or (86).
   Claim (b).   By Fubini, $E(|u + V|^p) < \infty$ for $\mathscr{P}U^{-1}$-almost all u. Choose one good u. Then

$$V = (u + V) + (-u)$$

is in L^p by (a). $\bigstar$

For (89), let M, W_1, W_2, ... be independent random variables on $(\Omega, \mathscr{F}, \mathscr{P})$. Suppose the W's are identically distributed. Suppose M is non-negative integer-valued, and $\mathscr{P}\{M = 1\} > 0$. Make the convention $\Sigma_1^0 = 0$.

(89) Lemma.   *Let* $p \geqq 1$ *and* $M \in L^p$, *or* $0 < p < 1$ *and* $M \in L^1$.

  (a) $\Sigma_{m=1}^M W_m \in L^p$ *implies* $W_1 \in L^p$.

  (b) $W_1 \in L^p$ *implies* $\Sigma_{m=1}^M W_m \in L^p$.

PROOF.   *Claim (a).*

$$\mathscr{P}\{M = 1\} \cdot E\{|W_1|^p\} = \int_{\{M=1\}} |W_1|^p \, d\mathscr{P}$$

$$= \int_{\{M=1\}} |\Sigma_{m=1}^M W_m|^p \, d\mathscr{P}$$

$$\leqq \int |\Sigma_{m=1}^M W_m|^p \, d\mathscr{P}$$

$$< \infty..$$

Claim (b).   Suppose $p \geqq 1$. Check this computation, using (85).

$$E\{|\Sigma_{n=1}^M W_n|^p\} = \Sigma_{m=1}^\infty E\{|\Sigma_{n=1}^m W_n|^p\} \cdot \mathscr{P}\{M = m\}$$

$$\leqq \Sigma_{m=1}^\infty m^{p-1} \Sigma_{n=1}^m E\{|W_n|^p\} \cdot \mathscr{P}\{M = m\}$$

$$= \{E\{|W_1|^p\} \cdot \Sigma_{m=1}^\infty m^p \, \mathscr{P}\{M = m\}$$

$$= E\{|W_1|^p\} \cdot E\{M^p\}$$

$$< \infty.$$

The argument for $0 < p < 1$ is similar, using (86).   ★

(90) Lemma.   *Suppose U and V are independent random variables, and suppose $\mathscr{P}\{U + V = 0\} = 1$. Then $\mathscr{P}\{U = -K\} = \mathscr{P}\{V = K\} = 1$ for some constant K.*

PROOF.   Fubini will produce a constant v with

$$\mathscr{P}\{U + v = 0\} = 1.$$ ★

For (91), suppose M, W_1, W_2, ... are independent random variables on $(\Omega, \mathscr{F}, \mathscr{P})$, the W's being identically distributed. Suppose M is nonnegative integer valued, and $\mathscr{P}\{M = 1\} > 0$. As before, $\Sigma_1^0 = 0$.

(91) Lemma.   *Suppose K is a constant and*

$$\mathscr{P}\{\Sigma_{m=1}^M W_m = K\} = 1.$$

(a) $\mathscr{P}\{W_1 = K\} = 1$.

(b) *If $K \neq 0$, then $\mathscr{P}\{M = 1\} = 1$.*

PROOF.   *Claim (a).*

$$\mathscr{P}\{W_1 = K\} = \mathscr{P}\{W_1 = K \mid M = 1\}$$
$$= \mathscr{P}\{\Sigma_{m=1}^{M} W_m = K \mid M = 1\}$$
$$= 1.$$

Claim (b).   Clearly $\Sigma_{m=1}^{M} W_m = KM$, so $\mathscr{P}\{KM = K\} = 1$.   ★

Return to Markov chains; I is a recurrent class for the stochastic matrix P. Confine ω to the set where $\xi_n = j$ for infinitely many n, for all $j \in I$. This set has P_j-probability 1 by (1.73). Let $0 \leq \tau_1(j) < \tau_2(j) < \cdots$ be the n with $\xi_n = j$. The nth j-block is the sample sequence on $[\tau_n(j), \tau_{n+1}(j))$, shifted so as to start at time 0. Here, $n = 1, 2, \ldots$ Fix a function h on I, and let

$$Y_n(j) = \Sigma \{h(\xi_m) : \tau_n(j) \leq m < \tau_{n+1}(j)\}.$$

For any particular j, the variables $\{Y_n(j) : n = 1, 2, \ldots\}$ are independent and identically distributed relative to P_i. The distribution depends on j, but not on i.

(92) Theorem.   (a) *If $Y_n(j)$ is in L^p relative to P_i for some i and j, then $Y_n(j)$ is in L^p relative to P_i for all i and j.*

(b) *If $P_i\{Y_n(j) = 0\} = 1$ for some i and j, then $P_i\{Y_n(j) = 0\} = 1$ for all i and j.*

PROOF.   There is no interest in varying i, so fix it. Fix $j \neq k$. I will interchange j and k. Look at Figure 4. Let $N(j, k)$ be the least n such that the nth j-block contains a k. Abbreviate

$$V(j, k) = Y_1(j) + \cdots + Y_{N(j,k)}(j).$$

I claim that with respect to P_i,

(93) $V(j, k)$ is distributed like $V(k, j)$.

Indeed, let $A(j, k)$ be the sample sequence from the first j until just before the first k after the first j. Let $B(k, j)$ be the sample sequence from this k until just before the next j. Then $A(j, k)B(k, j)$ is the sample sequence from the first j until just before the first j after the first k after the first j. So $V(j, k)$ is the sum of $h(\eta)$ as η moves through the state sequence $A(j, k)B(k, j)$. Equally, $V(k, j)$ is the sum of $h(\eta)$ as η moves through the state sequence

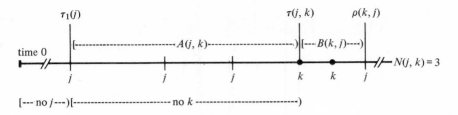

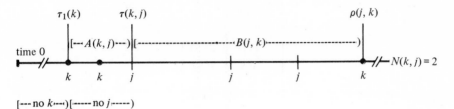

Figure 4.

$A(k, j)B(j, k)$. As strong Markov (1.22) implies, $A(j, k)$ is independent of $B(k, j)$, and is distributed like $B(j, k)$. So the rearranged sample sequence $B(k, j)A(j, k)$ is distributed like $A(k, j)B(j, k)$. Addition being commutative, (93) is proved.

Let $c(j, k)$ be the P_i-probability that the first j-block contains a k, so

$$P_i\{N(j, k) = n\} = c(j, k)[1 - c(j, k)]^{n-1} \quad \text{for } n = 1, 2, \ldots.$$

NOTE. $0 < c(j, k) \leqq 1$.

On some convenient probability triple $(\Omega, \mathcal{F}, \mathcal{P})$, construct independent random variables $M(j, k)$, $C(j, k)$, $D_1(j, k)$, $D_2(j, k)$, $\ldots$ with the following three properties:

(a) $\mathcal{P}\{M(j, k) = m\} = c(j, k)[1 - c(j, k)]^m$ for $m = 0, 1, \ldots$;

(b) the $\mathcal{P}$-distribution of $C(j, k)$ is the conditional P_i-distribution of $Y_1(j)$ given the first j-block contains a k;

(c) the $\mathcal{P}$-distribution of $D_n(j, k)$ is the conditional P_i-distribution of $Y_1(j)$ given that the first j-block does not contain a k, and does not depend on n.

In particular,

(94)   the P_i-distribution of $Y_1(j)$ is $c(j, k)$ times the $\mathcal{P}$-distribution of $C(j, k)$, plus $[1 - c(j, k)]$ times the $\mathcal{P}$-distribution of $D_n(j, k)$.

Abbreviate

$$U(j, k) = \Sigma_{m=1}^{M(j,k)} D_m(j, k),$$

with the usual convention that $\Sigma_{m=1}^{0} = 0$. Blocks (1.31) and (4.48) show

(95)   the $\mathscr{P}$-distribution of $C(j, k) + U(j, k)$ coincides with the P_i-distribution of $V(j, k)$.

Claim (a). Assume $Y_1(j) \in L^p$ for P_i. I have to show $Y_1(k) \in L^p$ for P_i. I claim

(96) $V(j, k) \in L^p$ for P_i.

Suppose $c(j, k) < 1$, the other case being easier. Now $C(j, k)$ and $D_1(j, k)$ are in L^p, using (94) and (87b). So $U(j, k) \in L^p$ by (89b). This and (88a) force $C(j, k) + U(j, k) \in L^p$. Now (95) proves (96).

As (93, 96) imply, $V(k, j) \in L^p$ for P_i. So $C(k, j) + U(k, j) \in L^p$ by (95). Consequently, $C(k, j) \in L^p$ and $U(k, j) \in L^p$ by (88b). In particular, $D_1(k, j) \in L^p$ by (89a). This gets $Y_1(k) \in L^p$ by (94, 87a).

Claim (b). Assume $P_i\{Y_1(j) = 0\} = 1$. I have to show $P_i\{Y_1(k) = 0\} = 1$. Clearly, $P_i\{V(j, k) = 0\} = 1$. So (93) implies $P_i\{V(k, j) = 0\} = 1$. By (95),

$$\mathscr{P}\{C(k, j) + U(k, j) = 0\} = 1.$$

By (90), there is a constant K with

$$\mathscr{P}\{C(k, j) = -K\} = \mathscr{P}\{U(k, j) = K\} = 1.$$

So (91a) makes $\mathscr{P}\{D_1(k, j) = K\} = 1$. But $c(1 - c) < 1$ for $0 \leq c \leq 1$, so $\mathscr{P}\{M(k, j) = 1\} < 1$. This and (91b) force $K = 0$. Now (94) gets

$$P_i\{Y_1(k) = 0\} = 1. \qquad \bigstar$$

Remember $\tau_1(j)$ is the least n with $\xi_n = j$. Let $\tau(j, k)$ be the least $n > \tau_1(j)$ with $\xi_n = k$. Let $\rho(k, j)$ be the least $n > \tau(j, k)$ with $\xi_n = j$. See Figure 4.

(97) Corollary. *If $Y_1(j) \in L^p$ relative to P_i for some i and j, then*

$$\Sigma_n \{h(\xi_n): \tau_1(j) \leq n < \tau(j, k)\} \in L^p$$

relative to P_i for all i, j, k.

PROOF.   Suppose $j \neq k$. Let

$$S(j, k) = \Sigma_n \{h(\xi_n): \tau_1(j) \leq n < \tau(j, k)\}$$
$$T(k, j) = \Sigma_n \{h(\xi_n): \tau(j, k) \leq n < \rho(k, j)\}.$$

Then $S(j, k)$ is the sum of $h(\eta)$ as η moves through $A(j, k)$. And $T(k, j)$ is the sum of $h(\eta)$ as η moves through $B(k, j)$. So $S(j, k)$ is independent of $T(k, j)$.

But
$$S(j, k) + T(k, j) = V(j, k) \in L^p$$
by (96). So $S(j, k) \in L^p$ by (88b). Use (92) to vary j.   ★

The next result generalizes (1.76).

(98) Corollary. *If $\tau_2(j) - \tau_1(j) \in L^p$ relative to P_i for some i and j, then $\tau_2(j) - \tau_1(j) \in L^p$ and $\tau(j, k) - \tau_1(j) \in L^p$ relative to P_i for all i, j, k.*

PROOF. Put $h \equiv 1$ in (92, 97).   ★

3

SOME INVARIANCE PRINCIPLES

1. INTRODUCTION

This chapter deals with the asymptotic behavior of the partial sums of functionals of a Markov chain, and in part is an explanation of the central limit theorem for these processes. Markov (1906) introduced his chains in order to extend the central limit theorem; this chapter continues his program. Section 3 contains an arcsine law for functional processes. The invariance principles of Donsker (1951) and Strassen (1964), to be discussed in *B & D*, are extended to functional processes in Section 4. For an alternative treatment of some of these results, see (Chung, 1960, Section I.16).

Throughout this chapter, let I be a finite or countably infinite set, with at least two elements. Let P be a stochastic matrix on I, for which I is one positive recurrent class. Let π be the unique invariant probability; see (1.81). Recall that the probability P_i on sequence space I^∞ makes the coordinate process $\{\xi_n\}$ Markov with stationary transitions P and starting state i. Fix a reference state $s \in I$. Confine ω to the set where $\xi_n = s$ for infinitely many n. Let $0 \leqq \tau_1 < \tau_2 < \cdots$ be the times n with $\xi_n = s$. Let f be a real-valued function on I. Let

$$Y_j = \Sigma_n \{f(\xi_n) : \tau_j \leqq n < \tau_{j+1}\}$$

and

$$U_j = \Sigma_n \{|f(\xi_n)| : \tau_j \leqq n < \tau_{j+1}\}.$$

Here and elsewhere in this chapter, j is used as a running index with values $1, 2, \ldots$; and not as a generic state in I. Let $V_0 = 0$ and $V_m = \Sigma_{j=1}^m Y_j$ and

---

I want to thank Pedro Fernandez and S.-T. Koay for checking the final draft of this chapter.

$S_n = \Sigma_{j=0}^n f(\xi_j)$. For (3) and (4) below, assume:

(1) $\qquad\qquad \Sigma_{i \in I} \pi(i) |f(i)| < \infty \quad$ and $\quad \Sigma_{i \in I} \pi(i) f(i) = 0,$

and

(2) $\qquad\qquad\qquad U_j^2$ has finite P_s-expectation.

NOTE. If (2) holds for one reference state s, it holds for all s: the dependence of U_j on s is implicit. This follows from (2.92), and can be used to select the reference state equal to the starting state. I will not take advantage of this. Theorems (3) and (4) hold if $[x]$ is interpreted as any nonnegative integer m with $|m - x| \leq 2$. The max can be taken over all values of [ ]. And i is a typical element of I.

(3) Theorem. $\quad n^{-\frac{1}{2}} \max \{|S_j - V_{[j\pi(s)]}| : 0 \leq j \leq n\} \to 0$ *in P_i-probability.*

(4) Theorem. *With P_i-probability* 1,

$$(n \log \log n)^{-\frac{1}{2}} \max \{|S_j - V_{[j\pi(s)]}| : 0 \leq j \leq n\} \to 0.$$

The idea of comparing S_j with $V_{[j\pi(s)]}$ is in (Chung, 1960, p. 78).

For (6), do not assume (1) and (2), but assume

(5) $\qquad\qquad Y_j$ differs from 0 with positive P_s-probability.

Let v_m be 1 or 0 according as V_m is positive or nonpositive. Similarly, let s_n be 1 or 0 according as S_n is positive or nonpositive.

NOTE. s is a fixed state, and s_n is a random variable.

(6) Theorem. *With P_i-probability* 1,

$$\frac{1}{n} \Sigma_j \{s_j : 1 \leq j \leq n\} - \frac{1}{n\pi(s)} \Sigma_j \{v_j : 1 \leq j \leq n\pi(s)\} \to 0.$$

NOTE. The quantities τ_n, Y_j, U_j, V_m, v_m depend on the reference state s. This dependence is not displayed in the notation. The quantities s_m and S_m do not depend on the reference state. In (3), the convergence may be a.e. I doubt it, but have no example.

2. ESTIMATING THE PARTIAL SUMS

Blocks (1.31) imply: Y_1, Y_2, ... are P_i-independent and identically distributed. So are U_1, U_2, ... The joint P_i-distribution of the Y's and U's does not depend on the state i, but does of course depend on the reference state s. Introduce E_i for expectation relative to P_i. Assumption (1) implies $E_i(Y_1) = 0$: it is enough to check this when $i = s$. Let ζ_k be the number of k's in the first s-block. Then

$$\Sigma_n \{f(\xi_n):\tau_1 \leqq n < \tau_2\} = \Sigma_{k \in I} f(k)\zeta_k.$$

Using (2.24),

$$E_s(Y_1) = \Sigma_{k \in I} f(k)E_s(\zeta_k) = \Sigma_{k \in I} f(k)\pi(k)/\pi(s) = 0.$$

Assumption (2) implies $E_i(Y_1^2) < \infty$.

On $\{\tau_1 \leqq n\}$, let:

$$l(n) \text{ be the largest } j \text{ with } \tau_j \leqq n;$$

$$Y'(n) = \Sigma_j \{f(\xi_j):0 \leqq j \leqq \tau_1 - 1\};$$

$$Y''(n) = \Sigma_j \{f(\xi_j):\tau_{l(n)} \leqq j \leqq n\}.$$

On $\{\tau_1 > n\}$, let $l(n) = 0$ and $Y'(n) = S_n$ and $Y''(n) = 0$. Let $V_{-1} = 0$. As in Figure 1,

(7) $$S_n = Y'(n) + Y''(n) + V_{l(n)-1}.$$

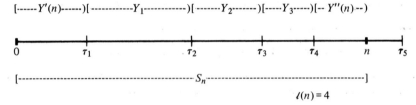

Figure 1.

Then (3) follows from (10), (11), and (13), to which (8) and (9) are preliminary.

(8) Lemma. *Let $a_1, a_2, \ldots$ and $0 < b_1 \leqq b_2 \leqq \cdots$ be real numbers with $b_j \to \infty$ and $a_j/b_j \to 0$. Then*

$$\max \{a_1, \ldots, a_n\}/b_n \to 0.$$

PROOF. Easy.   ★

(9) Lemma. *Let $a > 0$. Let $Z_1, Z_2, \ldots$ be identically distributed, not necessarily independent, with $E(|Z_1|^a) < \infty$. Then*

$$n^{-1/a} \max_j \{|Z_j|:1 \leqq j \leqq n\} \to 0 \quad \text{a.e.}$$

PROOF. It is enough to do the case $a = 1$ and $Z_j \geqq 0$: to get the general case, replace Z_j by $|Z_j|^a$. By (8), it is enough to show

$$Z_n/n \to 0 \quad \text{a.e.}$$

Let $\varepsilon > 0$, abbreviate $A_m = \{\varepsilon m \leqq Z_1 < \varepsilon(m+1)\}$, and check this computation.

$$\Sigma_{n=1}^{\infty} \text{Prob} \{Z_n \geqq \varepsilon n\} = \Sigma_{n=1}^{\infty} \Sigma_{m=n}^{\infty} \text{Prob } A_m$$

$$= \Sigma_{m=1}^{\infty} \Sigma_{n=1}^{m} \text{Prob } A_m$$

$$= \frac{1}{\varepsilon} \Sigma_{m=1}^{\infty} \varepsilon m \text{ Prob } A_m$$

$$\leqq \frac{1}{\varepsilon} \Sigma_{m=1}^{\infty} \int_{A_m} Z_1$$

$$\leqq \frac{1}{\varepsilon} E(Z_1)$$

$$< \infty.$$

Borel-Cantilli implies that almost everywhere, $Z_n \geqq \varepsilon n$ for only finitely many n. Let $\varepsilon \downarrow 0$ through a sequence.   ★

(10) Lemma. $n^{-\frac{1}{2}} \max_j \{|Y'(j)| : 0 \leqq j \leqq n\} \to 0$ *with P_i-probability* 1.

PROOF. $|Y'(n)| \leqq \Sigma_j \{|f(\xi_j)| : 0 \leqq j \leqq \tau_1 - 1\}$, for all n.   ★

(11) Lemma. $n^{-\frac{1}{2}} \max_j \{|Y''(j)| : 0 \leqq j \leqq n\} \to 0$ *with P_i-probability* 1.

PROOF.   Plainly, $|Y''(j)| \leqq U_{l(j)}$, where U_0 is temporarily 0. But $l(n) \leqq n + 1$, so

$$\max_j \{|Y''(j)| : 0 \leqq j \leqq n\} \leqq \max_j \{U_j : 1 \leqq j \leqq n + 1\}.$$

Use (2) and (9).   ★

(12) Lemma.   **(a)** $l(n)/n \to \pi(s)$ *with P_i-probability* 1.

          (b) $\tau_{l(n)}/n \to 1$ *with P_i-probability* 1.

PROOF.   Let $r_j = \tau_{j+1} - \tau_j$. By blocks (1.31), the P_i-distribution of $\{r_j\}$ does not depend on i; and with respect to P_i, the random variables $\{r_j\}$ are independent and identically distributed. Moreover, $E_i(r_1) = 1/\pi(s)$ by (1.81). The strong law implies that with P_i-probability 1,

$$(r_1 + \cdots + r_m)/m \to 1/\pi(s).$$

Put $m = l(n)$ and take reciprocals to see that with P_i-probability 1,

(a) $l(n)/[r_1 + \cdots + r_{l(n)}] \to \pi(s).$

Remember $l(n) \leqq n + 1$ and look at Figure 2:

$$\tau_{l(n)+1} - n \leqq r_{l(n)} \leqq \max_j \{r_j : 1 \leqq j \leqq n + 1\}.$$

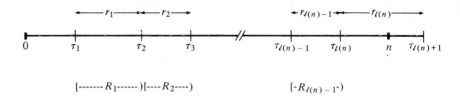

Figure 2.

Use (9) to deduce:

(b) $$r_{l(n)} = o(n) \quad \text{a.e.}$$

(c) $$\tau_{l(n)+1} - n = o(n) \quad \text{a.e.}$$

As Figure 2 shows,

$$r_1 + \cdots + r_{l(n)} = n + (\tau_{l(n)+1} - n) - \tau_1.$$

Clearly,

(d) $$\tau_1 = o(n).$$

Use (c) and (d) to deduce:

(e) $$r_1 + \cdots + r_{l(n)} = n + o(n) \quad \text{a.e.}$$

Combine this with (a) to prove (12a). Next,

$$\tau_{l(n)} = (r_1 + \cdots + r_{l(n)}) - r_{l(n)} + \tau_1;$$

combine this with (b), (d), and (e) to get (12b). ★

The proof of the next result involves (*B & D*, 1.118), which is quoted below.

(13) Lemma. $n^{-\frac{1}{2}} \max \{|V_{l(j)-1} - V_{[j\pi(s)]}| : 0 \leq j \leq n\} \to 0$ *in* P_i *probability.*

PROOF. Use (13.118) and (12a). Namely, fix $\varepsilon > 0$ and $\delta > 0$. For $r > 1$, let $\theta(r, n)$ be the P_i-probability—which does not depend on i—of the event

$$A_n = \{\varepsilon n^{\frac{1}{2}} > \max [|V_a - V_b| : 0 \leq a \leq b \leq ra \text{ and } 0 \leq a \leq n]\}.$$

Use (*B & D*, 1.118) to find one r for which there is an $n_0 = n_0(r, \varepsilon, \delta) < \infty$ so large that for all $n > n_0$,

$$\theta(r, n) > 1 - \delta.$$

Then use (12a) to find $N < \infty$ so large that $N\pi(s) > 2$ and the event

$$B = \left\{ \frac{1}{r} < \frac{l(j) - 1}{[j\pi(s)]} < r \text{ for all } j > N \right\}$$

has P_i-probability more than $1 - \delta$. Choose $n_1 > n_0$, finite but so large that for $n > n_1$, the event

$$C_n = \{\varepsilon n^{\frac{1}{2}} > \max \ [|V_{l(j)-1} - V_{[j\pi(s)]}|:0 \leqq j \leqq N]\}$$

has P_i-probability more than $1 - \delta$. Thus, $P_i(A_n \cap B \cap C_n) > 1 - 3\delta$ for $n > n_1$. If $n > n_1$ and $j = 0, \ldots, n$, I claim

$$|V_{l(j)-1} - V_{[j\pi(s)]}| < \varepsilon n^{\frac{1}{2}},$$

provided ω is confined to $A_n \cap B \cap C_n$. Indeed, if $j \leqq N$ the inequality holds because $\omega \in C_n$. Suppose $N < j \leqq n$. Then $l(j) - 1 \leqq n$, and $[j\pi(s)] \geqq 1$. This and $\omega \in B$ force $l(j) - 1 \geqq 1$. Of $l(j) - 1$ and $[j\pi(s)]$, the lesser is between 1 and n; and the greater is at most r times the lesser, because $\omega \in B$. The inequality is now visibly true, because $\omega \in A_n$. ★

This completes the proof of (3). The proof of (4) is similar, using $(B \& D, 1.119)$ instead of $(B \& D, 1.118)$. To quote $(B \& D, 1.118-119)$, let $Y_1, Y_2, \ldots$ be independent, identically distributed random variables on some probability triple $(\Omega, \mathscr{F}, \mathscr{P})$. Suppose Y_1 has mean 0 and finite variance. Let $V_0 = 0$, and for $n \geqq 1$ let

$$V_n = Y_1 + \cdots + Y_n.$$

(B & D, 1.118). *Let $\varepsilon > 0$ and $r > 1$. Let $p(r, n) = \mathscr{P}(A)$, where A is the event that*

$$n^{\frac{1}{2}} < \max \ \{|V_j - V_k|:0 \leqq j \leqq n \text{ and } j \leqq k \leqq rj\}.$$

Then

$$\lim_{r \downarrow 1} \lim \sup_{n \to \infty} p(r, n) = 0.$$

(B & D, 1.119). *Let $\varepsilon > 0$. There is an $r > 1$, which depends on ε but not on the distribution of Y_1, such that $\mathscr{P}(\lim \sup A_n) = 0$, where A_n is the event that*

$$\max \ \{|V_j - V_k|:0 \leqq j \leqq n \text{ and } j \leqq k \leqq rj\}$$

exceeds $\varepsilon(n \log \log n)^{\frac{1}{2}}$.

3. THE NUMBER OF POSITIVE SUMS

Assume (5). Suppose that $V_m > 0$ for infinitely many m along P_i-almost all paths. In the opposite case, $V_m \leqq 0$ for all large m, along P_i-almost all paths, by the Hewitt–Savage 0–1 Law (1.122). The argument below then establishes (6), with $1 - s_j$ for s_j and $1 - v_j$ for v_j. This modified (6) is equivalent to original (6).

As in (12), let $r_j = \tau_{j+1} - \tau_j$. With respect to P_i, the r_j are independent, identically distributed, and have mean $1/\pi(s)$. The P_i-distribution of r_j does not depend on i, but does depend on the reference state s. Theorem (6) will be proved by establishing (14) and (15). To state them, let

$$A_n = \frac{1}{n\pi(s)} \Sigma_j \{v_j : 1 \leqq j \leqq n\pi(s)\}$$

$$B_n = \frac{1}{n} \Sigma_j \{v_j r_{j+1} : 1 \leqq j \leqq l(n) - 2\}$$

$$C_n = \frac{1}{n} \Sigma_m \{s_m : 1 \leqq m \leqq n\}.$$

The two estimates are

(14) $A_n - B_n \to 0$ with P_i-probability 1

(15) $B_n - C_n \to 0$ with P_i-probability 1.

Add (14) and (15) to get (6) in the form $A_n - C_n \to 0$ a.e. Relation (14) is obtained by replacing r_{j+1} with its mean value $1/\pi(s)$. The error is negligible after dividing by n, in view of the strong law. Relation (15) will follow from the fact that essentially all $m \leqq n$ are in intervals $[\tau_{j+1}, \tau_{j+2})$ over which $s_m = v_j$, because $|V_j|$ is large by comparison with U_{j+1} and $Y'(m)$.

Making this precise requires lemmas (16) and (17). For (16), let $r_1, r_2, \ldots$ be any independent, identically distributed random variables, with finite expectation. Let $\mathscr{F}_1 \subset \mathscr{F}_2 \subset \cdots$ be σ-fields, such that r_n is $\mathscr{F}_{n+1}$-measurable, and $\mathscr{F}_n$ is independent of r_n. Let $z_1, z_2, \ldots$ be random variables, taking only the values 0 and 1, such that z_n is $\mathscr{F}_{n+1}$-measurable, and $\Sigma_1^\infty z_n = \infty$ a.e.

(16) Lemma. $(\Sigma_{j=1}^n z_j r_{j+1})/(\Sigma_{j=1}^n z_j)$ *converges to* $E(r_j)$ a.e.

PROOF.   Let $Z_n = \Sigma_1^n z_j$. For $m = 1, 2, \ldots$, let W_m be 1 plus the smallest n such that $Z_n = m$. I say that $\{W_m = j\} \in \mathscr{F}_j$. Indeed, for $j \geqq m + 1$,

$$\{W_m = j\} = \{z_1 + \cdots + z_{j-1} = m > z_1 + \cdots + z_{j-2}\}.$$

If $m' < m$ and A is a Borel subset of the line, I deduce

$$\{r_{W_{m'}} \in A \text{ and } W_m = j\} \in \mathscr{F}_j;$$

for this set is

$$\cup_{k=1}^{j-1} \{r_k \in A \text{ and } W_{m'} = k \text{ and } W_m = j\}$$

I conclude that $r_{W_1}, r_{W_2}, \ldots$ are independent and identically distributed, the distribution of r_{W_m} being that of r_j. Indeed, if $A_1, \ldots, A_m$ are Borel subsets of the line, then

Prob $\{r_{W_1} \in A_1, \ldots, r_{W_{m-1}} \in A_{m-1}, r_{W_m} \in A_m\}$

$= \Sigma_{j=1}^\infty$ Prob $\{r_{W_1} \in A_1, \ldots, r_{W_{m-1}} \in A_{m-1} \text{ and } W_m = j \text{ and } r_j \in A_m\}$

$= \Sigma_{j=1}^\infty$ Prob $\{r_{W_1} \in A_1, \ldots, r_{W_{m-1}} \in A_{m-1} \text{ and } W_m = j\} \cdot$ Prob $\{r_j \in A_m\}$

$=$ Prob $\{r_{W_1} \in A_1, \ldots, r_{W_{m-1}} \in A_{m-1}\} \cdot$ Prob $\{r_j \in A_m\}$,

because $\{r_{W_1} \in A_1, \ldots, r_{W_{m-1}} \in A_{m-1}$ and $W_m = j\} \in \mathscr{F}_j$, while r_j is independent of $\mathscr{F}_j$ and Prob $\{r_j \in A_m\}$ does not depend on j.

By the strong law,

$$\frac{1}{n}\Sigma_{k=1}^{n}\, r_{W_k} \to E(r_1) \quad \text{a.e.}$$

Because $Z_n \to \infty$ a.e.,

$$Z_n^{-1}\Sigma_k\, \{r_{W_k}\!:\! 1 \leqq k \leqq Z_n\} \to E(r_1) \quad \text{a.e.}$$

But

$$\Sigma_j\, \{z_j r_{j+1}\!:\! 1 \leqq j \leqq n\} = \Sigma_k\, \{r_{W_k}\!:\! 1 \leqq k \leqq Z_n\}:$$

for Z_n is the number of $j = 1, \ldots, n$ with $z_j = 1$; and W_k is 1 plus the kth ^{i} with $z_j = 1$.   ★

For (17), let $Y_1, Y_2, \ldots$ be any sequence of independent, identically distributed random variables. Put $V_n = \Sigma_1^n\, Y_j$. Make no assumptions about the moments of Y_j. Let M be a positive, finite number. Let d_n be 1 or 0 according as $|V_n| \leqq M$ or $|V_n| > M$.

(17) Lemma. *Suppose Y_j differs from 0 with positive probability. Then*

$$\frac{1}{n}\Sigma_1^n\, d_j \to 0 \quad \text{a.e.}$$

PROOF. Suppose Y_j differs from x with positive probability, for each x. Otherwise the result is easy. Let C_n be the concentration function of V_n, as defined in Section 5. Fix k at one of $1, 2, \ldots$ and fix r equal to one of $0, \ldots,$ $k - 1$. Let θ_n be the conditional probability that $|V_{nk+r}| \leqq M$ given $Y_1, \ldots, Y_{(n-1)k+r}$. I claim

(a) $\theta_n \leqq C_k(2M)$ a.e.

Indeed, let μ be the distribution of $Y = (Y_1, \ldots, Y_{(n-1)k+r})$, a probability on the set of $(n - 1)k + r$-vectors $y = (y_1, \ldots, y_{(n-1)k+r})$. Let A be a Borel subset of this vector space. Let $T = Y_{(n-1)k+r+1} + \cdots + Y_{nk+r}$, so T is independent of Y and distributed like V_k. Let $s(y) = y_1 + \cdots + y_{(n-1)k+r}$, so

$$V_{nk+r} = s(Y) + T.$$

By Fubini,

Prob $\{Y \in A \text{ and } |V_{nk+r}| \leqq M\}$

$$= \int_A \text{Prob }\{|s(y) + T| \leqq M\}\, \mu(dy)$$

$$= \int_A \text{Prob }\{-M - s(y) \leqq T \leqq M - s(y)\}\, \mu(dy)$$

$$\leqq \int_A C_k(2M)\, \mu(dy)$$

$$= C_k(2M) \cdot \mu(A)$$

$$= C_k(2M) \cdot \text{Prob }\{Y \in A\}.$$

This completes the proof of (a).

I claim

(b) $\lim \sup_{n \to \infty} \frac{1}{n} \Sigma_{j=1}^n d_{jk+r} \leq C_k(2M)$ a.e.

Claim (b) follows from (a) and this martingale fact:

$$\lim_{n \to \infty} \frac{1}{n} \Sigma_{j=1}^n (d_{jk+r} - \theta_j) = 0 \quad \text{a.e.}$$

This fact may not be in general circulation; two references are (Dubins and Freedman, 1965, Theorem (1)) and (Neveu, 1965, p. 147). Claim (b) can also be deduced from the strong law for coin tossing. Suppose without real loss that there is a uniform random variable independent of $Y_1, Y_2, \ldots$. Then you can construct independent 0–1 valued random variables $e_1, e_2, \ldots$ such that: e_n is 1 with probability $C_k(2M)$, and $e_n \geq d_{nk+r}$.

Sum out $r = 0, \ldots, k - 1$ in claim (b) and divide by k:

(c) $\lim \sup_{n \to \infty} \frac{1}{nk} \Sigma_j \{d_j : 1 \leq j < (n + 1)k\} \leq C_k(2M)$ a.e.

Let m and n tend to ∞, with $nk \leq m < (n + 1)k$. Then $m/(nk) \to 1$, so (c) implies

$$\lim \sup_{m \to \infty} \frac{1}{m} \Sigma_{j=1}^m d_j \leq C_k(2M) \quad \text{a.e.}$$

Now let $k \to \infty$, so $C_k(2M) \to 0$ by (36) below.

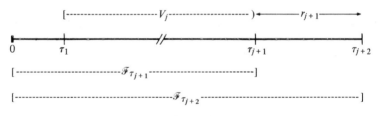

Figure 3.

PROOF OF (14).   Recall the definitions of A_n and B_n, made before (14). Introduce

$$D_n = \frac{1}{n\pi(s)} \Sigma_j \{v_j : 1 \leq j \leq l(n) - 2\}.$$

You should use (16) to see that with P_i-probability 1,

$$(\Sigma_{j=1}^m v_j r_{j+1})/(\Sigma_{j=1}^m v_j) \to 1/\pi(s).$$

The conditions of (16) are satisfied by strong Markov (1.21), with $\mathscr{F}_j = \mathscr{F}_{\tau_j}$: look at Figure 3. The condition $\Sigma\, v_j = \infty$ a.e. is the assumption made at the beginning of this section. Put $m = l(n) - 2$ and rearrange to get

$$(18) \qquad\qquad D_n/B_n \to 1 \quad \text{with } P_i\text{-probability } 1.$$

Use (12a) to see

$$(19) \qquad\qquad A_n - D_n \to 0 \quad \text{with } P_i\text{-probability } 1.$$

As Figure 2 shows, $B_n \leqq 1$. Thus

$$|D_n - B_n| \leqq |D_n - B_n|/B_n \to 0$$

with P_i-probability 1 by (18). Combine this with (19) to get (14). ★

PROOF OF (15). Temporarily, let S be a subset of the nonnegative integers. A *random subset R of S* assigns a subset $R(\omega)$ of S to each $\omega \in I^\infty$, so that $\{\omega : j \in R(\omega)\}$ is measurable for each $j \in S$. The cardinality $\#R$ of R is a random variable on I^∞, whose value at ω is the number of elements of $R(\omega)$. For $j = 1, 2, \ldots$, let

$$R_j(\omega) = \{m : m \text{ is a nonnegative integer and } \tau_j(\omega) \leqq m < \tau_{j+1}(\omega)\}.$$

Then R_j is a random subset of the nonnegative integers, and $\#R_j = r_j$, as in Figure 2.

Fix $\varepsilon > 0$. Choose M so large that

$$\int_{\{U_1 > M/2\}} r_1 \, dP_i < \varepsilon/3;$$

the integral does not depend on i. Let G_n be the following random subset of the positive integers: $j \in G_n$ iff $\tau_{j+2} \leqq n$ and $|V_j| > M$ and $U_{j+1} \leqq M/2$. In particular, $j \in G_n$ implies $1 \leqq j \leqq l(n) - 2$. See Figure 4. Of course, G_n depends on M, although this is not explicit in the notation. Let

$$H_n = \bigcup \{R_{j+1} : j \in G_n\},$$

a random subset of $\{1, \ldots, n\}$. In particular, $m \in H_n$ implies $\tau_2 \leqq m < \tau_{l(n)}$, as in Figure 4. The main part of the argument is designed to show

(20) For P_i-almost all sample sequences, $\#H_n \geqq (1 - \varepsilon)n$ for all large n.

How large depends on the sample sequence.

Before proving (20), I will derive (15) from (20). Let E_M be the subset of I^∞ where

$$Y' = \Sigma_n \{|f(\xi_n)| : 0 \leqq n < \tau_1\} \leqq M/2.$$

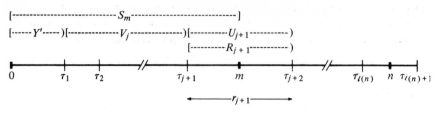

Figure 4.

By looking at Figure 2,

$$\Sigma_j \{r_{j+1}: 1 \leq j \leq l(n) - 2\} \leq n$$

and

$$\Sigma_j \{r_{j+1}: j \in G_n\} = \#H_n.$$

So

$$\Sigma_j \{r_{j+1}: 1 \leq j \leq l(n) - 2 \text{ and } j \notin G_n\} \leq n - \#H_n.$$

But $v_j = 0$ or 1, so

$$0 \leq \Sigma_j \{v_j r_{j+1}: 1 \leq j \leq l(n) - 2 \text{ and } j \notin G_n\} \leq n - \#H_n.$$

Similarly,

$$0 \leq \Sigma_m \{s_m: 1 \leq m \leq n \text{ and } m \notin H_n\} \leq n - \#H_n.$$

On E_M, I say that

$$\Sigma_j \{v_j r_{j+1}: j \in G_n\} = \Sigma_m \{s_m: m \in H_n\}.$$

Indeed, $m \in R_{j+1}$ and $j \in G_n$ and $Y' \leq M/2$ force $s_m = v_j$: because $\tau_{j+1} \leq m < \tau_{j+2} \leq n$ and $|V_j| > M$ and $U_{j+1} \leq M/2$, so

$$|S_m - V_j| \leq Y' + U_{j+1} \leq M < |V_j|,$$

and S_m is positive or negative with V_j. See Figure 4. Consequently, on E_M,

$$|\Sigma_j \{v_j r_{j+1}: 1 \leq j \leq l(n) - 2\} - \Sigma_m \{s_m: 1 \leq m \leq n\}| \leq n - \#H_n.$$

Recall the definitions of B_n and C_n given before (14). Relation (20) implies that P_i-almost everywhere on E_M,

$$|B_n - C_n| \leq \varepsilon \quad \text{for all large } n.$$

Let M increase to ∞ through a countable set, so E_M swells to the whole space. Then let ε decrease to 0 through a countable set, and get (15) from (20).

Turn now to the proof of (20). Let d_j be 1 or 0 according as $|V_j| \leq M$ or $|V_j| > M$. I claim that for P_i-almost all sample sequences,

(21) $\Sigma_{j=1}^n d_j r_{j+1} \leq \varepsilon n/3 \quad \text{for all large } n.$

If $\Sigma_{j=1}^{\infty} d_j < \infty$ with positive P_i-probability, Hewitt-Savage (1.122) implies $d_j = 0$ for all large j, with P_i-probability 1; and (21) holds. Suppose $\Sigma_{j=1}^{\infty} d_j = \infty$ with P_i-probability 1. Lemma (16) makes

$$(\Sigma_{j=1}^{n} d_j r_{j+1})/(\Sigma_{j=1}^{n} d_j) \to 1/\pi(s)$$

with P_i-probability 1. The conditions of (16) are satisfied by strong Markov (1.21), with $\mathscr{F}_j = \mathscr{F}_{r_j}$. See Figure 3. But (17) makes $\frac{1}{n}\Sigma_{j=1}^{\infty} d_j \to 0$ with P_i-probability 1. So (21) still holds. Put $l(n) - 2$ for n in (21) and remember $l(n) \leqq n + 1$:

(22) $\Sigma_j \{d_j r_{j+1} : 1 \leqq j \leqq l(n) - 2\} \leqq \varepsilon n/3$ for all large n, with P_i-probability 1.

Next, blocks (1.31) and the strong law imply that with P_i-probability 1,

$$\frac{1}{n-2}\Sigma_j \{r_{j+1} : 1 \leqq j \leqq n-2 \text{ and } U_{j+1} > M/2\} \to \int_{\{U_1 > M/2\}} r_1 \, dP_i.$$

Put $l(n)$ for n; remember that $l(n) \leqq n + 1$, and the integral is less than $\varepsilon/3$ by choice of M;

(23) $\Sigma_j \{r_{j+1} : 1 \leqq j \leqq l(n) - 2 \text{ and } U_{j+1} > M/2\} \leqq \varepsilon n/3$ for all large n, with P_i-probability 1.

Finally, use (12b) to see that

(24)   The number of $m \leqq n$ with $m < \tau_2$ or $m \geqq \tau_{l(n)}$ is at most $\varepsilon n/3$ for all large n, with P_i-probability 1.

Let $\bar{H}_n$ be the random set of $m \in \{1, \ldots, n\}$ which have property (a) or (b) or (c):

(a) $m \in R_{j+1}$ for some $j = 1, \ldots, l(n) - 2$ with $d_j = 1$.

(b) $m \in R_{j+1}$ for some $j = 1, \ldots, l(n) - 2$ with $U_{j+1} > M/2$.

(c) $m < \tau_2$ or $m \geqq \tau_{l(n)}$.

Combine (22–24) to get

(25) $\#\bar{H}_n \leqq \varepsilon n$ for all large d, with P_i-probability 1.

But $\bar{H}_n$ is the complement of H_n relative to $\{1, \ldots, n\}$, proving (20).   ★

To state the arcsine law (26), define F_a as follows. For $a = 0$ or 1, let F_a be the distribution function of point mass at a. For $0 < a < 1$, let F_a be the probability on $[0, 1]$, with density proportional to $y \to y^{a-1}(1 - y)^{-a}$.

(26) Corollary. *Suppose (5). The P_i-distribution of $\frac{1}{n}\Sigma_{j=1}^n s_j$ converges to F_a iff the P_i-mean of $\frac{1}{n}\Sigma_{j=1}^n s_j$ converges to a.*

PROOF. Use (6) and (Spitzer, 1956, Theorem 7.1).   ★

NOTE. Temporarily, let $M_n = \frac{1}{n}\Sigma_{j=1}^n s_j$. If the distribution of M_n converges to anything, say F, then the P_i-mean of M_n converges to the mean μ of F, because $0 \leq M_n \leq 1$. Thus $F = F_\mu$. This need not hold for subsequential limits of the distribution of M_n. Furthermore, if the convergence holds for one i, it holds for all i. More generally, (6) shows that the P_i-distribution of $\frac{1}{n}\Sigma_{j=1}^n s_j$ is asymptotically free of i; because the P_i-distribution of the Y's, so of the v's, does not depend on i.

The balance of this section concerns the exceptional case

(27) $Y_j = 0$ with P_i-probability 1.

(28) Theorem. *Let τ be the least positive n with $\xi_n = i$. If (27) holds, then with P_i-probability 1,*

(29) $\lim_{n\to\infty} \frac{1}{n}\Sigma_{m=1}^n s_m = \pi(i)\int\Sigma_m \{s_m : 0 \leq m < \tau\}\, dP_i.$

PROOF. If (27) holds for one reference state s, it holds for all s by (2.92); the dependence of Y_j on s is implicit. Consequently, in studying the asymptotic behavior of $\frac{1}{n}\Sigma_1^n s_m$ relative to P_i, it is legitimate to use i as the reference state s: condition (27) will still hold. The simplification is

$$P_i\{\tau_1 = 0 \text{ and } \tau_2 = \tau\} = 1.$$

For $j = 1, 2, \ldots$, let T_j be the random vector of random length $r_j = \tau_{j+1} - \tau_j$, whose mth term is S_{τ_j+m} for $m = 0, \ldots, r_j - 1$. Now

$$S_{\tau_j-1} = Y_1 + \cdots + Y_{j-1} = 0 \quad \text{a.e.,}$$

so the mth term in T_j is really

$$f(\xi_{\tau_j}) + \cdots + f(\xi_{\tau_j+m})$$

except on a P_i-null set. The first summand is $f(i)$, but this does not help. Blocks (1.31) imply that the vectors $T_1, T_2, \ldots$ are independent and identically distributed relative to P_i. By the strong law, with P_i-probability 1,

$$\lim_{j \to \infty} \frac{1}{j-1} \sum_{m=0}^{\tau_j - 1} s_m = \int \sum_m \{s_m : 0 \leq m < \tau_2\} \, dP_i.$$

Confine j to the sequence $l(n)$, and use (12).   ★

The limit in (29) depends on i. For example, let $I = \{1, 2, 3\}$ and let

$$P = \begin{pmatrix} 0 & 1 & 0 \\ 0 & 0 & 1 \\ 1 & 0 & 0 \end{pmatrix},$$

so $1 \to 2 \to 3 \to 1$. Then $\pi(i) = \frac{1}{3}$ for $i = 1, 2, 3$; condition (1) is equivalent to

$$f(1) + f(2) + f(3) = 0.$$

Condition (27) is automatic. When $i = 1$, the right side of (29) is $\frac{1}{3}$ times the number of positive sums

$$f(1), \quad f(1) + f(2), \quad f(1) + f(2) + f(3).$$

When $i = 2$, the right side of (29) is $\frac{1}{3}$ times the number of positive sums

$$f(2), \quad f(2) + f(3), \quad f(2) + f(3) + f(1).$$

4. SOME INVARIANCE PRINCIPLES

As in Section 1.7 of $B \& D$, let g be a real-valued function on the non-negative integers, whose value at n is g_n. Then $g_{(n)}$ is the continuous, real-valued function on $[0, 1]$ whose value at j/n is $g_j / n^{\frac{1}{2}}$, and which is linearly interpolated. Let $\{B(t) : 0 \leq t < \infty\}$ be normalized Brownian motion. That is: $B(0) = 0$; all the sample paths of B are continuous; and for

$$0 \leq t_0 \leq t_1 \leq \cdots \leq t_n,$$

the differences $B(t_0)$, $B(t_1) - B(t_0)$, ..., $B(t_n) - B(t_{n-1})$ are independent normal random variables, with means 0 and variances

$$t_0, t_1 - t_0, \ldots, t_n - t_{n-1}$$

respectively. Such a construct exists by ($B \& D$, 1.6). Let $C[0, 1]$ be the space of continuous, real-valued functions on $[0, 1]$, with the sup norm

$$\|f\| = \max \{|f(t)| : 0 \leq t \leq 1\}.$$

Give $C[0, 1]$ the metric

$$\text{distance } (f, g) = \|f - g\|,$$

and the σ-field generated by the sets which are open relative to this metric. The distribution Π_σ of $\{B(\sigma^2 t): 0 \leq t \leq 1\}$ is a probability on $C[0, 1]$. For more discussion of all these objects, see Sections 1.5 and 1.7 of $B \& D$. The next theorem, which is an extension of Donsker's invariance principle to functional processes, depends on some results from $B \& D$. These are quoted at the end of the section.

(30) Theorem. *Suppose* (1) *and* (2). *Then* $\sigma^2 = \Pi(s) \cdot \int Y_1^2 \, dP_i$ *depends neither on i nor on the reference state s. Let φ be a bounded, measurable, real-valued function on $C[0, 1]$, which is continuous with Π_σ-probability 1. Then*

$$\lim_{n \to \infty} \int_{I^\infty} \varphi(S_{(n)}) \, dP_i = \int_{C[0,1]} \varphi \, d\Pi_\sigma.$$

PROOF. Set $V_0 = 0$. Then $V_{(n)}$ is linear on $\left[\dfrac{j}{n}, \dfrac{j+1}{n}\right]$ and takes the value $V_j/n^{\frac{1}{2}}$ at j/n. Similarly for $S_{(n)}$, except that $S_0 = f(\xi_0)$. Let

$$F_n(t) = V_{(n)}(\pi(s)t) \quad \text{for } 0 \leq t \leq 1.$$

Thus, F_n is a measurable mapping from I^∞ to $C[0, 1]$. The values of F_n are piecewise linear functions, with corners at $\dfrac{m}{n\pi(s)}$ for $m = 0, 1, \dots, m_0$, where m_0 is the largest integer no more than $n\pi(s)$.

I claim

(31) $$\|S_{(n)} - F_n\| \to 0 \quad \text{in } P_i\text{-probability.}$$

Study two successive corners $\dfrac{m}{n\pi(s)}$ and $\dfrac{m+1}{n\pi(s)}$ of F_n; so

$$0 \leq \frac{m}{n\pi(s)} \leq \frac{m+1}{n\pi(s)} \leq 1;$$

and study the successive corners of $S_{(n)}$ from the greatest one $\dfrac{a}{n}$ no greater than $\dfrac{m}{n\pi(s)}$ to the least one $\dfrac{b}{n}$ no less than $\dfrac{m+1}{n\pi(s)}$. This is depicted in Figure 5. Analytically,

$$0 \leq \frac{a}{n} \leq \frac{m}{n\pi(s)} \leq \frac{a+1}{n} \leq \cdots \leq \frac{b-1}{n} \leq \frac{m+1}{n\pi(s)} \leq \frac{b}{n} \leq 1;$$

that is,

$$a\pi(s) \leq m \leq (a+1)\pi(s) \leq \cdots \leq (b-1)\pi(s) \leq m+1 \leq b\pi(s).$$

By the linearity,

$$\max_t \left\{ |S_{(n)}(t) - F_n(t)| : \frac{m}{n\pi(s)} \leq t \leq \frac{m+1}{n\pi(s)} \right\}$$

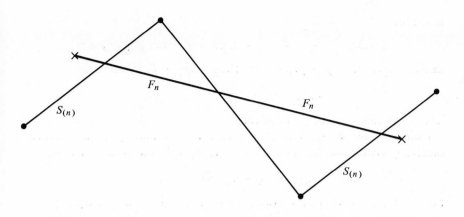

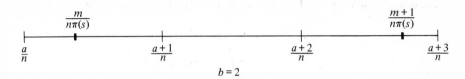

$$b = 2$$

Figure 5.

is at most

$$n^{-\frac{1}{2}} \max_{j,\mu} \{|S_j - V_\mu| : \mu = m \text{ or } m + 1 \text{ and } a \leqq j \leqq b\};$$

and $|\mu - j\pi(s)| \leqq 2$. Suppose, as is likely, that the last corner $\dfrac{m}{n\pi(s)}$ of F_n is less than 1. That is

$$\frac{m}{n\pi(s)} < 1 < \frac{m+1}{n\pi(s)}.$$

Again, let $\dfrac{a}{n}$ be the greatest corner of $S_{(n)}$ no greater than $\dfrac{m}{n\pi(s)}$, so

$$\frac{a}{n} \leqq \frac{m}{n\pi(s)} \leqq \frac{a+1}{n} \leqq \cdots \leqq \frac{n}{n} \leqq \frac{m+1}{n\pi(s)};$$

that is,

$$a\pi(s) \leqq m \leqq (a+1)\pi(s) \leqq \cdots \leqq n\pi(s) \leqq m + 1.$$

As before,

$$\max_t \left\{ |S_{(n)}(t) - F_n(t)| : \frac{m}{n\pi(s)} \leqq t \leqq 1 \right\}$$

is at most

$$n^{-\frac{1}{2}} \max_{j,\mu} \{|S_j - V_\mu| : \mu = m \text{ or } m + 1 \text{ and } a \leqq j \leqq n\};$$

in this display too, $|\mu - j\pi(s)| \leqq 2$. Thus, $\|S_{(n)} - F_n\|$ is at most

$$n^{-\frac{1}{2}} \max_j \{|S_j - V_{[j\pi(s)]}| : 0 \leqq j \leqq n\}.$$

Use (3) to complete the proof of (31).

I will now derive (30) from (31). Temporarily, confine φ to the bounded, uniformly continuous functions on $C[0, 1]$. By (31) and $(B \& D, 1.129)$,

$$E_i\{\varphi(S_{(n)})\} - E_i\{\varphi(F_n)\} \to 0.$$

For $f \in C[0, 1]$, let $\hat{f}(t) = f(\pi(s)t)$ for $0 \leqq t \leqq 1$, so $\hat{f} \in C[0, 1]$ and $\|\hat{f}\| \leqq \|f\|$. Let $\hat{\varphi}(f) = \varphi(\hat{f})$. Then $\hat{\varphi}$ is still bounded and uniformly continuous. Check $F_n = \hat{V}_{(n)}$. Let $\rho^2 = E_i(Y_1^2)$. By Donsker $(B \& D, 1.120)$,

$$E_i\{\varphi(\hat{V}_{(n)})\} = E_i\{\hat{\varphi}(V_{(n)})\}$$

$$\to \int \hat{\varphi}(f) \, \Pi_\rho(df)$$

$$= \int \varphi(\hat{f}) \, \Pi_\rho(df)$$

$$= \int \varphi(f) \, \Pi_\sigma(df).$$

This completes the proof of (30) for bounded and uniformly continuous φ. Now use $(B \& D, 1.127)$.

Finally, σ^2 does not depend on i because the P_i-distribution of Y_1 does not depend on i. And σ^2 does not depend on s because the approximants to $\int \varphi \, d\Pi_\sigma$ do not depend on s, and this integral for all bounded, uniformly continuous φ determines Π_σ, so σ.   ★

NOTE. In principle, σ^2 can be computed from P. See (Chung, 1960, p. 81) for an explicit formula.

Let K_σ be the set of absolutely continuous functions f on $[0, 1]$ which satisfy: $f(0) = 0$ and $\int_0^1 |f'(t)|^2 \, dt \leqq \sigma^2$. The extension of Strassen's invariance principle to functional processes is

(32) Theorem. *Suppose* (1) *and* (2). *For P_i-almost all sample sequences, the indexed subset*

$$\{(2 \log \log n)^{-\frac{1}{2}} S_{(n)} : n = 3, 4, \ldots\}$$

of $C[0, 1]$ is relatively compact, and its set of limit points is K_σ, where $\sigma^2 = \pi(s) \cdot E_i(Y_1^2)$.

PROOF.   Use (4) and Strassen's invariance principle (B & D, 1.133), as in the proof of (30).   ★

Some results from B & D

(B & D, 120). Let $Y_1, Y_2, \ldots$ be independent, identically distributed random variables on $(\Omega, \mathcal{F}, \mathcal{P})$, with mean 0 and finite variance ρ^2. Let $V_0 = 0$ and

$$V_n = Y_1 + \cdots + Y_n \quad for\ n \geqq 1.$$

Let φ be a bounded, real-valued, measurable function on $C[0, 1]$, which is continuous $\mathcal{P}$-almost everywhere. Then

$$\lim_{n \to \infty} \int_\Omega \varphi(V_{(n)})\, d\mathcal{P} = \int_{C[0,1]} \varphi(f)\, \Pi_\rho\, (df).$$

(B & D, 127). Let $\{\mathcal{P}_n, \mathcal{P}\}$ be probabilities on $C[0, 1]$. Suppose

$$\int \varphi\, d\mathcal{P}_n \to \int \varphi\, d\mathcal{P}$$

for all bounded, uniformly continuous φ. Then the convergence holds for all bounded, measurable φ which are continuous $\mathcal{P}$-almost everywhere.

(B & D, 129).   Let $\{Z_n, W_n\}$ be measurable maps from $(\Omega, \mathcal{F}, \mathcal{P})$ to $C[0, 1]$. Suppose

$$\|Z_n - W_n\| \to 0 \quad in\ \mathcal{P}\text{-probability.}$$

Let φ be a bounded and uniformly continuous function on $C[0, 1]$. Then

$$\int \varphi(Z_n)\, d\mathcal{P} - \int \varphi(W_n)\, d\mathcal{P} \to 0.$$

(B & D, 133).   Let $Y_1, Y_2, \ldots$ be independent, identically distributed random variables on $(\Omega, \mathcal{F}, \mathcal{P})$, with mean 0 and finite variance ρ^2. Let $V_0 = 0$ and

$$V_n = Y_1 + \cdots + Y_n \quad for\ n \geqq 1.$$

Let Ω^* be the set of ω such that the indexed subset

$$\{(2 \log \log n)^{-\frac{1}{2}}\, V_{(n)}(\cdot, \omega) : n = 3, 4, \ldots\}$$

of $C[0, 1]$ is relatively compact, with limit set K_ρ. Then $\Omega^* \in \mathcal{F}$ and $\mathcal{P}(\Omega^*) = 1$.

5.   THE CONCENTRATION FUNCTION

Let X be a random variable on $(\Omega, \mathcal{F}, \mathcal{P})$.

DEFINITION.   The concentration function C_X of X is this function of non-negative u:

$$C_X(u) = \sup_x \mathcal{P}\{x \leqq X \leqq x + u\}.$$

For (33), let $-\infty < a < b < \infty$. Let $a_n \to a$ and $b_n \to b$.

(33) Lemma. $\mathscr{P}\{a \leqq X \leqq b\} \geqq \limsup \mathscr{P}\{a_n \leqq X \leqq b_n\}$.

PROOF. Fix $\varepsilon > 0$. Choose $\delta > 0$ so small that

$$\mathscr{P}\{a - \delta \leqq X \leqq b + \delta\} \leqq \mathscr{P}\{a \leqq X \leqq b\} + \varepsilon.$$

For all large n,
$$[a_n, b_n] \subset [a - \delta, b + \delta].$$
So

$$\mathscr{P}\{a_n \leqq X \leqq b_n\} \leqq \mathscr{P}\{a - \delta \leqq X \leqq b + \delta\} \leqq \mathscr{P}\{a \leqq X \leqq b\} + \varepsilon. \quad \bigstar$$

(34) Lemma. (a) *If $\lambda > 0$, then $C_{\lambda X}(\lambda u) = C_X(u)$.*

    (b) C_X *is nondecreasing.*

    (c) C_X *is continuous from the right.*

    (d) $C_X(u + v) \leqq C_X(u) + C_X(v)$.

    (e) *If X and Y are independent, $C_{X+Y} \leqq C_X$.*

    (f) *If X and Y are independent and identically distributed, then*

$$C_X(u)^2 \leqq \mathscr{P}\{|X - Y| \leqq u\}.$$

PROOF. *Claims (a, b) are easy.*
Claim (c). Let $u \geqq 0$ and $u_n \downarrow u$. As (b) implies, $C_X(u_n)$ converges, and the limit is at least $C_X(u)$. Select positive ε_n tending to 0. Select a_n such that

$$C_X(u_n) \leqq \mathscr{P}\{a_n \leqq X \leqq a_n + u_n\} + \varepsilon_n.$$

Then
$$\lim_n C_X(u_n) \leqq \limsup_n \mathscr{P}\{a_n \leqq X \leqq a_n + u_n\}.$$

If $|a_n| \to \infty$, this forces
$$\lim_n C_X(u_n) = 0 \leqq C_X(u).$$

Otherwise, pass to a subsequence n^* with $a_{n*} \to a$ and use (33):

$$\lim_n C_X(u_n) \leqq \mathscr{P}\{a \leqq X \leqq a + u\} \leqq C_X(u).$$

Claim (d). Fix $\varepsilon > 0$. For suitable a,

$$C_X(u + v) \leqq \mathscr{P}\{a \leqq X \leqq a + u + v\} + \varepsilon$$
$$\leqq \mathscr{P}\{a \leqq X \leqq a + u\} + \mathscr{P}\{a + u \leqq X \leqq a + u + v\} + \varepsilon$$
$$\leqq C_X(u) + C_X(v) + \varepsilon.$$

Claim (e). Let F be the distribution function of Y. Use Fubini:

$$\mathscr{P}\{x \leqq X + Y \leqq x + u\} = \int \mathscr{P}\{x - y \leqq X \leqq x - y + u\} \, F(dy)$$

$$\leqq \int C_X(u) \, F(dy)$$

$$= C_X(u).$$

Claim (f). If $x \leqq X \leqq x + u$ and $x \leqq Y \leqq x + u$, then $|X - Y| \leqq u$.
So

$$\mathscr{P}\{x \leqq X \leqq x + u\}^2 \leqq \mathscr{P}\{|X - Y| \leqq u\}. \qquad\qquad ★$$

For (35–36), let $X_1, X_2, \ldots$ be independent, identically distributed random variables on the probability triple $(\Omega, \mathscr{F}, \mathscr{P})$. Suppose $\mathscr{P}\{X_i = 0\} < 1$. Let K be a positive number. Let $S_n = X_1 + \cdots + X_n$. Let τ be the least n if any with $|S_n| \geqq K$, and $\tau = \infty$ if none. Use E for expectation relative to $\mathscr{P}$.

(35) Lemma. (a) *There are $A < \infty$ and $\rho < 1$ such that $\mathscr{P}\{\tau > n\} \leqq A\rho^n$.*

(b) *$E\{\tau\} < \infty$ and $\mathscr{P}\{\tau < \infty\} = 1$.*

(c) *Either $\mathscr{P}\{\limsup S_n = \infty\} = 1$ or $\mathscr{P}\{\liminf S_n = -\infty\} = 1$.*

NOTE. In (a), the constants A and ρ do not depend on n; the inequality holds for all n.

PROOF. *Claim (a).* Suppose $\mathscr{P}\{X_i > 0\} > 0$; the case $\mathscr{P}\{X_i < 0\} > 0$ is symmetric. Find $\delta > 0$ so small that $\mathscr{P}\{X_1 \geqq \delta\} > 0$. If $\mathscr{P}\{X_1 \geqq \delta\} = 1$ the proof terminates; so suppose $\mathscr{P}\{X_1 \geqq \delta\} < 1$. Find a positive integer N so large that $N\delta > 2K$. Fix $k = 0, 1, \ldots$. Now $S_{Nk} > -K$ and $X_i \geqq \delta$ for $i = Nk + 1, \ldots, N(k + 1)$ imply $S_{N(k+1)} \geqq K$. So the relation $\tau > N(k + 1)$ implies $|S_n| < K$ for $1 \leqq n \leqq Nk$, and $X_i < \delta$ for at least one $i = Nk + 1, \ldots, N(k + 1)$. Consequently,

$$\mathscr{P}\{\tau > N(k + 1)\} \leqq (1 - \theta)\,\mathscr{P}\{\tau > Nk\},$$

where $\theta = \mathscr{P}\{X_1 \geqq \delta\}^N > 0$. By substituting,

$$\mathscr{P}\{\tau > N(k + 1)\} \leqq (1 - \theta)^{k+1}.$$

If $Nk \leqq n < N(k + 1)$,

$$\mathscr{P}\{\tau > n\} \leqq \mathscr{P}\{\tau > Nk\} \leqq (1 - \theta)^k \leqq \frac{1}{1 - \theta}\,\rho^n,$$

where $\rho = (1 - \theta)^{1/N}$.

Claim (b) is easy, using (a).

Claim (c). Suppose the claim is false. By Hewitt-Savage (1.122),

$$\mathscr{P}\{\limsup S_n < \infty\} = \mathscr{P}\{\liminf S_n > -\infty\} = 1.$$

So
$$\mathscr{P}\{\sup |S_n| < \infty\} = 1.$$

By countable additivity, there is a finite K with
$$\mathscr{P}\{\sup |S_n| < K\} > 0.$$

This contradicts (b).   ★

Recall that $X_1, X_2, \ldots$ are independent and identically distributed, and $S_n = X_1 + \cdots + X_n$.

(36) Theorem. *If $\mathscr{P}\{X_1 = c\} < 1$ for all c, then $C_{S_n}(u) \to 0$ as $n \to \infty$, for each $u \geq 0$.*

PROOF. If the theorem fails, use (34b, e) to find $u > 0$ and $\delta > 0$ with $C_{S_m}(u) \geq \delta$ for all m. Let $X_1, Y_1, X_2, Y_2, \ldots$ be independent and identically distributed. Let $U_i = X_i - Y_i$ and $T_n = U_1 + \cdots + U_n$. By (34f),

(37) $$\mathscr{P}\{|T_n| \leq u\} \geq \delta^2 \quad \text{for all } n.$$

Let F be the distribution function of X_1. By Fubini,
$$\mathscr{P}\{T_1 = 0\} = \mathscr{P}\{X_1 = Y_1\}$$
$$= \int \mathscr{P}\{X_1 = y\} \, F(dy)$$
$$< 1.$$

By (35c) and symmetry,

(38) $$\mathscr{P}\{\limsup T_n = \infty\} = 1.$$

Choose a positive integer N so large that $N\delta^2 \geq 2$. I will obtain a contradiction by constructing a positive integer M, and N disjoint intervals $I_1, I_2, \ldots, I_N$, so that
$$\mathscr{P}\{T_M \in I_j\} \geq \tfrac{3}{4}\delta^2 \quad \text{for } j = 1, \ldots, N.$$

Let $I_1 = [-u, u]$. By (37),
$$\mathscr{P}\{T_n \in I_1\} \geq \tfrac{3}{4}\delta^2 \quad \text{for all } n.$$

As in Figure 6, let τ_1 be the least n with $T_n > 3u$. As (38) implies, $\mathscr{P}\{\tau_1 < \infty\} = 1$. Choose positive integers M_1 and K_1 so large that
$$\mathscr{P}\{\tau_1 \leq M_1 \text{ and } T_{\tau_1} \leq K_1\} \geq \tfrac{3}{4}.$$

In particular, $K_1 > 3u$. Let $I_2 = [2u, K_1 + u]$. I claim
$$\mathscr{P}\{T_n \in I_2\} \geq \tfrac{3}{4}\delta^2 \quad \text{for all } n \geq M_1.$$

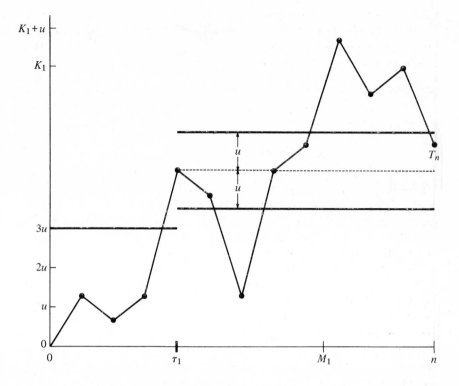

Figure 6.

Indeed, $\{T_n \in I_2\} \supset \bigcup_{j=1}^{M_1} A_j$ for $n \geqq M_1$, where

$$A_j = \{\tau_1 = j \text{ and } T_j \leqq K_1 \text{ and } |T_n - T_j| \leqq u\}.$$

Now $\{\tau_1 = j \text{ and } T_j \leqq K_1\}$ is measurable on $U_1, \ldots, U_j$; moreover, $T_n - T_j$ is measurable on $U_{j+1}, U_{j+2}, \ldots$ and is distributed like T_{n-j}. Using (37),

$$\mathscr{P}\{A_j\} \geqq \mathscr{P}\{\tau_1 = j \text{ and } T_j \leqq K_1\} \, \delta^2.$$

Sum out $j = 1, \ldots, M_1$:

$$\mathscr{P}\{T_n \in I_2\} \geqq \mathscr{P}\{\tau_1 \leqq M_1 \text{ and } T_{\tau_1} \leqq K_1\} \, \delta^2.$$

To proceed, let τ_2 be the least $n > \tau_1$ with $T_n > K_1 + 3u$; find $M_2 > M_1$ and K_2 with

$$\mathscr{P}\{\tau_2 \leqq M_2 \text{ and } T_{\tau_2} \leqq K_2\} \geqq \tfrac{3}{4}.$$

In particular, $K_2 > K_1 + 3u$. Let $I_2 = [K_1 + 2u, K_2 + u]$. Stages 3, 4, ..., N should be clear; M is any integer greater than M_N.   ★

The much more interesting inequality (39) of Kolmogorov shows $C_{S_n}(u) = O(n^{-\frac{1}{2}})$, instead of $o(1)$. This result is included because of its interest; it is not used in the book. I learned the proof from Lucien LeCam. A reference is (Rogozin, 1961).

(39) Theorem. *Let n be a positive integer. Let $X_1, \ldots, X_n$ be independent random variables, perhaps with different distributions. Let u and v be nonnegative numbers. Then*

$$C_{X_1 + \cdots + X_n}(u) \leq \tfrac{3}{2}(1 + \langle u/v \rangle) \{\Sigma_{i=1}^n [1 - C_{X_i}(v)]\}^{-\frac{1}{2}}.$$

In the formula, $0/0 = 1$ and $u/0 = \infty$ for $u > 0$; while $\langle x \rangle$ is the greatest integer which does not exceed x.

The main lemma (40) is combinatorial. It leads to an estimate (42) of the concentration function of a sum of symmetric, two point variables. General variables turn out (45) to have a symmetric part. This does the case $u = v = 2$ in (49). The case $u = v > 0$ follows by scaling, and $u = v = 0$ by continuity. Finally, the general case drops out of subadditivity.

Here are the preliminaries. Write $\#G$ for the cardinality of G, namely the number of elements of G. Let F be a finite set. If $A \subset F$ and $B \subset F$, say A and B are *incomparable* iff neither $A \subset B$ nor $B \subset A$. For example, A and B are incomparable provided $A \neq B$ and $\#A = \#B$. A family $\mathscr{F}$ of subsets of F is *incomparable* iff $A \in \mathscr{F}$ and $B \in \mathscr{F}$ and $A \neq B$ imply A and B are incomparable. Any family of subsets with common cardinality is incomparable. Let $n = \#F$. The family $\hat{\mathscr{F}}$ of subsets of F having cardinality $\langle n/2 \rangle$ is incomparable and has cardinality $\binom{n}{\langle n/2 \rangle}$.

(40) Lemma. $\#\mathscr{F} \leq \binom{n}{\langle n/2 \rangle}$ *for any incomparable family $\mathscr{F}$ of subsets of F: where $n = \#F$.*

PROOF. Suppose n is even. Let $\mathscr{F}$ be an incomparable family with maximal $\#\mathscr{F}$. I assert that $\#A = n/2$ for all $A \in \mathscr{F}$. Since $\mathscr{F}^* = \{F \backslash A : A \in \mathscr{F}\}$ is also incomparable, and $\#\mathscr{F}^* = \#\mathscr{F}$ is also maximal, I only have to show that $A \in \mathscr{F}$ implies $\#A \geq n/2$.

By way of contradiction, suppose

$$r = \min\{\#A : A \in \mathscr{F}\} < n/2,$$

and suppose $A_1, \ldots, A_j$ in $\mathscr{F}$ have cardinality r, while other $A \in \mathscr{F}$ have $\#A > r$. Consider the set Σ of all pairs (A_i, x), such that $i = 1, \ldots, j$, and for each i, the point $x \in F \backslash A_i$. Of course,

$$\#\Sigma = j(n - r).$$

Let $\mathscr{F}_0$ be the family of subsets of F of the form $A_i \cup \{x\}$ for $(A_i, x) \in \Sigma$. This representation is not unique, and I must now estimate $\#\mathscr{F}_0$. Consider the set Σ_0 of all pairs (B, y), where $B \in \mathscr{F}_0$ and $y \in B$. Plainly

$$\#\Sigma_0 = (\#\mathscr{F}_0)(r + 1).$$

Now $(A_i, x) \to (A_i \cup \{x\}, x)$ is a 1-1 mapping of Σ into Σ_0, so

$$(\#\mathscr{F}_0)(r + 1) = \#\Sigma_0 \geqq \#\Sigma = j(n - r).$$

But $r < n/2$ and n is even, so $r < (n - 1)/2$, and $n - r > r + 1$. Therefore,

$$\#\mathscr{F}_0 > j.$$

Let $\mathscr{F}' = \mathscr{F}_0 \cup (\mathscr{F}\setminus\{A_1, \ldots, A_j\})$. So $\#\mathscr{F}' > \#\mathscr{F}$. I will argue that $\mathscr{F}'$ is incomparable. This contradicts the maximality of $\#\mathscr{F}$, and settles the even case. First, $\mathscr{F}_0$ is incomparable since all $A \in \mathscr{F}_0$ have the same cardinality. Second, $\mathscr{F}\setminus\{A_1, \ldots, A_j\}$ is incomparable because $\mathscr{F}$ is. Third, if $(A_i, x) \in \Sigma$ and $B \in \mathscr{F}\setminus\{A_1, \ldots, A_j\}$, then

$$\#(A_i \cup \{x\}) \leqq \#B,$$

so $B \subset A_i \cup \{x\}$ entails $B = A_i \cup \{x\}$ and $A_i \subset B$. And $A_i \cup \{x\} \subset B$ also implies $A_i \subset B$. But $A_i \subset B$ contradicts the incomparability of $\mathscr{F}$. This completes the argument that $\mathscr{F}'$ is incomparable, so the proof for even n.

Suppose n is odd. Let $\mathscr{F}$ be incomparable with maximal $\#\mathscr{F}$. The argument for even n shows that $A \in \mathscr{F}$ has

$$\#A = (n - 1)/2 \text{ or } (n + 1)/2.$$

Suppose some $A \in \mathscr{F}$ have $\#A = (n - 1)/2$ and some $\#A = (n + 1)/2$. Let $A_1, \ldots, A_j$ have cardinality $(n - 1)/2$. Repeat the argument for even n to construct an incomparable family $\mathscr{F}'$ with $\#\mathscr{F}' \geqq \#\mathscr{F}$, and all $A \in \mathscr{F}$ having $\#A = (n + 1)/2$. ★

To state (41), let $x_1, \ldots, x_n$ be real numbers greater than 1. Let V be the set of n-tuples $v = (v_1, \ldots, v_n)$ of ± 1. Let a be a real number. Let U be the set of $v \in V$ with

$$a \leqq \Sigma_{i=1}^n v_i x_i \leqq a + 2.$$

(41) Lemma. $\#U \leqq \binom{n}{\langle n/2 \rangle}$.

PROOF. For $v \in V$, let $A(v) = \{i : i = 1, \ldots, n \text{ and } v_i = 1\}$. For $u \neq v$ in U, the sets $A(u)$ and $A(v)$ are incomparable, because all $x_i > 1$. Use (40). ★

To state (42), let $X_1, \ldots, X_n$ be independent random variables on $(\Omega, \mathscr{F}, \mathscr{P})$. Suppose for each i there is a nonnegative real number x_i with

$$\mathscr{P}\{X_i = x_i\} = \mathscr{P}\{X_i = -x_i\} = \tfrac{1}{2}.$$

Let m be the number of $x_i > 1$.

(42) Lemma. $C_{X_1+\cdots+X_n}(2) \leq 2^{-m} \begin{pmatrix} m \\ \langle m/2 \rangle \end{pmatrix}.$

PROOF. By (34e), it is enough to do the case $m = n$. This case is immediate from (41). ★

(43) Lemma. $2^{-n} \begin{pmatrix} n \\ \langle n/2 \rangle \end{pmatrix} \leq (1 + n)^{-\frac{1}{2}}.$

PROOF. Suppose first that n is even. Abbreviate

$$t_m = 2^{-2m} \begin{pmatrix} 2m \\ m \end{pmatrix}.$$

By algebra,

$$t_{m+1} = \frac{2m + 1}{2m + 2} t_m;$$

so

$$(1 + 2m + 2)^{\frac{1}{2}} t_{m+1} = \frac{(2m + 1)^{\frac{1}{2}}(2m + 3)^{\frac{1}{2}}}{2m + 2}(1 + 2m)^{\frac{1}{2}} t_m.$$

Geometric means are less than arithmetic means, so

$(1 + 2m)^{\frac{1}{2}} t_m$ decreases as m increases.

This proves the even n case. Parenthetically, $(2m)^{\frac{1}{2}} t_m$ increases with m.
Suppose n is odd. By algebra,

$$2^{-2m-1} \begin{pmatrix} 2m + 1 \\ m \end{pmatrix} = \frac{2m + 1}{2m + 2} 2^{-2m} \begin{pmatrix} 2m \\ m \end{pmatrix},$$

and by inspection,

$$\frac{2m + 1}{2m + 2}(1 + 2m)^{-\frac{1}{2}} \leq (1 + 2m + 1)^{-\frac{1}{2}};$$

so the lemma holds for odd n because it does for even n. ★

To state (44), let W be a nonnegative random variable, and k a positive real number.

(44) Lemma. *If* $\operatorname{Var} W \leq E(W) = k$, *then* $E[(1 + W)^{-\frac{1}{2}}] \leq \frac{3}{2}k^{-\frac{1}{2}}.$

PROOF. Define the function g on $[0, \infty)$ by $g(x) = (1 + x)^{-\frac{1}{2}}$. Verify that g is convex and decreasing. Let $f(0) = g(0)$; let f be linear on $[0, k]$; and let $f(x) = g(k)$ for $x \geq k$. See Figure 7. Algebraically,

$$f(x) = k^{-1}[1 - g(k)](x - k)^- + g(k),$$

where $y^- = \max\{-y, 0\}$. Now $g \leqq f$, so $E[g(W)] \leqq E[f(W)]$. But

$$E[(W - k)^-] = \tfrac{1}{2}E[|W - k|] \leqq \tfrac{1}{2}\{E[(W - k)^2]\}^{\frac{1}{2}} \leqq \tfrac{1}{2}k^{\frac{1}{2}},$$

by Schwarz. Consequently,

$$\begin{aligned} E[g(W)] &\leqq k^{-1}\,[1 - g(k)]\,\tfrac{1}{2}\,k^{\frac{1}{2}} + g(k) \\ &\leqq \tfrac{1}{2}k^{-\frac{1}{2}} + k^{-\frac{1}{2}} \\ &= \tfrac{3}{2}k^{-\frac{1}{2}}. \end{aligned}$$

★

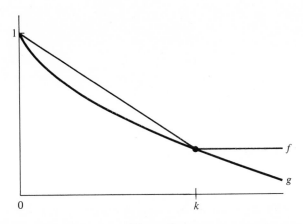

Figure 7

For (45–50), let $X_1, \ldots, X_n$ be independent random variables, with distribution functions $F_1, \ldots, F_n$ and concentration functions $C_1, \ldots, C_n$, respectively. The best case to think about is continuous, strictly increasing F_i. Do not assume the F_i are equal. There are nondecreasing functions $f_1, \ldots, f_n$ on $(0, 1)$, whose distribution functions with respect to Lebesgue measure are $F_1, \ldots, F_n$: for example $f_i(y) = \inf\{x : F_i(x) > y\}$. Let

$$g_i(y) = \frac{f_i(1 - y) + f_i(y)}{2} \quad \text{and} \quad h_i(y) = \frac{f_i(1 - y) - f_i(y)}{2}.$$

On a convenient probability triple $(\Omega, \mathscr{F}, \mathscr{P})$, construct independent random variables $Y_1, \ldots, Y_n, \delta_1, \ldots, \delta_n$, where each Y_i is uniformly distributed on $[0, \tfrac{1}{2}]$, and each δ_i is ± 1 with probability $\tfrac{1}{2}$. Let

$$Z_i = g_i(Y_i) + \delta_i h_i(Y_i).$$

(45) Lemma. $\{Z_i : i = 1, \ldots, n\}$ *is distributed like* $\{X_i : i = 1, \ldots, n\}$.

PROOF. Plainly, the Z_i are $\mathscr{P}$-independent. Let φ be a bounded, measurable function on $(-\infty, \infty)$. Check this computation, where E is expectation relative to $\mathscr{P}$.

$$E\{\varphi[g_i(Y_i) + \delta_i h_i(Y_i)]\} = \tfrac{1}{2}E\{\varphi[g_i(Y_i) - h_i(Y_i)]\} + \tfrac{1}{2}E\{\varphi[g_i(Y_i) + h_i(Y_i)]\}$$
$$= \tfrac{1}{2}E\{\varphi[f_i(Y_i)]\} + \tfrac{1}{2}E\{\varphi[f_i(1 - Y_i)]\}$$
$$= \int_0^{\frac{1}{2}} \varphi[f_i(y)] \, dy + \int_0^{\frac{1}{2}} \varphi[f_i(1 - y)] \, dy$$
$$= \int_0^{\frac{1}{2}} \varphi[f_i(y)] \, dy + \int_{\frac{1}{2}}^1 \varphi[f_i(y)] \, dy$$
$$= \int_0^1 \varphi[f_i(y)] \, dy$$
$$= \int_{-\infty}^{\infty} \varphi(x) \, F_i \, (dx). \qquad \qquad \bigstar$$

(46) Lemma. $\mathscr{P}\{h_i(Y_i) > 1\} \geqq 1 - C_i(2)$.

PROOF. Fix y with

(47) $$0 < y < \tfrac{1}{2}[1 - C_i(2)].$$

Then $y < \tfrac{1}{2}$, so

(48) $$f_i(y) \leqq f_i(z) \leqq f_i(1 - y) \quad \text{for } y \leqq z \leqq 1 - y.$$

I say

$$C_i(2) < 1 - 2y$$
$$\leqq \text{Lebesgue } \{z : f_i(y) \leqq f_i(z) \leqq f_i(1 - y)\}$$
$$= \text{Prob } \{f_i(y) \leqq X_i \leqq f_i(1 - y)\}:$$

the first line follows from (47), the second from (48), and the third from the fact that the Lebesgue distribution of f_i coincides with the distribution of X_i. From the definition of the concentration function,

$$f_i(1 - y) - f_i(y) > 2;$$

so

(47)   implies $h_i(y) > 1$.

Therefore,

$$\{Y_i < \tfrac{1}{2}[1 - C_i(2)]\} \subset \{h_i(Y_i) > 1\}.$$

But Y_i is uniform on $[0, \tfrac{1}{2}]$. $\qquad \qquad \bigstar$

(49) Proposition. *Let n be a positive integer. Let $X_1, \ldots, X_n$ be independent random variables, perhaps with different distributions. Then*

$$C_{X_1 + \cdots + X_n}(2) \leqq \tfrac{3}{2} \{\Sigma_{i=1}^n [1 - C_{X_i}(2)]\}^{-\frac{1}{2}}.$$

PROOF.   Remember that X_i has distribution function F_i and concentration function C_i. Remember that f_i is a nondecreasing function on $(0, 1)$ whose Lebesgue distribution is F_i. Remember

$$g_i(y) = \frac{f_i(1 - y) + f_i(y)}{2} \quad \text{and} \quad h_i(y) = \frac{f_i(1 - y) - f_i(y)}{2}.$$

Remember $Y_1, \ldots, Y_n, \delta_1, \ldots, \delta_n$ are independent random variables on $(\Omega, \mathscr{F}, \mathscr{P})$; the Y_i are uniform over $[0, \frac{1}{2}]$; the δ_i are ± 1 with probability $\frac{1}{2}$. Remember

$$Z_i = g_i(Y_i) + \delta_i h_i(Y_i).$$

Remember that E is expectation relative to $\mathscr{P}$.

Let $\varphi_i(y) = 1$ or 0 according as $h_i(y) > 1$ or $h_i(y) \leq 1$, for $0 \leq y \leq \frac{1}{2}$. Let Q be the $\mathscr{P}$-distribution of $Y = (Y_1, \ldots, Y_n)$, a probability on S, the set of all n-tuples $y = (y_1, \ldots, y_n)$ with $0 \leq y_i \leq \frac{1}{2}$. For $y \in S$, let

$$\varphi(y) = \varphi_1(y_1) + \cdots + \varphi_n(y_n).$$

Let a be a real number. Let

$$\pi = \text{Prob} \{a \leq \Sigma_{i=1}^{n} X_i \leq a + 2\}.$$

By (45) and Fubini,

$$\pi = \mathscr{P}\{a \leq \Sigma_{i=1}^{n} Z_i \leq a + 2\}$$

$$= \int_{y \in S} \mathscr{P}\{a \leq \Sigma_{i=1}^{n} g_i(y_i) + \Sigma_{i=1}^{n} h_i(y_i)\delta_i \leq a + 2\} \, Q(dy).$$

For each vector y, the integrand is at most

$$2^{-\varphi(y)} \binom{\varphi(y)}{\langle \varphi(y)/2 \rangle} \leq [1 + \varphi(y)]^{-\frac{1}{2}}$$

by (42–43). Therefore,

$$\pi \leq E\{[1 + \varphi(Y)]^{-\frac{1}{2}}\}.$$

Abbreviate $p_i = \mathscr{P}\{h_i(Y_i) > 1\}$ and verify this computation.

$$E[\varphi(Y)] = \Sigma_{i=1}^{n} p_i.$$

$$\text{Var} [\varphi(Y)] = \Sigma_{i=1}^{n} \text{Var} [\varphi_i(Y_i)]$$
$$\leq \Sigma_{i=1}^{n} E\{[\varphi_i(Y_i)]^2\}$$
$$= \Sigma_{i=1}^{n} E\{\varphi_i(Y_i)\}$$
$$= \Sigma_{i=1}^{n} p_i.$$

By (44),

$$\pi \leq \tfrac{3}{2}(\Sigma_{i=1}^{n} p_i)^{-\frac{1}{2}}.$$

Finally, use (46).   ★

(50) Proposition.   *Let n be a positive integer. Let $X_1, \ldots, X_n$ be independent random variables, perhaps with different distributions. Let v be a nonnegative number. Then*

$$C_{X_1 + \cdots + X_n}(v) \leqq \tfrac{3}{2} \left\{ \Sigma_{i=1}^{n} \left[1 - C_{X_i}(v) \right] \right\}^{-\frac{1}{2}}.$$

PROOF.   For $v > 0$, use (49) and (34a). Then let $v \downarrow 0$ and use (34c).   ★

PROOF OF (39).   It is harmless to suppose $v > 0$. Abbreviate $C = C_{X_1 + \cdots + X_n}$. If $u < v$, then $C(u) \leqq C(v)$ by (34b), and the inequality (39) holds by (50). Let m be a positive integer and $mv \leqq u < (m + 1)v$. Then $\langle u/v \rangle = m$. By (34b, d)

$$\begin{aligned}
C(u) &\leqq C[(m + 1)v] \\
&\leqq (m + 1)C(v) \\
&= [1 + \langle u/v \rangle]C(v).
\end{aligned}$$

Use (50).   ★

4

THE BOUNDARY

1. INTRODUCTION

This chapter is based on work of Blackwell (1955, 1962), Doob (1959), Feller (1956), and Hunt (1960). Let P be a substochastic matrix on I: that is, $P(i,j) \geqq 0$ and $\Sigma_{j \in I} P(i,j) \leqq 1$. Let P^0 be the identity matrix, and $G = \Sigma_{n=0}^{\infty} P^n$. Suppose $G < \infty$. By (1.51), this is equivalent to saying that all $i \in I$ are transient. Let p be a probability on I such that $pG(i) > 0$ for all $i \in I$. Here $pG(i)$ means $\Sigma_{j \in I} p(j)G(j,i)$. A function h on I is *excessive* iff:

(1) $$h \geqq 0;$$

(2) $$\Sigma_{i \in I} p(i)h(i) = 1;$$

and

(3) $$\Sigma_{j \in I} P(i,j)h(j) \leqq h(i) \quad \text{for all } i \in I.$$

Check, $h(i) < \infty$ for all $i \in I$. If equality holds in (3), then h is *harmonic*. Because of (2), these definitions are relative to the *reference probability p*. Throughout, i, j are used for generic elements of I, and h for a generic excessive function.

The set of h is convex, and compact in the coordinatewise topology. One object of this chapter is to identify the extreme h, and prove that any h can be represented as a unique integral average of extreme h. This is equivalent to constructing a regular conditional distribution for the Markov chain given the invariant σ-field.

Give I the σ-field of all its subsets. Let Ω^* be the set of all finite, nonempty I-sequences, with the σ-field of all its subsets. Let I^{∞} be the set of infinite

I want to thank Isaac Meilijson for checking the final draft of this chapter.

111

I-sequences, with the product σ-field. Give $\Omega^* \cup I^\infty$ the σ-field generated by all the subsets of Ω^* and all the measurable subsets of I^∞. Let $\xi_0, \ldots$ be the coordinate processes on $\Omega^* \cup I^\infty$. Let Ω be the union of Ω^* and the set Ω^∞ of $\omega \in I^\infty$ such that $\xi_n(\omega) = i$ for only finitely many n, for each $i \in I$. Retract $\xi_0, \ldots$ to Ω, and give Ω the relative σ-field.

For any probability q on I, let P_q be the probability on Ω for which the coordinate process is Markov with starting probability q and stationary transitions P. Of course, $P_q(\Omega) = 1$ because I is transient and q is a probability. If $q(i) = 1$, write P_i for P_q. Let $I^h = \{i : i \in I \text{ and } h(i) > 0\}$. Let $(ph)(i) = p(i)h(i)$, and

$$P^h(i, j) = h(i)^{-1}P(i, j)h(j)$$

for i and j in I^h. Plainly, P^h is substochastic on I^h,

$$(P^h)^n = (P^n)^h \quad \text{and} \quad G^h(i, j) = h(i)^{-1}G(i, j)h(j) = \Sigma_{n=0}^\infty (P^h)^n(i, j) < \infty.$$

Abbreviate Q_h for P^h_{ph}, the probability on Ω such that the coordinate process is Markov with starting distribution ph and stationary transitions P^h. That is,

$$Q_h\{\xi_0 = i_0, \ldots, \xi_n = i_n\} = 0$$

unless $i_0, \ldots, i_n$ are all in I^h, in which case

$$Q_h\{\xi_0 = i_0, \ldots, \xi_n = i_n\} = (ph)(i_0) \Pi_{m=0}^{n-1} P^h(i_m, i_{m+1}).$$

Let $\mathscr{I}$ be the invariant σ-field of Ω; the definition will be given later. The main result will now be summarized. There is one subset $H \in \mathscr{I}$, of Q_h-measure 1 for all h; and for each $i \in I$, there is an $\mathscr{I}$-measurable function $g(i)$ from H to the real line, such that for each $\omega \in H$, the function $g(\cdot)(\omega)$ on I is excessive; moreover, abbreviating Q_ω for $Q_{g(\cdot)(\omega)}$, the mapping $\omega \to Q_\omega$ is a regular conditional Q_h-probability given $\mathscr{I}$ for all h.

This result, and general reasoning, give the following: Let E be the set of $\omega \in H$ satisfying:

$$Q_\omega\{\omega' : \omega' \in H \text{ and } Q_{\omega'} = Q_\omega\} = 1;$$

and

$$Q_\omega(\Omega^\infty) = 1 \quad \text{for } \omega \in \Omega^\infty.$$

Then $E \in \mathscr{I}$ and $Q_h(E) = 1$. Let $\mathscr{E}$ be the σ-field of subsets of E measurable on $\omega \to Q_\omega$. Then there is one and only one probability on $\mathscr{E}$, namely Q_h, integrating g to h. Thus, as ω runs through E, the function $g(\cdot)(\omega)$ on I runs through the extreme, excessive h. Finally, the extreme, excessive h which are not harmonic are precisely the functions $G(\cdot, j)/pG(j)$ as j varies over I.

The function $g(i)$ is a version of the Radon–Nikodym derivative of P_i with respect to P_p, retracted to $\mathscr{I}$. The main difficulty is to choose this version properly from the point of view of the Q_h, for as h varies through the extreme excessive functions, the Q_h are mutually orthogonal. There are two properties $g(i)$ must have from the point of view of Q_h: it must vanish a.e. when $i \notin I^h$; and $g(i)h(i)$ must be a version of the Radon–Nikodym derivative of P_i^h with respect to Q_h, when retracted to $\mathscr{I}$, for $i \in I^h$. Perhaps the leading special case is the following: p concentrates on one state, P is stochastic, $G(j, k) > 0$ for all $j, k \in I$, and h is harmonic. Then h is positive everywhere and Ω^* is not needed. Moreover, Ω_n has measure 1 if $p(I_n) = 1$; these quantities will be defined later. Section 2 contains proofs. Section 3 contains the following theorem:

$$G(i, \xi_n)/p(\xi_n) \text{ converges to a finite limit } P_p\text{-almost surely.}$$

Section 4 contains examples. Section 5 contains related material, which will be referred to in *ACM*.

NOTATION.   (a) G used to be called eP in (1.49).

(b) Ω^* does not include the empty sequence, thereby differing from I^* in Section 1.3.

(c) If S is a set, $\mathscr{F}$ is a σ-field of subsets of S, and $F \in \mathscr{F}$, then $F\mathscr{F}$ is the σ-field of subsets of F of the form FA, with $A \in \mathscr{F}$. This σ-field is called $\mathscr{F}$ *relativized to* F, or the *relative σ-field* if F and $\mathscr{F}$ are understood. This notation is only used when $F \in \mathscr{F}$.

2.   PROOFS

Recall that $\Omega = \Omega^* \cup \Omega^\infty$, where Ω^* is the set of nonempty, finite I-sequences, and Ω^∞ is the set of infinite I-sequences which visit each state only finitely often. And $\{\xi_n\}$ is the coordinate process on Ω. The *shift* T is this mapping of Ω into Ω. If $\omega \in \Omega^\infty$, then $T\omega \in \Omega^\infty$ and

$$\xi_n(T\omega) = \xi_{n+1}(\omega) \quad \text{for } n = 0, 1, \dots.$$

If $\omega \in \Omega^*$ has length $m \geq 2$, then $T\omega \in \Omega^*$ has length $m - 1$ and

$$\xi_n(T\omega) = \xi_{n+1}(\omega) \quad \text{for } n = 0, \dots, m - 2.$$

If $\omega \in \Omega^*$ has length 1, then $T\omega = \omega$. For all ω, let $T^0\omega = \omega$, and $T^{n+1}\omega = TT^n\omega$. The *invariant σ-field* $\mathscr{I}$ of Ω is the σ-field of measurable subsets A of Ω which are *invariant*: $T^{-1}A = A$. Let $\mathscr{I}^* = \Omega^*\mathscr{I}$ and $\mathscr{I}^\infty = \Omega^\infty\mathscr{I}$.

The first lemma below gives a more constructive definition of $\mathscr{I}$. To state it, and for use throughout the section, fix a sequence I_n of finite subsets of I which increase to I; let Ω_n be the set of $\omega \in \Omega$ with $\xi_m(\omega) \in I_n$ for some $m = 0, 1, \dots$. On Ω_n, let τ_n be the largest m with $\xi_m \in I_n$, and let $Y_n = \xi_{\tau_n}$.

Let T_n be this mapping from Ω_n to Ω:

$$T_n(\omega) = T^{\tau_n(\omega)}(\omega).$$

Verify the measurability of T_n. Let $\mathscr{I}_n$ be the σ-field generated by the subsets of Ω_n measurable on T_n, and all measurable subsets of $\Omega\backslash\Omega_n$. For $\omega \in \Omega^*$, let $L(\omega)$ be the last defined coordinate of ω.

(4) Lemma.   (a) $\mathscr{I}_n \downarrow \mathscr{I}$.

   (b) $\mathscr{I}^*$ is the σ-field of subsets of Ω^* generated by L.

PROOF.   *Assertion* (a). I claim that $\mathscr{I}_n$ decreases. To begin with, $T_{n+1} = T_{n+1} \circ T_n$ on Ω_n. If $A \in \Omega_{n+1}\mathscr{I}_{n+1}$, then $A = T_{n+1}^{-1}B$ for some measurable B. So

$$A = (T_n^{-1} T_{n+1}^{-1} B) \cup (A\backslash\Omega_n) \in \mathscr{I}_n.$$

And Ω_n increases with n, completing the proof that $\mathscr{I}_n$ decreases. I claim that $\bigcap_n \mathscr{I}_n \subset \mathscr{I}$. Indeed, let $A \in \bigcap_n \mathscr{I}_n$, and fix ω. I have to show $\omega \in A$ iff $T\omega \in A$. Suppose ω has length at least 2, for otherwise $\omega = T\omega$. Fix a positive integer n so large that $\xi_1(\omega) \in I_n$. Then ω and $T\omega$ are in Ω_n, and $T_n(\omega) = T_n(T\omega)$. But $A \in \mathscr{I}_n$, so

$$A = (T_n^{-1}B) \cup (A\backslash\Omega_n)$$

for some measurable B. This shows $\omega \in A$ iff $T\omega \in A$. Finally, I claim $\mathscr{I} \subset \mathscr{I}_n$. Indeed, let $A \in \mathscr{I}$. The problem is to show $A \cap \Omega_n \in \mathscr{I}_n$. But

$$A \cap \Omega_n = T_n^{-1}A \in \mathscr{I}_n.$$

Assertion (b) is immediate from (a) and

$$(5) \qquad\qquad L(\omega) = \xi_0(T^n\omega) \quad \text{for all large } n$$

$$= \xi_0(T_n\omega) \quad \text{for all large } n. \qquad\qquad \bigstar$$

WARNING. $\Omega_n \notin \mathscr{I}_{n+1}$.

(6) Lemma.   *Let $i \in I_N$. Then P_i is absolutely continuous with respect to P_p when both probabilities are retracted to $\Omega_N \mathscr{I}_N$, a version of the Radon-Nikodym derivative on Ω_N being $G(i, Y_N)/pG(Y_N)$.*

PROOF.   Fix M and $j_0, \ldots, j_M$, with $j_0 \in I_N$ and $j_1 \notin I_N, \ldots, j_M \notin I_N$. Let

$$\pi = P(j_0, j_1) \cdots P(j_{M-1}, j_M) \cdot P_{j_M}\{\xi_n \in I_N \text{ for no } n > 0\}.$$

The event

$$A_m = \{\Omega_N \text{ and } \xi_{\tau_N} = j_0, \ldots, \xi_{\tau_N + M} = j_M \text{ and } \tau_N = m\}$$

is the same as

$$\{\xi_m = j_0, \ldots, \xi_{m+M} = j_M \text{ and } \xi_n \in I_N \text{ for no } n > m + N\}.$$

Let q be a generic probability on I. Then Markov (1.15) makes

$$P_q(A_m) = P_q\{\xi_m = j_0\} \cdot \pi.$$

Let $A = \bigcup_{m=0}^{\infty} A_m$, so

$$A = \{Q_N \text{ and } \xi_{\tau_N} = j_0, \ldots, \xi_{\tau_N+M} = j_M\}.$$

Sum out m:

$$P_q\{A\} = qG(j_0) \cdot \pi.$$

Let q assign mass 1 to i or let $q = p$; remember $Y_N = j_0$ on A:

$$P_i\{A\} = \int_A \frac{G(i, Y_N)}{pG(Y_N)} dP_p.$$

Use (10.16) to vary A.   ★

Let C_i be the set where $G(i, Y_n)/pG(Y_n)$ converges to a finite limit as $n \to \infty$. Call the limit $g(i)$. Of course, $g(i)$ may be 0. Plainly, $C_i \in \mathscr{I}$, and $g(i)$ is $\mathscr{I}$-measurable.

(7) Lemma. P_i *is absolutely continuous with respect to* P_p, *when both probabilities are retracted to* $\mathscr{I}$. *Moreover,* $P_p(C_i) = 1$, *and* $g(i)$ *is a version of the Radon–Nikodym derivative of* P_i *with respect to* P_p, *when both probabilities are retracted to* $\mathscr{I}$.

PROOF.   As (4) and (6) imply, P_i is absolutely continuous with respect to P_p on $\Omega_N \cdot \mathscr{I}$. Since $\Omega_N \uparrow \Omega$ as $N \uparrow \infty$, the absolute continuity on $\mathscr{I}$ follows. Use (10.35) on (6) for convergence.   ★

Remember that the probabilities P_i^h and Q_h are concentrated in the part of Ω where all coordinates are in I^h.

(8) Lemma.   *If* $i \notin I^h$, *then* $Q_h(C_i) = 1$ *and* $g(i) = 0$ *with* Q_h-*probability* 1.

PROOF.   Clearly, $Q_h\{Y_m \in I^h\} = 1$. Moreover,

$$h(i) \geqq \sum_j P^n(i, j)h(j) \geqq P^n(i, j)h(j),$$

proving

(9) $i \notin I^h$ and $j \in I^h$ imply $G(i, j) = 0$.

So $i \notin I^h$ makes

$$Q_h\{G(i, Y_n)/pG(Y_n) = 0\} = 1.$$ ★

A helpful consequence of (9) is that for all $i_0, \ldots, i_n$ in I, whether in I^h or not,

(10) $Q_h\{\xi_0 = i_0, \ldots, \xi_n = i_n\} = p(i_0) \, [\Pi_{m=0}^{n-1} P(i_m, i_{m+1})] \, h(i_n).$

Suppose $j \in I^h$. Remember ph acts on I^h. So

(11)
$$phG^h(j) = \sum_{i \in I^h} p(i)h(i)h(i)^{-1}G(i,j)h(j)$$
$$= \sum_{i \in I^h} p(i)G(i,j)h(j)$$
$$= \sum_{i \in I} p(i)G(i,j)h(j) \qquad \text{by (9)}$$
$$= pG(j)h(j)$$
$$> 0.$$

(12) Lemma. *Let $i \in I^h$. Then $Q_h(C_i) = 1$. Moreover, P_i^h is absolutely continuous with respect to Q_h, with Radon-Nikodym derivative $g(i)/h(i)$, provided both probabilities are retracted to $\mathscr{I}$.*

PROOF. Keep n so large that $i \in I_n$. Then P_i^n is absolutely continuous with respect to Q_h when both are retracted to $\Omega_n \mathscr{I}_n$, a Radon-Nikodym derivative being

$$G^h(i, Y_n)/phG^h(Y_n) = h(i)^{-1}[G(i, Y_n)/pG(Y_n)]$$

when $Y_n \in I^h$. This follows from (6) on P^h and (11), in the part of Ω where all coordinates are in I^h. But this part has probability 1, both for P_i^h and Q_h. Now (7) on P^h makes this derivative sequence converge Q_h-almost surely in the part of Ω where all coordinates are in I^h, which is Q_h-almost all of Ω. This makes $Q_h(C_i) = 1$. Lemma (7) also identifies the limit of the derivative sequence, namely $g(i)/h(i)$, as the required Radon-Nikodym derivative.   ★

Let $C = \bigcap_i C_i$. So $C \in \mathscr{I}$, and $Q_h(C) = 1$ by (8) and (12).

(13) Lemma. $h(i) = \int_C g(i) \, dQ_h$.

PROOF. For $i \notin I^h$, use (8). For $i \in I^h$, use (12).   ★

The next step is to calculate the conditional distribution given $\mathscr{I}$.

(14) Lemma. *Let $i_0, \ldots, i_n \in I$. A version of the conditional Q_h-probability that $\xi_0 = i_0, \ldots, \xi_n = i_n$, given $\mathscr{I}$, is*

$$p(i_0)P(i_0, i_1) \cdots P(i_{n-1}, i_n)g(i_n).$$

PROOF. Let B be the event that $\xi_0 = i_0, \ldots, \xi_n = i_n$. Let $A \in \mathscr{I}$. Check

$$T^{-n}\{\xi_0 = i_n \text{ and } A\} = \{\xi_n = i_n \text{ and } A\}.$$

By Markov (1.15),
$$Q_h(B \cap A) = Q_h(B) P_{i_n}^h(A).$$

By (10),
$$Q_h(B) = P_p(B) h(i_n).$$

By (8) and (12),
$$h(i_n) P_{i_n}^h(A) = \int_A g(i_n) \, dQ_h.$$

That is,

$$Q_h(B \cap A) = \int_A P_p(B)\, g(i_n)\, dQ_h. \qquad \bigstar$$

Let $H = \{\omega : \omega \in C$ and $g(\cdot)(\omega)$ is excessive$\}$. Check $H \in \mathscr{I}$.

(15) Lemma. $Q_h(H) = 1$.

PROOF. Clearly, (1) is satisfied: $g(\cdot)(\omega) \geqq 0$ for all $\omega \in C$. Next, I will work on (2): the set of $\omega \in C$ such that

$$\Sigma_{i \in I}\, p(i)g(i)(\omega) = 1$$

has to have Q_h-probability 1 for all h. But $Q_h\{\xi_0 \in I^h\} = 1$. So with Q_h-probability 1,

$$\begin{aligned}
1 &= Q_h\{\xi_0 \in I^h \mid \mathscr{I}\} \\
&= \Sigma_{i \in I^h}\, Q_h\{\xi_0 = i \mid \mathscr{I}\} \\
&= \Sigma_{i \in I^h}\, p(i)g(i) \qquad\qquad \text{by (14).}
\end{aligned}$$

By (8), the sum can be extended over all of I, at the expense of changing the exceptional null set.

Finally, I will work on (3): for each $i \in I$, the set of ω such that

$$\Sigma_{j \in I}\, P(i, j)g(j)(\omega) \leqq g(i)(\omega)$$

has to have Q_h-probability 1 for all h. Suppose first $i \notin I^h$. Then $Q_h\{g(i) = 0\} = 1$ by (8). If $j \notin I^h$, use (8); if $j \in I^h$, use (9); either way

$$Q_h\{P(i, j)g(j) = 0\} = 1.$$

Suppose next $i \in I^h$. Use (9) and $pG(i) > 0$ to find n and $i_0, \ldots, i_{n-2}$ in I^h with

$$\pi = p(i_0)P(i_0, i_1) \cdots P(i_{n-2}, i) > 0.$$

Let

$$A = \{\xi_0 = i_0,\ \xi_1 = i_1, \ldots,\ \xi_{n-2} = i_{n-2},\ \xi_{n-1} = i\}.$$

Then $A \supset \{A$ and $\xi_n \in I^h\}$. Use (14): with Q_h-probability 1,

$$\begin{aligned}
\pi g(i) &= Q_h\{A \mid \mathscr{I}\} \\
&\geqq Q_h\{A \text{ and } \xi_n \in I^h \mid \mathscr{I}\} \\
&= \Sigma_{j \in I^h}\, Q_h\{A \text{ and } \xi_n = j \mid \mathscr{I}\} \\
&= \pi\, \Sigma_{j \in I^h}\, P(i, j)g(j).
\end{aligned}$$

Divide out π: with Q_h-probability 1,

$$g(i) \geqq \Sigma_{j \in I^h}\, P(i, j)g(j).$$

In view of (9), the sum can be extended over all of I, by changing the exceptional null set.   ★

For $\omega \in H$, let Q_ω be the probability on Ω making the coordinate process a Markov chain with starting probability $pg(\cdot)(\omega)$ and stationary transitions $P^{g(\cdot)(\omega)}$.

(16) Theorem. $\omega \to Q_\omega$ *is a regular conditional Q_h-probability, given $\mathscr{I}$.*

PROOF.   This follows from (14), (15), and (10.16).   ★

Let E_0 be the set of $\omega \in H$ such that: $Q_\omega(\Omega^*)$ is 0 or 1, according as $\omega \in \Omega^*$ or $\omega \in \Omega^\infty$. Let E_1 be the set of $\omega \in H$ such that $g(\cdot)(\omega') = g(\cdot)(\omega)$ for Q_ω-almost all ω'. Let $E = E_0 \cap E_1$. Let $\mathscr{E}$ be the σ-field of subsets of E measurable on g. In particular, $\mathscr{E}$ is countably generated. Verify that

$$Q_h \quad \text{determines} \quad h.$$

Consequently, E_1 is the set of $\omega \in H$ such that $Q_{\omega'} = Q_\omega$ for Q_ω-almost all $\omega' \in H$. And $\mathscr{E}$ is the σ-field generated by $\omega \to Q_\omega$. The atoms of $\mathscr{E}$ are precisely the sets of constancy of $\omega \to g(\cdot)(\omega)$, namely the sets of constancy of $\omega \to Q_\omega$: use (10.18).

(17) Theorem.   (a) $E \in \mathscr{I}$ *and* $Q_h(E) = 1$.

(b) $Q_\omega(A) = 1_A(\omega)$ *for* $\omega \in E$ *and* $A \in \mathscr{E}$.

PROOF.   The set $\Omega^* \in \mathscr{I}$, and E_0 is the set where $\omega \to Q_\omega(\Omega^*)$ is equal to the indicator function of Ω^*. Thus $E_0 \in \mathscr{I}$ and $Q_h(E_0) = 1$. Moreover, $E_1 \in \mathscr{I}$ and $Q_h(E_1) = 1$ by (10.52). Finally, $\omega \in E_1$ makes Q_ω concentrate on the $\mathscr{E}$-atom containing ω.

(18) Theorem.   *There is one and only one probability m on $\mathscr{E}$, such that*

$$\int_E g \, dm = h.$$

Namely, m is Q_h retracted to $\mathscr{E}$.

PROOF.   The retraction of Q_h to $\mathscr{E}$ works by (13) and (17). For the uniqueness, suppose m on $\mathscr{E}$ satisfies

$$\int_E g \, dm = h.$$

As (10) and (10.16) imply,

$$\int_E Q_\omega \, m(d\omega) = Q_h.$$

Use (17b):

$$Q_h(A) = \int_E Q_\omega(A) \, m(d\omega) = m(A).   ★$$

(19) Corollary. *As ω runs through E, the function $g(\cdot)(\omega)$ on I runs through the set of extreme, excessive h. Moreover, h is extreme iff $Q_h\{g = h\} = 1$.*

PROOF.   Using (18), the mapping $m \to \int g\,dm$ is 1–1 and affine, from the set of probabilities on $\mathscr{E}$ onto the set of h. The inverse image of h is the retraction of Q_h to $\mathscr{E}$. Now m is extreme iff it is 0–1, by (10.17a). Thus, h is extreme iff Q_h is 0–1 on $\mathscr{E}$. But Q_h is 0–1 on $\mathscr{E}$ iff

$$Q_h\{g = h'\} = 1$$

for some h', using (10.17b, -18).   Necessarily, $h' = h$ by (18). And there is an $\omega \in E$ with $g(\cdot)(\omega) = h$. Next, let $\omega \in E$ and put $h = g(\cdot)(\omega)$. Let m assign mass 1 to the $\mathscr{E}$-atom containing ω, so $m\{g = h\} = 1$. Then m is 0–1 and $\int g\,dm = h$, so h is extreme.   ★

(20) Lemma. *If $\omega \in E\Omega^\infty$, then $g(\cdot)(\omega)$ is harmonic.*

PROOF.   If h is excessive but not harmonic, then $Q_h(\Omega^*) > 0$. But $Q_\omega(\Omega^*) = 0$, because $\omega \in E_0$.   ★

(21) Lemma. *Fix $k \in I$, and let $h = G(\cdot, k)/pG(k)$. Then h is excessive, and equality holds in (3) iff $i \neq k$. In particular, h determines k.*

PROOF. $\Sigma_{j \in I} P(i, j)G(j, k)$ is the P_i-mean number of visits to k in positive time.   ★

NOTE.   This observation and Fatou afford another proof of (15).

(22) Lemma. *$\Omega^* \subset E$. And for $\omega \in \Omega^*$,*

$$g(\cdot)(\omega) = G(\cdot, L(\omega))/pG(L(\omega)).$$

Moreover, $\mathscr{I}^* = \Omega^*\mathscr{E}$.

PROOF.   Clearly, $\Omega^* \subset C$ and for $\omega \in \Omega^*$,

$$g(\cdot)(\omega) = G(\cdot, L(\omega))/pG(L(\omega)).$$

As (21) implies, $\Omega^* \subset H$. Fix $k \in I$. Let $A = \{\omega : \omega \in \Omega^* \text{ and } L(\omega) = k\}$, and let $h = G(\cdot, k)/pG(k)$.
   I say $Q_h(A) > 0$. Indeed, find n and $i_0, \ldots, i_n$ in I with

$$\pi = p(i_0)P(i_0, i_1) \cdots P(i_{n-1}, i_n)P(i_n, k) > 0.$$

Let

$$\rho = h(k) - \Sigma_j P(k, j)h(j),$$

which is positive by (21). Let

$$B = \{\xi_0 = i_0, \ldots, \xi_n = i_n \text{ and } \xi_{n+1} = k\}$$

and let

$$B_j = \{B \text{ and } \xi_{n+2} = j\}.$$

Use (10):

$$Q_h(B) = \pi h(k) \quad \text{and} \quad Q_h(B_j) = \pi P(k, j) h(j).$$

Then

$$
\begin{aligned}
Q_h(A) &\geq Q_h(B \backslash \cup_j B_j) \\
&= Q_h(B) - \Sigma_j\, Q_h(B_j) \\
&= \pi \rho.
\end{aligned}
$$

Now $Q_h(E) = 1$ by (17), so $E \cap A$ is nonempty. But (4b) and (21) make A an atom of $\mathscr{I}$; and $E \in \mathscr{I}$: therefore $A \subset E$. Consequently, $\Omega^* \subset E$. For $\omega \in \Omega^*$, the function $g(\cdot)(\omega)$ determines $L(\omega)$ by (21). Consequently, $\mathscr{I}^* = \Omega^* \mathscr{E}$.   ★

Incidentally, the argument shows $Q_h(A) = 1$. For a more direct proof, see (33).

(23) Theorem. *As k ranges over I, the function $G(\cdot, k)/pG(k)$ ranges over the extreme, excessive functions which are not harmonic.*

PROOF.   From (19) and (20), the only candidates for the role of extreme, excessive, non-harmonic functions are $g(\cdot)(\omega)$ for $\omega \in E\Omega^*$. As (22) implies, $\Omega^* \subset E$ and each candidate succeeds.   ★

(24) Remark. $\mathscr{E}$ is a countably generated sub σ-field of E. If $A \in \mathscr{I}$, then $\{\omega : \omega \in E \text{ and } Q_\omega(A) = 1\} \in \mathscr{E}$ differs from A by a Q_h-null set. In particular, $\mathscr{I}$ and $\mathscr{E}$ are equivalent σ-fields for Q_h. However, the σ-field $\mathscr{I}$ is inseparable. Each of its atoms is countable. In general, for $\omega \in E$ the probability Q_w is continuous, and therefore assigns measure 0 to the $\mathscr{I}$-atom containing ω.

(25) Remark. h is extreme iff Q_h is 0–1 on $\mathscr{I}$.

PROOF.   Proved in (19).   ★

(26) Remark.   Suppose P is stochastic. Then 1 is extreme iff there are no further bounded harmonic h.

PROOF.   For "if," use an argument based on (1). For "only if" suppose $h \neq 1$ is bounded harmonic and $\varepsilon > 0$ is small. Then

$$1 = \tfrac{1}{2}(1 - \varepsilon)\,\frac{1 - \varepsilon h}{1 - \varepsilon} + \tfrac{1}{2}(1 + \varepsilon)\,\frac{1 + \varepsilon h}{1 + \varepsilon}$$

displays 1 as a convex combination of distinct harmonic functions.   ★

(27) Theorem. *Suppose P is stochastic. Then P_p is 0–1 on $\mathscr{I}$ iff 1 is the only bounded harmonic h.*

PROOF. Use (25) and (26). ★

On a first reading of this chapter, skip to Section 3. It is possible to study the bounded, excessive functions in a little more detail, and in parentheses. To begin with, (10) implies that Q_h is absolutely continuous with respect to P_p on the first $n + 1$ coordinates, and has derivative $h(\xi_n)$. This martingale converges to dQ_h/dP_p by (10.35). Of course, Q_h need not be absolutely continuous with respect to P_p on the full σ-field. However, if h is bounded by K then $Q_h \leqq KP_p$ by what precedes, and $h(\xi_n)$ converges even in L^1. Conversely, if $Q_h \leqq KP_p$, even on $\mathscr{E}$, then h is bounded by K in view of (13). If $h^* = \lim h(\xi_n)$, then $h(i) = E(h^* \mid \xi_n = i)$.

Turn now to extreme, excessive functions which are bounded. The characterization is simple: h is bounded and extreme iff $P_p\{g = h\} > 0$. For (13) implies $1 = \int g \, dP_p$. If $P_p\{g = h\} = \alpha > 0$, then $\alpha h \leqq 1$; while h is extreme by (19). If h is bounded and extreme, then $Q_h\{g = h\} = 1$ by (19), and $Q_h \leqq KP_p$ by the previous paragraph. There are at most countably many such h, say $h_1, h_2, \ldots$. General h can be represented as

$$\Sigma_n q_n h_n + q_c h_c + q_s h_s.$$

Here the q's are nonnegative numbers which sum to 1. As usual, the h's are excessive. Retracting Q_h and P_p to $\mathscr{E}$,

$$q_n = Q_h\{g = h_n\} \quad \text{and} \quad q_c h_c = \int g \, dm_c \quad \text{and} \quad q_s h_s = \int g \, dm_s,$$

where: m_c is the part of Q_h which is absolutely continuous with respect to the continuous part of P_p; and m_s is the part of Q_h singular with respect to P_p. In particular, $q_s Q_{h_s} = m_s$ is singular with respect to P_p. As (10.35) implies, $h_s(\xi_n) \to 0$ with P_p-probability 1 and $h_s(\xi_n) \to \infty$ with Q_{h_s}-probability 1.

3.   A CONVERGENCE THEOREM

Let $e(j) = pG(j)$. Let

$$R(i, j) = e(j)P(j, i)/e(i),$$

so R is substochastic: $\Sigma_j e(j)P(j, i)$ is the P_p-mean number of visits to i in positive time, and is at most $e(i)$. Let $S = \Sigma_{n=0}^\infty R^n$, so

$$S(i, j) = e(j)G(j, i)/e(i) < \infty.$$

I remind you that Ω_N is the set where I_N is visited, and τ_N is the time of the last visit to I_N. On Ω_N, let $\zeta_m = \xi_{\tau_N - m}$ for $0 \leqq m \leqq \tau_N$. Of course, even ζ_0 is only partially defined, namely on Ω_N.

(28) Lemma. ζ *is a partially defined Markov chain with stationary transitions* R, *relative to* P_p.

PROOF.   Let $i_0 \in I_N$, and let $i_1, \ldots, i_M \in I$. Let

$$\pi = \Pi_{m=0}^{M-1} P(i_{M-m}, i_{M-m-1}).$$

For $i \in I_N$, let

$$u(i) = P_i\{\xi_\nu \in I_N \text{ for no } \nu > 0\}.$$

Let

$$A = \{\Omega_N \text{ and } \tau_N \geqq M \text{ and } \zeta_0 = i_0, \ldots, \zeta_M = i_M\}.$$

Then

$$A = \bigcup_{n=0}^\infty A_n,$$

where

$$A_n = \{\xi_n = i_M, \ldots, \xi_{n+M} = i_0, \text{ and } \xi_\nu \in I_N \text{ for no } \nu > n + M\}.$$

By Markov (1.15),

$$P_p(A_n) = P_p\{\xi_n = i_M\} \cdot \pi \cdot u(i_0).$$

Sum out n and manipulate:

$$P_p(A) = e(i_M) \cdot \pi \cdot u(i_0)$$

$$= \frac{u(i_0)}{e(i_0)} \Pi_{m=0}^{M-1} R(i_m, i_{m+1}). \qquad \qquad \bigstar$$

(29) Theorem.   *As* $n \to \infty$, *the ratio* $G(i, \xi_n)/pG(\xi_n)$ *tends to a limit* P_p-*almost surely.*

PROOF.   Abbreviate $\rho_n = G(i, \xi_n)/pG(\xi_n)$.   Let $0 \leqq a < b < \infty$.   On Ω_N: let β_N be the number of downcrossings, as defined for (10.33), of $[a, b]$ by ρ_n as n decreases from τ_N to 0. I will eventually prove inequality (β):

$$(\beta) \qquad \qquad \int_{\Omega_N} \beta_N \, dP_p \leqq 1/(b - a).$$

For the moment, take (β) on faith. Check that Ω_N and β_N are nondecreasing with N. So (β) and monotone convergence imply

$$\int \lim_{N \to \infty} \beta_N \, dP_p \leqq 1/(b - a).$$

By (10.10b),

$$P_p\{\lim_N \beta_N < \infty\} = 1.$$

Let Ω_g be the intersection of $\{\lim_N \beta_N < \infty\}$ as a, b vary over the rationals. Then Ω_g is measurable and $P_p\{\Omega_g\} = 1$. You can check that ρ_n converges as $n \to \infty$ everywhere on Ω_g, because $\tau_N \to \infty$ as $N \to \infty$. The limit is finite by (31) below.

I will now prove inequality (β). Define R, S, and ζ_m as for (28). On Ω_N, let

$$X_m = G(i, \xi_{\tau_N - m})/pG(\xi_{\tau_N - m}) = S(\zeta_m, i)/pG(i)$$

for $0 \leq m \leq \tau_N$, and let $X_m = 0$ for $m > \tau_N$. Let $X_m = 0$ off Ω_N. Let $\mathscr{F}_N$ be the σ-field spanned by all measurable subsets of

$$\Omega \backslash \{\Omega_N \text{ and } \tau_N \geq m\},$$

and by $\zeta_0, \ldots, \zeta_m$ on $\{\Omega_N \text{ and } \tau_N \geq m\}$. You should check that the $\mathscr{F}_m$ are nondecreasing, and X_m is $\mathscr{F}_m$-measurable. For $A \in \mathscr{F}_m$, I claim

$$\int_A X_{m+1} \, dP_p \leq \int_A X_m \, dP_p.$$

This can be checked separately for $A \subset \Omega \backslash \{\Omega_N \text{ and } \tau_N \geq m\}$ and for A of the form

$$\{\Omega_N \text{ and } \tau_N \geq m \text{ and } \zeta_0 = j_0, \ldots, \zeta_m = j_m\}.$$

You should do the first check. Here is the second. I say

$$\begin{aligned}
\int_A X_{m+1} \, dP_p &= \int_{\{A \text{ and } \tau_N \geq m+1\}} X_{m+1} \, dP_p \\
&= \Sigma_{k \in I} \, P_p\{A \text{ and } \zeta_{m+1} = k \text{ and } \tau_N \geq m + 1\} \cdot S(k, i)/pG(i) \\
&= \Sigma_{k \in I} \, P_p(A) \cdot R(j_m, k) \cdot S(k, i)/pG(i) \\
&\leq P_p(A) \cdot S(j_m, i)/pG(i) \\
&= \int_A X_m \, dP_p,
\end{aligned}$$

for these reasons: $X_{m+1} = 0$ on $\{\tau_N = m\}$ in the first line; split over the sets $\{\zeta_{m+1} = k\}$ and use the definition of X in the second; use (28) in the third; use (21) on R in the fourth; use the definition of X in the last. Consequently, the sequence $X_0, X_1, \ldots$ is an expectation-decreasing martingale. Plainly, β_N is at most the number of downcrossings of $[a, b]$ by $X_0, X_1, \ldots$. By (10.33),

$$\int_{\Omega_N} \beta_N \, dP_p \leq (b - a)^{-1} \int_{\Omega_N} X_0 \, dP_p \leq (b - a)^{-1} pG(i)^{-1} \int_{\Omega_N} S(\zeta_0, i) \, dP_p.$$

Use (28): the last integral is the mean number of visits to i by $\zeta_0, \zeta_1, \ldots$, that is, the mean number of visits to i by $\xi_0, \ldots, \xi_{\tau_N}$, and is no more than $pG(i)$. This proves (β).   ★

(30) Corollary. *For all h, the sequence* $G(i, \xi_n)/pG(\xi_n)$ *converges* Q_h*-almost surely. Moreover, h is extreme iff*

$$Q_h\left\{\lim_n \frac{G(\cdot, \xi_n)}{pG(\xi_n)} = h\right\} = 1.$$

Proof. $Q_h\{\xi_n \in I^h\} = 1$, so the case $i \notin I^h$ is easy by (9). For $i \in I^h$ and $\xi_n \in I^h$,

$$\frac{G^h(i, \xi_n)}{phG^h(\xi_n)} = \frac{G(i, \xi_n)/h(i)}{pG(\xi_n)}$$

converges Q_h-almost surely by (11), and (29) on P^h. The last assertion now follows from (19).   ★

(31) Remark. $G(i, \cdot)/pG(\cdot)$ is bounded, because (1.51d) makes

$$G(i, j)/pG(j) = S(j, i)pG(i) \leqq S(i, i)/pG(i) = G(i, i)/pG(i).$$

4.   EXAMPLES

Let $\phi(i, j) = P_i\{\xi_n = j \text{ for some } n \geqq 0\}$. By (1.51d),

(32) $G(i, j) = \phi(i, j)G(j, j).$

If $i_0 = i, \ldots, i_n = j$ are in I, and $i \in I^h$, then (9) shows

$$P_i^h\{\xi_0 = i_0, \ldots, \xi_n = i_n\} = h(i)^{-1}P_i\{\xi_0 = i_0, \ldots, \xi_n = i_n\}h(j).$$

Sum over all such sequences with $i_1, \ldots, i_{n-1}$ not equal to j to get

(33) $P_i^h\{\xi_n = j \text{ for some } n \geqq 0\} = h(i)^{-1}\phi(i, j)h(j)$ for $i \in I^h$.

Let $\{i_n\}$ be a sequence in I. Say i_n *converges with limit* h iff for each j in I

there are only finitely many n with $i_n = j$, and

$$\lim_{n \to \infty} G(j, i_n)/pG(i_n) = h(j).$$

In the conventional treatment, I is compactified by adding all these limits; the extra points form the *boundary*.

For given reference probability p and substochastic matrix P, the set of extreme harmonic h was identified by (19) as some of the limits of convergent sequences i_n. In this section, the extreme harmonic h are found in seven examples. The first four are artificial, and are introduced to clarify certain points in the theory. The last three present some well known processes.

(34) Example.   Let N be a positive integer. There will be precisely N extreme harmonic h. The state space I consists of 1 and all pairs (n, m), with $n = 1, \ldots, N$ and $m = 1, 2, \ldots$. The reference probability concentrates on 1. The transition probabilities P are subject to the following conditions, as in Figure 1:

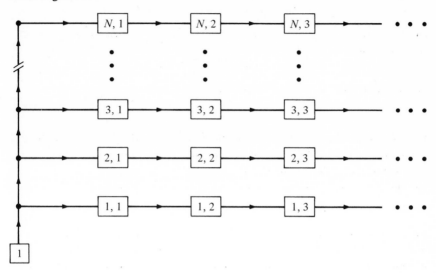

From any state other than 1, it is possible to jump to 1. This transition is not shown.

Figure 1.

P is transient;

$P[1, (n, 1)] > 0$ and $\Sigma_n P[1, (n, 1)] = 1$;

$0 < P[(n, m), (n, m + 1)] < 1$;

$P[(n, m), 1] = 1 - P[(n, m), (n, m + 1)]$.

In the presence of the other conditions, the first condition is equivalent to

$$\Pi_{m=1}^{\infty} P[(n, m), (n, m + 1)] > 0 \quad \text{for some } n;$$

use (1.16). For $a = 1, \ldots, N$, let h_a be this function on I:

$h_a(1) = 1$;

$h_a(n, m) = 1/\phi[1, (n, m)]$ for $n = a$;

$h_a(n, m) = \phi[(n, m), 1]$ for $n \neq a$.

Then $h_1, \ldots, h_N$ are the extreme harmonic h.

PROOF.   To see this, let $j \in I$, and let i_n be a sequence in I. By (32),

(35) $G(j, i_n)/G(1, i_n) = \phi(j, i_n)/\phi(1, i_n)$.

Suppose i_n converges. Let $i_n = (a_n, b_n)$. Clearly, $b_n \to \infty$. Let $j = (c, d)$. Suppose $b_n \geq d$. If $a_n = c$, then 1 leads to i_n only through j. By (1.16) and strong Markov (1.22),

$$\phi(1, i_n) = \phi(1, j) \cdot \phi(j, i_n).$$

The right side of (35) is therefore $1/\phi(1, j)$. If $a_n \neq c$, then j leads to i_n only through 1. So

$$\phi(j, i_n) = \phi(j, 1) \cdot \phi(1, i_n),$$

and the right side of (35) is $\phi(j, 1)$. But

$$\phi(j, 1) < 1/\phi(1, j),$$

for otherwise $\phi(1, j) = \phi(j, 1) = 1$ and P is recurrent by (1.54). Because i_n converges, a_n is eventually constant, say at a. The limit of i_n is then h_a. By (19), any extreme harmonic h is an h_a.

I will now check that h_a is harmonic. To begin with, I say

$$\phi[1, (a, 1)] = P[1, (a, 1)] + \Sigma_{n \neq a} P[1, (n, 1)] \cdot \phi[(n, 1), 1] \cdot \phi[1, (a, 1)].$$

Indeed, consider the chain starting from 1. How can it reach $(a, 1)$? The first move can be to $(a, 1)$. Or the first move can be to $(n, 1)$ with $n \neq a$: the chain has then to get back to 1, and from 1 must make it to $(a, 1)$. This argument can be rigorized using (1.16, 1.15, 1.22). Divide the equality by $\phi[1, (a, 1)]$:

$$1 = \Sigma_n P[1, (n, 1)] \cdot h_a(n, 1).$$

That is,

$$h_a(1) = \Sigma_j P(1, j) \cdot h_a(j).$$

Next, I say

$$\phi[1, (a, b + 1)] = \phi[1, (a, b)] \cdot P[(a, b), (a, b + 1)]$$
$$+ \phi[1, (a, b)] \cdot P[(a, b), 1] \cdot \phi[1, (a, b + 1)].$$

Indeed, a chain reaches $(a, b + 1)$ from 1 by first hitting (a, b). It then makes $(a, b + 1)$ on one move, or returns to 1 and must try again. Rearranging,

$$h_a(a, b) = P[(a, b), (a, b + 1)] \cdot h_a(a, b + 1) + P[(a, b), 1] \cdot h_a(1)$$
$$= \Sigma_j P[(a, b), j] \cdot h_a(j)$$

Finally, let $n \neq a$:

$$\phi[(n, m), 1] = P[(n, m), (n, m + 1)] \cdot \phi[(n, m + 1), 1] + P[(n, m), 1];$$

so

$$h_a(n, m) = \Sigma_j P[(n, m), j] \cdot h_a(j).$$

I will now check that h_a is extreme. Abbreviate

$$\pi_a = P_1^{h_a}.$$

As (33) implies, π_a-almost all sample sequences reach (a, b). Therefore, with π_a-probability 1 the first coordinate of ξ_n is a for infinitely many n. But ξ_n converges with π_a-probability 1 by (30). So $\pi_a\{\xi_n \to h_a\} = 1$, and h_a is extreme by (30).   ★

(36) Example. There are countably many extreme harmonic h. There is a sequence i_n in I which converges to an extreme excessive h which is not harmonic. This example is obtained by modifying (34) as follows: The state space consists of 1 and all pairs (n, m) of positive integers. The reference probability concentrates on 1. The transitions are constrained as in (34). The new convergence is (n, m) with $n \to \infty$ and m free. The limit is h_∞, where

$$h_\infty(j) = \phi(j, 1) = G(j, 1)/G(1, 1).$$

Use (21) to see h_∞ is not harmonic. The rest of the argument is like (34). ★

(37) Example. There are c extreme harmonic h. The state space I consists of all finite sequences of 0's and 1's, including the empty sequence $\varnothing$. The reference probability concentrates on $\varnothing$. The transition probabilities P are subject to the following conditions, as in Figure 2:

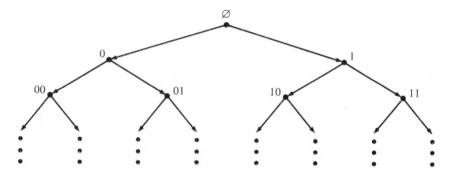

From any state other than $\varnothing$, it is possible to jump to $\varnothing$. This transition is not shown.

Figure 2.

P is transient;

$0 < P(\varnothing, 0) < 1$ and $P(\varnothing, 1) = 1 - P(\varnothing, 0)$;

for each $j \neq \varnothing$ in I, the three numbers $P(j, j0)$ and $P(j, j1)$ and $P(j, 0)$ are all positive and sum to 1.

For each infinite sequence s of 0's and 1's, let h_s be this function on I:

$$h_s(j) = 1/\phi(\varnothing, j) \quad \text{if } s \text{ extends } j;$$
$$h_s(j) = \phi(j, \varnothing) \qquad \text{otherwise.}$$

Then $\{h_s\}$ are the extreme harmonic h. The argument is like (34): a sequence i_n in I converges iff the length of i_n tends to ∞, and the mth component of i_n is eventually constant, say at s_m, for each m. Then $i_n \rightarrow h_s$.

Now suppose that $P(j,j0) = P(j,j1)$ and depends only on the length of j, for all j. I claim each h_s is unbounded. Indeed, suppose for a moment that j has length N. Let $\theta(j)$ be the $P_\varnothing$-probability that $\{\xi_n\}$ visits j before any other k of length N. By symmetry, $\theta(j) = 2^{-N}$. If ξ_n visits j after visiting some k other than j of length N, there is a return from k to $\varnothing$, except for miracles. Thus,

$$\phi(\varnothing, j) = \theta(j) + \delta(j),$$

where $\delta(j)$ is at most $P_\varnothing(A_N)$, and

$$A_N = \{\xi \text{ returns to } \varnothing \text{ after first having length } N\}.$$

But $\lim_{N \to \infty} P_\varnothing(A_N) = 0$, because ξ_n visits $\varnothing$ infinitely often on $\bigcap_N A_N$. Consequently,

$$\lim_{N \to \infty} \phi(\varnothing, j) = 0.$$

However, there are many bounded, harmonic, nonextreme h: here is an example:
$h(i)$ is twice the P_i-probability that the first coordinate of ξ_n is 0 for all large n. ★

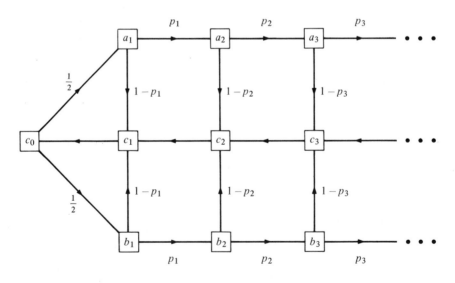

Figure 3.

(38) Example. A sequence i_n converges to an h which is harmonic but not extreme. The state space consists of $a_1, a_2, \ldots$ and $b_1, b_2, \ldots$ and

$c_0, c_1, \ldots$. The reference probability concentrates on c_0. Choose a sequence p_n with $0 < p_n < 1$ and

$$\Pi_{n=1}^{\infty} p_n > 0.$$

Define the transition probabilities P as in Figure 3:

$$P(c_0, a_1) = P(c_0, b_1) = \tfrac{1}{2};$$
$$P(a_n, a_{n+1}) = P(b_n, b_{n+1}) = p_n \quad \text{for } n = 1, 2, \ldots;$$
$$P(a_n, c_n) = P(b_n, c_n) = 1 - p_n \quad \text{for } n = 1, 2, \ldots;$$
$$P(c_{n+1}, c_n) = 1 \quad\quad\quad\quad \text{for } n = 0, 1, \ldots.$$

Then a_n converges to an extreme harmonic function, as does b_n; and this exhausts the extreme harmonic functions. But c_n converges to the constant function 1, which is not extreme.

PROOF. I will argue that c_n converges to 1, which isn't extreme. The rest is like (34). By symmetry,

$$\phi(a_j, c_n) = \phi(b_j, c_n).$$

Suppose $n \geqq j$. Then c_0 leads to c_n only through a_j or b_j. And P_{c_0}-almost all sample paths hit a_j or b_j. So

$$\phi(a_j, c_n) = \phi(b_j, c_n) = \phi(c_0, c_n).$$

Now (32) implies

$$G(a_j, c_n)/G(c_0, c_n) = G(b_j, c_n)/G(c_0, c_n) = 1.$$

For $n > j$,

$$\phi(c_j, c_n) = \phi(c_0, c_n)$$

because c_j leads to c_n only through 1, and P_{c_j}-almost all sample paths reach 1. Now use (32) again:

$$G(c_j, c_n)/G(c_0, c_n) = 1.$$

Thus, c_n converges to 1. The event $\{\xi_n$ is an a for all large $n\}$ is invariant and has P_{c_0}-probability $\tfrac{1}{2}$ by symmetry, so the invariant σ-field is not P_{c_0}-trivial. Now (25) prevents 1 from being extreme.   ★

(39) Example. *The random walk.* The state space I consists of the integers. The reference probability concentrates on 0. Let $\tfrac{1}{2} < p < 1$. The transition probabilities are given by:

$$P(n, n+1) = p \quad \text{and} \quad P(n, n-1) = 1 - p.$$

As (1.95, 1.96) show,

$$\phi(i, j) = 1 \ \text{ if } j \geqq i \quad \text{and} \quad \phi(i, j) = [(1-p)/p]^{i-j} \ \text{ if } j < i.$$

So i_v converges, to h_+ and h_- respectively, iff $i_v \to \infty$ or $i_v \to -\infty$; where

$$h_+(i) = 1 \quad \text{and} \quad h_-(i) = [(1-p)/p]^i.$$

Now h_+ is extreme because the invariant σ-field is P_1-trivial; and h_- is extreme for a similar reason, Ph_- being the random walk with $1-p$ replacing p. Triviality follows from Hewitt-Savage (1.122).   ★

Write $x_n \approx y_n$, or x_n is *asymptotic* to y_n, iff $x_n/y_n \to 1$. Suppose d is a nonnegative integer, and $v \to \infty$ through the integers. Then

$$v! = v(v-1) \cdots (v-d+1)(v-d)!;$$

so

(40) $$v! \approx v^d (v-d)! \quad \text{as } v \to \infty.$$

(41) Example. *The random walk in space-time.* The state space I consists of pairs (n, m) of integers with $0 \le m \le n$. The reference probability concentrates on $(0, 0)$. Let $0 < p < 1$. The situation with $p = 0$ or 1 is easier. The transition probabilities P are given by

$$P[(n, m), (n+1, m+1)] = p \quad \text{and} \quad P[(n, m), (n+1, m)] = 1 - p.$$

You should check that $G[(a, b), (n, m)] = 0$ unless $n \ge a$, and $m \ge b$, and $n - a \ge m - b$, in which case

$$G[(a, b), (n, m)] = \binom{n-a}{m-b} \cdot p^m (1-p)^{n-m} \cdot p^{-b} (1-p)^{b-a}.$$

Suppose (n, m) converges. Then, $n \to \infty$. If m is bounded, by passing to a subsequence suppose m is eventually constant, say at M. Then (n, m) converges to h_M, where

$$h_M(a, b) = 0 \qquad \text{for } b > 0$$
$$= (1-p)^{-a} \quad \text{for } b = 0.$$

This function is not harmonic. Similarly, (n, m) does not converge to a harmonic function if $n - m$ is bounded. So, suppose that $m \to \infty$ and $n - m \to \infty$. By passing to a subsequence if necessary, suppose m/n converges, say to q, with $0 \le q \le 1$. Then

$$\binom{n-a}{m-b} \bigg/ \binom{n}{m} = \frac{(n-a)!}{n!} \cdot \frac{m!}{(m-b)!} \cdot \frac{(n-m)!}{[n-m-(a-b)]!}$$
$$\approx n^{-a} m^b (n-m)^{a-b} \quad \text{by (40)}$$
$$= \left(\frac{m}{n}\right)^b \left(1 - \frac{m}{n}\right)^{a-b}$$
$$\to q^b (1-q)^{a-b}.$$

So (n, m) converges to h_q, where

$$h_q(a, b) = \left(\frac{q}{p}\right)^b \left[\frac{(1 - q)}{(1 - p)}\right]^{a-b}$$

Now, P^{h_q} is again a random walk in space-time, with q replacing p, so that h_q is extreme harmonic by (25) and Hewitt-Savage (1.122).	★

The next example, the Polya urn, has been studied recently by (Blackwell and Kendall, 1964).

(42) Example. An urn contains u balls at time 0, of which w are white and $u - w$ black. Assume $w > 0$ and $u - w > 0$. At time n, a ball is drawn at random, and replaced. A ball of the same color is added. Then time moves on to $n + 1$. Let U_n be the number of balls in the urn at time n, namely $n + u$. Let W_n be the number of white balls in the urn at time n. Then $\{(U_n, W_n) : n = 0, \ldots\}$ is a Markov chain starting from (u, w), with state space I consisting of pairs (t, v) of integers having $0 < v < t$. The chain has stationary transitions P, where

$$P[(t, v), (t + 1, v + 1)] = v/t$$
$$P[(t, v), (t + 1, v)] \quad = (t - v)/t.$$

I claim that $\{W_{n+1} - W_n : n = 0, 1, \ldots\}$ is *exchangeable*. In the present notation, suppose $a \geqq c$ and

$$v_0 = c \leqq v_1 \leqq v_2 \leqq \cdots \leqq v_n$$

and $v_{m+1} = v_m$ or $v_m + 1$. Then the $P_{(a,c)}$ probability that

$$\xi_0 = (a, v_0), \ \xi_1 = (a + 1, v_1), \ \xi_2 = (a + 2, v_2), \ldots, \ \xi_n = (a + n, v_n)$$

is alleged to depend only on v_n. The easiest method is to argue, inductively on n, that this probability is equal to the product

$$\frac{c}{a} \cdot \frac{c + 1}{a + 1} \cdots \frac{w - 1}{a + w - c - 1} \cdot \frac{a - c}{a + w - c} \cdot \frac{a - c + 1}{a + w - c + 1} \cdots \frac{u - w - 1}{u - 1},$$

where $u = a + n$ and $w = v_n$. Call this product $\theta(a, c, u, w)$.

Keep $a \geqq c$. Then $G[(a, c), (u, w)] = 0$ unless $a \leqq u$, and $c \leqq w$, and $a - c \leqq u - w$. In the latter case, by exchangeability, $G[(a, c), (u, w)]$ is the product of two factors: the first is the number of sequences of $u - a$ balls, of which $w - c$ are white and the others black; the second is the common probability $\theta(a, c, u, w)$ that an urn with c white balls and $a - c$ black balls will produce some specified sequence of $u - a$ draws, of which $w - c$ are white and $(u - w) - (a - c)$ are black. That is,

$$G[(a, c), (u, w)] = \binom{u - a}{w - c} \cdot \theta(a, c, u, w).$$

Let the reference probability concentrate on $(2, 1)$. Suppose

$$u \geq a \qquad w \geq c \qquad u - w \geq a - c$$

and

$$u \geq 2 \qquad w \geq 1 \qquad u - w \geq 1 \qquad a > c \geq 1.$$

Then

$$\frac{G[(a, c), (u, w)]}{G[(2, 1), (u, w)]} = (a - 1)\binom{a - 2}{c - 1}\binom{u - a}{w - c} \Big/ \binom{u - 2}{w - 1}.$$

Suppose (u_n, w_n) converges. Then $u_n \to \infty$. By compactness, suppose w_n/u_n converges, say to π. If $w_n \to \infty$ and $u_n - w_n \to \infty$, then $(u_n, w_n) \to h_\pi$, where

$$h_\pi(a, c) = (a - 1)\binom{a - 2}{c - 1} \cdot \pi^{c-1}(1 - \pi)^{a-c-1}.$$

This follows by (40):

$$\frac{(u_n - a)!}{(u_n - 2)!} \approx u_n^{2-a}$$

$$\frac{(w_n - 1)!}{(w_n - c)!} \approx w_n^{c-1}$$

$$\frac{(u_n - w_n - 1)}{(u_n - w_n - a + c)!} \approx (u_n - w_n)^{a-c-1};$$

so the product, namely $\binom{u_n - a}{w_n - c} \Big/ \binom{u_n - 2}{w_n - 1}$, is asymptotic to

$$\left(\frac{w_n}{u_n}\right)^{c-1} \cdot \left(\frac{u_n - w_n}{u_n}\right)^{a-c-1} \to \pi^{c-1}(1 - \pi)^{a-c-1}.$$

If $u_n \to \infty$ and w_n is bounded, then $(u_n, w_n) \to f$, where $f(a, c) = 0$ for $c > 1$, and $f(a, 1) = a - 1$. If $w_n \to \infty$ and $u_n - w_n$ is bounded, then $(u_n, w_n) \to g$, where $g(a, c) = 0$ for $a - c > 1$, and $g(a, c) = a - 1$ for $a - c = 1$. Now f and g are not harmonic. Therefore, $\{h_\pi : 0 \leq \pi \leq 1\}$ contains all extreme harmonic h by (19). But by algebra, P^{h_π} corresponds to a random walk in space-time with parameter π, so h_π is extreme harmonic by (25) and Hewitt-Savage (1.122). ★

5. THE LAST VISIT TO i BEFORE THE FIRST VISIT TO $J \backslash \{i\}$

This section is somewhat apart from the rest of the chapter, but uses similar technology. The results will be referred to in *ACM*. Let P be a

$$I = \{1, 2, 3, 4, 5\} \quad J = \{1, 2, 3\} \quad i = 1$$

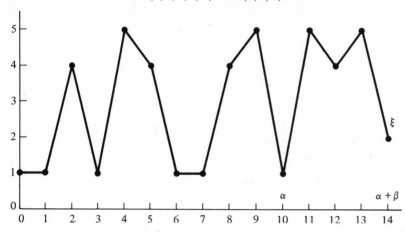

Figure 4.

stochastic matrix on the countable set I; suppose I forms one recurrent class relative to P. This stands in violent contrast to Sections 1–4. The coordinate process $\xi_0, \xi_1, \ldots$ on (I^∞, P_j) is Markov with starting state j and stationary transitions P. Fix $J \subset I$ and $i \in J$. Let $(\alpha + \beta)$ be the least n with $\xi_n \in J \backslash \{i\}$. Let α be the greatest $n < (\alpha + \beta)$ with $\xi_n = i$. See Figure 4. For $j \in I$, let

$$h(j) = P_j\{\xi \text{ visits } i \text{ before hitting } J \backslash \{i\}\}.$$

So $h(i) = 1$, while $h(j) = 0$ for $j \in J \backslash \{i\}$. Check

(43a) $$h(j) = \Sigma_{k \in I} P(j, k) h(k) \quad \text{for } j \in I \backslash J.$$

On the other hand,

(43b) $$\Sigma_{k \in I} P(i, k) h(k) = 1 - \theta,$$

where

$$\theta = P_i\{\xi \text{ hits } J \backslash \{i\} \text{ before returning to } i\} > 0.$$

Let

$$H = \{j : j \in I \text{ and } h(j) > 0\}.$$

Then $i \in H$, and $H \backslash \{i\} \subset I \backslash J$. Let

$$M(j, k) = \frac{1}{h(j)} P(j, k) h(k) \quad \text{for } j, k \in H.$$

Using (43), check that M is a substochastic matrix on H, whose rows sum to 1, except that row i sums to $1 - \theta$.

Let

$$H^* = \{j : j \in I \text{ and } h(j) < 1\} \cup \{i\}.$$

So $H^* \supset J$. Define a matrix M^* on H^* as follows.

$$M^*(j, k) = \frac{1}{1 - h(j)} P(j, k)[1 - h(k)] \quad \text{for } j \notin J.$$

$$M^*(j, k) = 0 \qquad\qquad\qquad\qquad\quad \text{for } j \in J\backslash\{i\}.$$

$$M^*(i, k) = \frac{1}{\theta} P(i, k)[1 - h(k)].$$

Using (43), check that M^* is a substochastic matrix on H^*, whose jth row sums to 1 or 0, according as $j \notin J\backslash\{i\}$ or $j \in J\backslash\{i\}$.

Let τ be the number of $n \leq \alpha$ with $\xi_n = i$.

(44) Theorem. *With respect to P_i:*

   (a) $\{\xi_n : 0 \leq n \leq \alpha\}$ *is independent of* $\{\xi_{\alpha+n} : 0 \leq n < \infty\}$;
   (b) $\{\xi_n : 0 \leq n \leq \alpha\}$ *is Markov with stationary transitions M;*
   (c) $\{\xi_{\alpha+n} : 0 \leq n \leq \beta\}$ *is Markov with stationary transitions M^*;*
   (d) $P_i\{\tau = \nu\} = \theta(1 - \theta)^{\nu-1}$ *for $\nu = 1, 2, \ldots.$*

PROOF. *Claim (a).* I say

   (e) $\{\xi_n : 0 \leq n \leq \alpha\}$ *is P_i-independent of* $\{\xi_{\alpha+n} : 0 \leq n \leq \beta\}$.

To check (e), stop ξ when it first enters $J\backslash\{i\}$, and reverse time. That is, look at $\zeta_n = \xi_{\alpha+\beta-n}$ for $0 \leq n \leq \alpha + \beta$. Then ζ is a Markov chain with stationary substochastic transitions, as in (28). Now $\beta = (\alpha + \beta) - \alpha$ is the least n with $\zeta_n = i$, and is Markov for ζ. Use strong Markov (1.21) on the process ζ and the time β. This proof of (e) was suggested by Aryeh Dvoretzky.

You can also derive (e) from blocks (1.31) and lemma (48) below. Use the successive i-blocks for $X_1, X_2, \ldots$. Let V be the set of i-blocks free of $J\backslash\{i\}$. Now L is the least n with $X_n \notin V$. Check that $\{\xi_n : 0 \leq n \leq \alpha\}$ is measurable on $(X_1, \ldots, X_{L-1})$, while $\{\xi_{\alpha+n} : 0 \leq n \leq \beta\}$ is measurable on X_L. But X_L is independent of $(X_1, \ldots, X_{L-1})$.

I will now argue (a) from (e) and strong Markov. Abbreviate

$$X = \{\xi_n : 0 \leq n \leq \alpha\}$$
$$Y = \{\xi_{\alpha+n} : 0 \leq n \leq \beta\}$$
$$Y^* = \xi_{\alpha+\beta}$$
$$Z = \{\xi_{\alpha+\beta+n} : 0 \leq n\}.$$

Check that $\xi_{\alpha+\cdot}$ is measurable on (Y, Z), and Y^* is measurable on Y. Let

A, B, C be measurable sets. I say

$$P_i\{X \in A \text{ and } Y \in B \text{ and } Y^* = j \text{ and } Z \in C\}$$
$$= P_i\{X \in A \text{ and } Y \in B \text{ and } Y^* = j\} \cdot P_j\{C\}$$
$$= P_i\{X \in A\} \cdot P_i\{Y \in B \text{ and } Y^* = j\} \cdot P_j\{C\}$$
$$= P_i\{X \in A\} \cdot P_i\{Y \in B \text{ and } Y^* = j \text{ and } Z \in C\};$$

the first and third equalities come from strong Markov (1.22) on the time $\alpha + \beta$; the second equality comes from (e). This proves (a).

Claim (b). Fix $i_0 = i$. Fix $n \geq 1$. Fix $i_1, \ldots, i_{n-1}$ in H and i_n in I. Then

$$\{\xi_m = i_m \text{ for } 0 \leq m \leq n \text{ and } \alpha \geq n\}$$
$$= \{\xi_m = i_m \text{ for } 0 \leq m \leq n \text{ and } \xi_{n+\cdot} \text{ visits } i \text{ before } J\backslash\{i\}\}.$$

By Markov (1.15),

$$P_i\{\xi_m = i_m \text{ for } 0 \leq m \leq n \text{ and } \alpha \geq n\} = [\Pi_{m=0}^{n-1} P(i_m, i_{m+1})] \, h(i_n)$$
$$= \Pi_{m=0}^{n-1} M(i_m, i_{m+1}) \quad \text{if } i_n \in H$$
$$= 0 \qquad\qquad\qquad \text{if } i_n \notin H.$$

Claims (c–d). Fix $i_0 = i$. Fix $n \geq 1$. Fix $i_1, \ldots, i_{n-2}$ in $H^*\backslash J$. Fix i_{n-1} in H^* and i_n in I. Let

$$A = \{\xi_{\alpha+m} = i_m \text{ for } 0 \leq m \leq n \text{ and } \beta \geq n\}.$$

To get (c), I have to compute $P_i\{A\}$. Clearly, $P_i\{A\} = 0$ if $i_n = i$, or if $n \geq 2$ and $i_{n-1} \in J$. Exclude these two cases. Let B_ν be the event that ξ visits i at least ν times before the first visit to $J\backslash\{i\}$. So $B_\nu = \{\tau \geq \nu\}$. To get (d), I have to compute $P_i\{B_\nu\}$. Let σ_ν be the time of the νth visit to i for $\nu = 1, 2, \ldots$; so

$$P_i\{\sigma_1 = 0\} = 1.$$

Let $\zeta(\nu)$ be the post-σ_ν process. Check that σ_ν is Markov, B_ν is in the pre-σ_ν sigma field, and $\zeta(\nu)_0 = i$. Check

$$B_{\nu+1} = B_\nu \cap \{\zeta(\nu) \in B_2\} \quad \text{for } \nu \geq 1.$$

By definition,

$$P_i\{B_0\} = P_i\{B_1\} = 1 \quad \text{and} \quad P_i\{B_2\} = 1 - \theta.$$

By strong Markov (1.22) and induction,

(45a) $P_i\{B_{\nu+1}\} = P_i\{B_\nu\} \cdot (1 - \theta) = (1 - \theta)^\nu \quad \text{for } \nu = 1, 2, \ldots$

This settles (d).

Let

$$C = \{\xi_m = i_m \text{ for } 0 \leq m \leq n \text{ and } \xi_{n+\cdot} \text{ visits } J\backslash\{i\} \text{ before } i\}.$$

By Markov (1.15),

(45b) $\qquad P_i\{C\} = [\Pi_{m=0}^{n-1} P(i_m, i_{m+1})] [1 - h(i_n)].$

Check that

$$\{B_\nu \text{ and } \zeta(\nu) \in C\} \quad \text{for } \nu = 1, 2, \ldots$$

are pairwise disjoint and their union is A. So

$$
\begin{aligned}
P_i\{A\} &= \Sigma_{\nu=1}^{\infty} P_i\{B_\nu \text{ and } \zeta(\nu) \in C\} \\
&= \Sigma_{\nu=1}^{\infty} P_i\{B_\nu\} \cdot P_i\{C\} && \text{by strong Markov (1.22)} \\
&= \Sigma_{\nu=1}^{\infty} (1 - \theta)^{\nu-1} \cdot P_i\{C\} && \text{by (45a)} \\
&= \frac{1}{\theta} [\Pi_{m=0}^{n-1} P(i_m, i_{m+1})] [1 - h(i_n)] && \text{by (45b)} \\
&= \Pi_{m=0}^{n-1} M^*(i_m, i_{m+1}) && \text{if } i_n \in H^* \\
&= 0 && \text{if } i_n \notin H^*. \qquad \bigstar
\end{aligned}
$$

Suppose $\theta < 1$. Define a new matrix $\hat{M}$ on H as follows:

$$\hat{M}(j, k) = M(j, k) \quad \text{for } j \neq i;$$

$$\hat{M}(i, k) = \frac{1}{1 - \theta} M(i, k).$$

This $\hat{M}$ is stochastic. Let λ be the least $n > 0$ with $\xi_n = i$. Let D be the conditional P_i-distribution of $\{\xi_n : 0 \leq n < \lambda\}$, given that $\xi_n \in J\backslash\{i\}$ for no $n < \lambda$. Define a new process $\{\zeta_n : 0 \leq n\}$ with state space I, starting from i, visiting i an infinite number of times, such that the i-blocks of ζ are independent and have common distribution D.

(46) Theorem. ζ *is Markov with stationary transitions* $\hat{M}$.

PROOF. Let $i_0 = i$. Let $i_1, \ldots, i_n$ be in I/J. Let

$$A = \{\xi_m = i_m \text{ for } 0 \leq m \leq n\}$$
$$B = \{\xi \text{ returns to } i \text{ before visiting } J\backslash\{i\}\}.$$

The D-probability of starting off $(i_0, i_1, \ldots, i_n)$ is

$$
\begin{aligned}
P_i\{A \mid B\} &= P_i\{A \cap B\}/P_i\{B\} \\
&= \frac{1}{1 - \theta} [\Pi_{m=0}^{n-1} P(i_m, i_{m+1})] h(i_n),
\end{aligned}
$$

by Markov (1.15). The D-probability of starting off $(i_0, i_1, \ldots, i_n)$, and terminating at time n, is

$$
\begin{aligned}
P_i\{A \text{ and } \xi_{n+1} = i \mid B\} &= P_i\{A \text{ and } \xi_{n+1} = i\}/P_i\{B\} \\
&= \frac{1}{1 - \theta} [\Pi_{m=0}^{n-1} P(i_m, i_{m+1})] P(i_n, i)h(i).
\end{aligned}
$$

Let $i_0 = i$ and $N \geqq 1$. Let $i_1, \ldots, i_{N-1}$ be in H. Let i_N be in I. Suppose m of $i_0, \ldots, i_{N-1}$ are equal to i. Then

$$\text{Prob}\,\{\zeta_n = i_n \text{ for } 0 \leqq n \leqq N\} = (1 - \theta)^{-m}\,[\Pi_{n=0}^{N-1}\,P(i_n, i_{n+1})]\,h(i_N)$$
$$= \Pi_{n=0}^{N-1}\,\hat{M}(i_n, i_{n+1}) \quad \text{if } i_N \in H$$
$$= 0 \qquad\qquad\qquad \text{if } i_N \notin H. \qquad \bigstar$$

Let T be independent of ζ. Let the distribution of T coincide with the distribution of τ relative to P_i, as described in (44d). Let T^* be the time of the Tth visit to i in ζ.

(47) Theorem. *The joint distribution of T and $\{\zeta_n : 0 \leqq n \leqq T^*\}$ coincides with the joint P_i-distribution of τ and $\{\xi_n : 0 \leqq n \leqq \alpha\}$.*

PROOF. Use (48) below. Let X_n be the nth i-block in ξ. Let V be the set of i-blocks free of $J\backslash\{i\}$. So $L = \tau$, and the two θ's coincide. $\qquad \bigstar$

To state (48), let $X_1, X_2, \ldots$ be a sequence of independent and identically distributed random objects. Let V be a measurable set of values, such that $X_1 \notin V$ has positive probability θ less than 1. Let L be the least n with $X_n \notin V$. Next, let $T, Z, Y_1, Y_2, \ldots$ be independent random objects. Suppose T is n with probability $\theta(1 - \theta)^{n-1}$ for $n = 1, 2, \ldots$. Suppose the distribution of Z coincides with the conditional distribution of X_1 given $X_1 \notin V$. Suppose the distribution of Y_n coincides with the conditional distribution of X_1 given $X_1 \in V$ for $n = 1, 2, \ldots$.

(48) Lemma. $(X_1, \ldots, X_{L-1}, X_L, L)$ *is distributed like* $(Y_1, \ldots, Y_{T-1}, Z, T)$.

PROOF. Let n be a positive integer. Let $A_1, \ldots, A_{n-1}$ be measurable subsets of V. Let B be a measurable set disjoint from V. Then

$$\text{Prob}\,\{X_m \in A_m \text{ for } 1 \leqq m \leqq n - 1 \text{ and } X_n \in B \text{ and } L = n\}$$
$$= \text{Prob}\,\{X_m \in A_m \text{ for } 1 \leqq m \leqq n - 1 \text{ and } X_n \in B\}$$
$$= [\Pi_{m=1}^{n-1}\,\text{Prob}\,\{X_1 \in A_m\}] \cdot \text{Prob}\,\{X_1 \in B\}$$
$$= [\Pi_{m=1}^{n-1}\,(1 - \theta)\,\text{Prob}\,\{Y_1 \in A_m\}] \cdot \theta\,\text{Prob}\,\{Z \in B\}$$
$$= [\Pi_{m=1}^{n-1}\,\text{Prob}\,\{Y_1 \in A_m\}] \cdot \text{Prob}\,\{Z \in B\} \cdot \text{Prob}\,\{T = n\}$$
$$= \text{Prob}\,\{Y_m \in A_m \text{ for } 1 \leqq m \leqq n - 1 \text{ and } Z \in B \text{ and } T = n\}. \quad \bigstar$$

NOTE. The first i-block in ξ, conditioned on missing $J\backslash\{i\}$, is Markov relative to P_i. If you condition the first i-block on hitting $J\backslash\{i\}$, however, it stops being Markov. The process $\xi_{\alpha+}$ isn't Markov: its first i-block hits $J\backslash\{i\}$; the second one could miss.

5

INTRODUCTION TO CONTINUOUS TIME

1. SEMIGROUPS AND PROCESSES

Let I be a finite or countably infinite set. A *matrix* M on I is a function $(i, j) \to M(i, j)$ from $I \times I$ to the real line. Call M *stochastic* iff $M(i, j) \geqq 0$ for all i and j, while $\Sigma_j M(i, j) = 1$ for all i. Call M *substochastic* iff $M(i, j) \geqq 0$ for all i and j, while $\Sigma_j M(i, j) \leqq 1$ for all i. Matrix multiplication is defined as usual:

$$MN(i, j) = \Sigma_{k \in I} \quad M(i, k)N(k, j).$$

(1) Definition. $P = \{P(t): 0 \leqq t < \infty\}$ *is a* stochastic semigroup *on I iff*

   (a) $P(t)$ *is a stochastic matrix on I for each $t \geqq 0$;*

and

   (b) $P(t + s) = P(t)P(s)$ *for all $t \geqq 0$ and $s \geqq 0$;*

and

   (c) $P(0)(i, j) = 1$ *or* 0 *according as $i = j$ or $i \neq j$.*

Call P a substochastic semigroup *iff (b) and (c) hold, but $P(t)$ is substochastic for each $t \geqq 0$. Call P* standard *iff*

   (d) $\lim_{t \to 0} P(t)(i, i) = 1$ *for all i.*

Usually, $P(t)(i, j)$ is abbreviated to $P(t, i, j)$.

I want to thank Richard Olshen for checking the final draft of this chapter.

NOTE. Let P be a standard substochastic semigroup. Then $P(t, i, j)$ is continuous at $t = 0$; this is an immediate consequence of the definition. In fact, $P(\cdot, i, j)$ is continuous on $[0, \infty)$; this is proved as (9). Finally, if $P(t)$ is stochastic for any $t > 0$, then $P(t)$ is stochastic for all t; see (6).

For finite I, let $\bar{I} = I$. For infinite I, endow I with the discrete topology, and let $\bar{I} = I \cup \{\varphi\}$ be the one-point compactification of I. Let $(\Omega, \mathcal{F}, \mathcal{P})$ be a probability triple. For each $t \geq 0$, let $X(t)$ be an $\mathcal{F}$-measurable function from Ω to $\bar{I}$. Let p be a probability on I and let P be a stochastic semigroup on I.

(2) Definition. $\{X(t) : 0 \leq t < \infty\}$ *is a* Markov chain *with* starting distribution p *and* stationary transitions P *iff*

$$\mathcal{P}\{X(t_m) = i_m \text{ for } m = 0, \ldots, n\} = p(i_0) \prod_{m=0}^{n-1} P(t_{m+1} - t_m, i_m, i_{m+1})$$

when

$$0 = t_0 < t_1 < \ldots < t_n < \infty \quad \text{and} \quad i_0, \ldots, i_n \in I.$$

If $p(i) = 1$, say the chain starts *from i. By convention, an empty product is 1.*

It is worth noting that $\mathcal{P}\{X(t) \in I\} = 1$ for all $t \geq 0$; it is impossible to prove, for it is usually false or meaningless, that $\mathcal{P}\{X(t) \in I \text{ for all } t \geq 0\} = 1$.

Does there always exist such a Markov chain? The answer is yes. This will be proved as a by-product of some rather difficult arguments and only for standard P. In other treatments, an affirmative answer is obtained immediately. Here is a digression which sketches the usual procedure. For simplicity, suppose I is finite. Let Ω be the set of all functions ω from $[0, \infty)$ to I; for $t \geq 0$, let $X(t, \omega) = \omega(t)$; let $\mathcal{F}$ be the usual product σ-field on Ω, namely, the smallest σ-field over which all the $X(t)$ are measurable. From the Kolmogorov consistency theorem (10.53), there is a unique probability $\mathcal{P}$ on $\mathcal{F}$ making $\{X(t) : 0 \leq t < \infty\}$ a Markov chain with the specified distribution. It is commonly held that, for a reasonable Markov chain, almost all the sample functions are step functions. How does X behave? Let S be the set of $\omega \in \Omega$ which are step functions. As everybody knows,

$$S \text{ includes no nonempty } \mathcal{F}\text{-set.}$$

Indeed, the monotone class argument associates to each $A \in \mathcal{F}$ a countable subset $C(A)$ of $[0, \infty)$, such that:

$$\omega \in A \text{ and } \omega' = \omega \text{ on } C(A) \text{ entails } \omega' \in A.$$

If $A \in \mathcal{F}$ and $A \subset S$ and $\omega \in A$, there is an $\omega' \in \Omega \backslash S$ with $\omega' = \omega$ on $C(A)$: a palpable contradiction. Consequently, S has inner $\mathcal{P}$-measure 0. Therefore, X is badly behaved; the most you can hope for is that S has outer $\mathcal{P}$-measure 1, which is in fact the case for standard P and finite I. The usual method of proof is to complete $\mathscr{F}$ under $\mathscr{P}$ and modify each $X(t)$ on a set of $\mathscr{P}$-measure 0, so the resulting process X^* is separable; this uses an uncountable axiom of choice. Then, you prove that for almost all ω, the function $t \to X^*(t, \omega)$ for rational t is a step function. As a function of real t, it is separable, and therefore a step function. The method here is to construct one process which not only has the required distribution, but also has step functions for all its sample functions.

Most results in this chapter are standard. References are usually given for proofs appropriated from others, but not for results. This section concludes with lemma (4), which will be used in most of the constructions in the rest of the book. Section 2 establishes the basic analytic properties of standard stochastic semigroups; these results will also be used many times. Sections 3 and 4 are independent of Section 2, and cover a special topic: the analytic properties of uniformly continuous semigroups. The results in this case are simpler and more complete. Given a uniformly continuous stochastic semigroup P, Section 7 constructs a Markov chain with stationary transitions P, all of whose sample functions are step functions. This construction depends on the construction in Section 6 of the general Markov chain whose sample functions are step functions up to the first bad discontinuity, and are then constant. Section 5 contains preliminary material on the exponential distribution. I will refer to Section 5 repeatedly, and Section 6 occasionally, in later constructions.

Finite I

Here is a summary of the results for finite I. Let P be a standard stochastic semigroup on I. Then
$$P'(0) = Q$$
exists and is finite; $Q(i, i) \leq 0$; while $Q(i, j) \geq 0$ for $i \neq j$; and
$$\Sigma_j \, Q(i, j) = 0.$$
Any finite matrix Q which satisfies these three conditions is the derivative at 0 of some standard stochastic P. Furthermore, Q determines P; in fact,
$$P(t) = e^{Qt}.$$
Fix a standard stochastic P, and let $Q = P'(0)$. Let
$$q(i) = -Q(i, i).$$
Let
$$\Gamma(i, j) = Q(i, j)/q(i) \quad \text{when } i \neq j \text{ and } q(i) > 0$$
$$= 0 \qquad\qquad \text{otherwise.}$$

Construct a process X as follows. The sample functions of X are I-valued right continuous step functions. The length of each visit to j is exponential with parameter $q(j)$, independent of everything else. The sequence of states visited by X is a discrete time Markov chain, with stationary transitions Γ and starting state i. Then X is Markov with stationary transitions P and starting state i.

Let Y be any Markov chain with stationary transitions P and starting state i. Then there is a process Y^* such that:

(a) $Y^*(t) = Y(t)$ almost surely, for each fixed t;

(b) the Y^* sample functions are right continuous I-valued step functions.

The jumps and holding times in Y^* are automatically distributed like the jumps and holding times in X. To be explicit, the sequence of states visited by Y^* is a discrete time Markov chain, with stationary transitions Γ and starting state i. The length of each visit to j is exponential with parameter $q(j)$, independent of everything else.

A lemma for general I

To state (4), let $R(t)$ be a matrix on I for each $t \in [0, \infty)$. Remember that $\bar{I} = I$ for finite I, and $\bar{I}$ is the one-point compactification of discrete I for countably infinite I. For each i, let X_i be an $\bar{I}$-valued process on the probability triple $(\Omega_i, \mathscr{F}_i, P_i)$. Let $\mathscr{F}_i(t)$ be the σ-field in Ω_i spanned by $X_i(s)$ for $0 \leq s \leq t$. Suppose

(3a) $P_i\{X_i(0) = i\} = 1$ for all i

(3b) $P_i\{X_i(t) \in I\} = 1$ for all $t \geq 0$ and all i

(3c) $P_i\{A \text{ and } X_i(t + s) = k\} = P_i\{A\} \cdot R(s, j, k)$ for all i, j, k in I, all nonnegative s and t, and all $A \in \mathscr{F}(t)$ with
$$A \subset \{X_i(0) = i \text{ and } X_i(t) = j\}.$$

(4) Lemma.   *Suppose conditions (3). Then R is a stochastic semigroup. Relative to P_i, the process X_i is Markov with stationary transitions R and starting state i.*

PROOF.   In (3c), put $t = 0$, $i = j$, $A = \{X_i(0) = i\}$, and use (3a):

(5) $R(s, i, k) = P_i\{X_i(s) = k\} \geq 0$.

Sum out $k \in I$ and use (3b):

$$\Sigma_{k \in I} R(s, i, k) = P_i\{X_i(s) \in I\} = 1.$$

So $R(s)$ is a stochastic matrix, taking care of (1a).

In (3c), put $A = \{X_i(0) = i$ and $X_i(t) = j\}$, and use (3a, 5):

$$R(t, i, j) \cdot R(s, j, k) = P_i\{X_i(t) = j \text{ and } X_i(t + s) = k\}.$$

Sum out $j \in I$ and use (3b, 5):

$$\begin{aligned}
\Sigma_{j \in I} R(t, i, j) \cdot R(s, j, k) &= P_i\{X_i(t) \in I \text{ and } X_i(t + s) = k\} \\
&= P_i\{X_i(t + s) = k\} \\
&= R(t + s, i, k).
\end{aligned}$$

So R satisfies (1b). Condition (1c) is taken care of by (3a) and (5). This makes R a stochastic semigroup.

Let $i_0, \ldots, i_n, i_{n+1} \in I$. Let $0 = t_0 < \cdots < t_n < t_{n+1} < \infty$. In (3c), put $i = i_0$, $j = i_n$, $k = i_{n+1}$, $t = t_n$, and $s = t_{n+1} - t_n$. Put

$$A = \{X_i(t_m) = i_m \text{ for } m = 0, \ldots, n\}.$$

Then

$$\begin{aligned}
P_{i_0}\{A \text{ and } X_i(t_{n+1}) = i_{n+1}\} &= P_{i_0}\{A\} \cdot R(t_{n+1} - t_n, i_n, i_{n+1}) \\
&= \Pi_{m=0}^{n} R(t_{m+1} - t_m, i_m, i_{m+1})
\end{aligned}$$

by induction: the case $n = 0$ is (5). This and (3a) make X_i Markov with stationary transitions R and starting state i, relative to P_i.   ★

2. ANALYTIC PROPERTIES

Let P be a substochastic semigroup on the finite or countably infinite set I. Except for (6), suppose

$$P \text{ is standard.}$$

(6) **Lemma.** *Even if P is not standard,*

     (a) $\Sigma_j P(t, i, j)$ *is nonincreasing with t.*

     (b) *If $P(t)$ is stochastic for some $t > 0$, then $P(t)$ is stochastic for all t.*

PROOF. *Claim (a).* Compute.

$$\begin{aligned}
\Sigma_j P(t + s, i, j) &= \Sigma_j \Sigma_k P(t, i, k) P(s, k, j) \\
&= \Sigma_k \Sigma_j P(t, i, k) P(s, k, j) \\
&\leqq \Sigma_k P(t, i, k).
\end{aligned}$$

Claim (b). Claim (a) shows that $P(s)$ is stochastic for $0 \leqq s \leqq t$. Now $P(u) = P(u/n)^n$ is visibly stochastic when $u/n \leqq t$.   ★

NOTE. Fix $i \in I$. If $\Sigma_{j \in I} P(t, i, j) = 1$ for some $t > 0$, then equality holds for all t. This harder fact follows from Lévy's dichotomy (*ACM*, 2.8).

(7) Lemma. *For each i,*

$$P(t, i, i) > 0 \quad \text{for all } t.$$

PROOF. $P(t, i, i) \geq P(t/n, i, i)^n$, and $P(t/n, i, i) \to 1$ as $n \to \infty$.   ★

(8) Lemma.   *Fix $i \in I$. If $P(t, i, i) = 1$ for some $t > 0$, then $P(t, i, i) = 1$ for all t.*

PROOF.   Let $0 < s < t$. Then

$$0 = 1 - P(t, i, i) \geq \Sigma_{j \neq i} P(s, i, j) P(t - s, j, j).$$

But $P(t - s, j, j) > 0$ by (7), forcing $P(s, i, j) = 0$ for all $j \neq i$. Using (6a),

$$P(s, i, i) = 1 - \Sigma_{j \neq i} P(s, i, j) = 1.$$

For general s,

$$P(s, i, i) \geq P(s/n, i, i)^n = 1$$

when $s/n < t$. That is,

$$P(s, i, i) = 1.$$ ★

(9) Lemma.   *For each i and j in I, the function $t \to P(t, i, j)$ is continuous. In fact,*

$$|P(t + \varepsilon, i, j) - P(t, i, j)| \leq 1 - P(|\varepsilon|, i, i).$$

PROOF.   It is enough to prove this for $\varepsilon > 0$: replace t by $t - \varepsilon$ to get $\varepsilon < 0$. Now

$$P(t + \varepsilon, i, j) = \Sigma_k P(\varepsilon, i, k) P(t, k, j);$$

so

$$P(t + \varepsilon, i, j) - P(t, i, j) = [P(\varepsilon, i, i) - 1]P(t, i, j) + \Sigma_{k \neq i} P(\varepsilon, i, k) P(t, k, j).$$

But $0 \leq P(t, k, j) \leq 1$ and $\Sigma_{k \neq i} P(\varepsilon, i, k) \leq 1 - P(\varepsilon, i, i)$.   ★

(10) Lemma. $P'(0, i, i)$ *exists and is nonpositive.*

PROOF.   Let $f(t) = -\log P(t, i, i)$. Then $0 \leq f(t) < \infty$ for all $t > 0$ by (7), and $f(0) = 0$, and f is subadditive, and f is continuous by (9). Let

(11) $q = \sup_{t>0} t^{-1} f(t).$

If $q = 0$, then $f \equiv 0$ and $P(t, i, i) \equiv 1$. So assume $q > 0$.

Fix a with $0 \leq a < q$. Fix $t > 0$ so that $t^{-1} f(t) \geq a$. Think of s as small and positive. Of course, $t = ns + \delta$ for a unique $n = 0, 1, \dots$ and δ with $0 \leq \delta < s$; both n and δ depend on s. So,

$$a \leq t^{-1} f(t) \leq t^{-1}[nf(s) + f(\delta)] = (t^{-1}ns)s^{-1}f(s) + t^{-1}f(\delta).$$

Let $s \to 0$. Then $t^{-1}ns \to 1$ and $\delta \to 0$, so

$$a \leq \lim \inf_{s \to 0} s^{-1} f(s).$$

Let a increase to q, proving

$$\lim_{s\to 0} s^{-1} f(s) = q.$$

In particular, $f(s) > 0$ and in consequence

$$s(d) = 1 - P(s, i, i) > 0$$

for small positive s. Of course, $d(s) \to 0$ as $s \to 0$. For $x > 0$,

$$\lim_{x\to 0} \frac{1}{x}[-\log(1 - x)] = 1,$$

so

$$\lim_{s\to 0} \frac{f(s)}{d(s)} = \lim_{s\to 0} \frac{-\log[1 - d(s)]}{d(s)} = 1.$$

Consequently,

$$\lim_{s\to 0} \frac{d(s)}{s} = q. \qquad \bigstar$$

Let

$$q(i) = -P'(0, i, i).$$

WARNING. $q(i) = \infty$ for all $i \in I$ is a distinct possibility. For examples, see Sections 9.1, (ACM, Sec. 3.3), (B & D, Sec. 2.12).

A state i with $q(i) < \infty$ is called *stable*. If all i are stable, the semigroup or process is called stable. A state i with $q(i) = \infty$ is called *instantaneous*. A state i with $q(i) = 0$ is called *absorbing*; this is equivalent to $P(t, i, i) = 1$ for some or all positive t, by (8) or (12), below.

In view of (11),

(12) $$P(t, i, i) \geqq e^{-q(i)t}.$$

This proves:

(13)   If $\sup_i q(i) < \infty$, then $\lim_{t\to 0} P(t, i, i) = 1$ uniformly in i.

The converse of (13) is also true: see (29).

(14) Proposition. *Fix $i \in I$, with $q(i) < \infty$. Fix $j \neq i$. Then $P'(0, i, j) = Q(i, j)$ exists and is finite. Moreover, $\sum_{j\in I} Q(i, j) \leqq 0$.*

PROOF. I say

(15) $$P(n\delta, i, j) \geqq \sum_{m=0}^{n-1} P(\delta, i, i)^m P(\delta, i, j) P[(n - m - 1)\delta, j, j].$$

Indeed, the mth term on the right side of (15) is the probability that a discrete time chain with transitions $P(\delta)$ and starting state i stays in i for m moves, then jumps to j, and is again in j at the nth move. Fix $\varepsilon > 0$. Using (12), choose $t > 0$ so that $m\delta < t$ implies $P(\delta, i, i)^m > 1 - \varepsilon$, and $s < t$

implies $P(s, j, j) > 1 - \varepsilon$. For $n\delta < t$, relation (15) implies

$$P(n\delta, i, j) > (1 - \varepsilon)^2 n P(\delta, i, j).$$

That is,

(16) $(n\delta)^{-1} P(n\delta, i, j) > (1 - \varepsilon)^2 \delta^{-1} P(\delta, i, j).$

Let $Q(i, j) = \limsup_{d \to 0} d^{-1} P(d, i, j)$. Let $\delta \to 0$ in such a way that $\delta^{-1} P(\delta, i, j) \to Q(i, j)$, and let $n \to \infty$ in such a way that $n\delta \to s < t$. From (16),

$$s^{-1} P(s, i, j) \geq (1 - \varepsilon)^2 Q(i, j),$$

so $\liminf_{s \to 0} s^{-1} P(s, i, j) \geq Q(i, j)$. This proves the first claim.

For the second claim, rearrange $\Sigma_j P(t, i, j) \leq 1$ to get

$$\frac{P(t, i, i) - 1}{t} + \Sigma_{j \neq i} \frac{P(t, i, j)}{t} \leq 0.$$

Use (10) on the first term. Use the first claim and Fatou on the sum.   ★

This proof is based on (Doob, 1953, p. 246).

(17) Lemma. *Fix $i \in I$, and $j \neq i$. Then $P'(0, i, j) = Q(i, j)$ exists and is finite.*

This differs from (13) in that $q(i)$ may be infinite. The proof is similar but harder, and is taken from (Chung, 1960, II.2).

PROOF. Let $\delta > 0$. Of course, $P(\delta)$ is a substochastic matrix on I, and $P(\delta)^n = P(n\delta)$. From Section 1.3, recall that $\{\xi_n\}$ is Markov with stationary transitions $P(\delta)$ and starting state i, relative to $P(\delta)_i$. Let

$$g(n) = P(\delta)_i \{\xi_n = i \text{ but } \xi_m \neq j \text{ for } 0 < m < n\}.$$

Then

(18) $P(n\delta, i, j) \geq \Sigma_{m=0}^{n-1} g(m) P(\delta, i, j) P[(n - m - 1)\delta, j, j],$

since the mth term on the right is the $P(\delta)_i$-probability that $\xi_n = j$, and the first j among $\xi_0, \ldots, \xi_n$ occurs at the $m + 1$st place, and is preceded by an i. Let

$$f(n) = P(\delta)_i \{\xi_n = j \text{ but } \xi_m \neq j \text{ for } 0 < m < n\}.$$

Then

(19) $P(m\delta, i, i) = g(m) + \Sigma_{\nu=1}^{m-1} f(\nu) P[(m - \nu)\delta, j, i]:$

for

$$\{\xi_m = i\} = A \cup (\bigcup_{\nu=1}^{m-1} B_\nu)$$

where

$$A = \{\xi_m = i \text{ but } \xi_\mu \neq j \text{ for } 0 < \mu < m\}$$

$$B_v = \{\xi_m = i \text{ and } \xi_v = j \text{ but } \xi_\mu \neq j \text{ for } 0 < \mu < v\}.$$

Since $\sum_{v=1}^{m-1} f(v) \leq 1$, relation (19) shows

(20) $$g(m) \geq P(m\delta, i, i) - \max\{P(s, j, i): 0 \leq s \leq m\delta\}.$$

Fix $\varepsilon > 0$. Find $t = t(i, j, \varepsilon) > 0$ so small that $P(s, i, i) > 1 - \varepsilon$ and $P(s, j, j) > 1 - \varepsilon$ for $0 \leq s \leq t$; then $P(s, j, i) < \varepsilon$. If $n\delta < t$, and $m \leq n$, then $g(m) \geq 1 - 2\varepsilon$ by (20). Combine this with (18);

$$P(n\delta, i, j) \geq (1 - \varepsilon)(1 - 2\varepsilon)nP(\delta, i, j).$$

Complete the argument as in (14).   ★

The main result, due to (Doob, 1942) and (Kolmogorov, 1951), is

(21) Theorem. *If P is a standard substochastic semigroup on the finite or countably infinite set I, then $P'(0, i, j) = Q(i, j)$ exists for all i, j and is finite for $i \neq j$.*

PROOF.   Use (10) and (17).   ★

The matrix $Q = P'(0)$ will be called the *generator* of P.

WARNING. Q is not the infinitesimal generator of P, and in fact does not determine P. For examples, see Sections 6.3 and 6.5.

The following theorem (Ornstein, 1960) and (Chung, 1960, p. 269) will not be used later, but is stated for its interest. A special case will be proved later, in Section 7.6.

(22) Theorem.   *$P(t, i, j)$ is continuously differentiable on $(0, \infty)$. For $i \neq j$ or $i = j$ and $q(i) < \infty$, the derivative is continuous at 0. For $0 < t < \infty$,*

$$\Sigma_j |P'(t, i, j)| < \infty \quad \text{and} \quad \Sigma_j P'(t, i, j) = 0.$$

For positive finite s and t,

$$P'(s + t) = P'(s)P(t) = P(s)P'(t).$$

An example of Smith (1964) shows that $P'(t, i, i)$ may oscillate as $t \to 0$ if $q(i) = \infty$. A similar example is presented in Section 3.6 of *ACM*.

NOTE.   Let Q be a matrix on I. When does there exist a standard stochastic or substochastic semigroup P with $P'(0) = Q$? This is one of the most interesting open questions in the Markov business. It is particularly intriguing when Q is allowed to take the value $-\infty$ on the diagonal. For some recent work on this question, see (Williams, 1967); one of his results is discussed in

Section 3.5 of *ACM*. When all entries in Q are finite, a partial solution to the question is known; this is discussed in Section 7.5. But the solution is not complete. When all entries in Q are finite, but $\Sigma_j Q(i,j) < 0$ for some i, it is not known when Q generates a standard stochastic semigroup. For uniform semigroups, the question is settled by (29).

3.   UNIFORM SEMIGROUPS

Let $\mathscr{A}$ be a *Banach algebra* with norm $\| \quad \|$ and identity Δ. That is, $\mathscr{A}$ is a noncommutative algebra over the reals, with identity Δ. By convention, $\alpha A = A\alpha$ for real α and $A \in \mathscr{A}$. For $A, B \in \mathscr{A}$ and real α,

$$\|A + B\| \leq \|A\| + \|B\|, \quad \|AB\| \leq \|A\| \cdot \|B\|, \quad \|\alpha A\| = |\alpha| \cdot \|A\|,$$

$$A \neq 0 \quad \text{implies} \quad \|A\| > 0, \quad \text{and} \quad \|\Delta\| = 1.$$

Finally, $\mathscr{A}$ is complete in the metric $\rho(A, B) = \|A - B\|$. In this section, convergence and continuity are with respect to ρ. Here are the local high points.

For $A \in \mathscr{A}$, by definition

$$e^A = \Delta + A + \frac{A^2}{2!} + \frac{A^3}{3!} + \cdots.$$

The series is Cauchy because

$$\left\| \frac{A^n}{n!} + \cdots + \frac{A^{n+m}}{(n+m)!} \right\| \leq \frac{\|A\|^n}{n!} + \cdots + \frac{\|A\|^{n+m}}{(n+m)!},$$

so converges (completeness) to an element of $\mathscr{A}$. The function $A \to e^A$ is uniformly continuous on $\{A : \|A\| \leq K < \infty\}$. If A and B *commute*, that is $AB = BA$, then

$$e^{A+B} = e^A e^B$$

by power series manipulation. If α is real,

$$e^{\alpha \Delta} = e^\alpha \Delta.$$

If $\|A\| < 1$, then by definition,

$$i(A) = \Delta + A + A^2 + \cdots$$

and by manipulation,

$$i(A)(\Delta - A) = (\Delta - A)i(A) = \Delta.$$

Consequently, if $B \in \mathscr{A}$ and β is real and $\|\Delta - \beta B\| < 1$, then B is *invertible*: there is a unique $C \in \mathscr{A}$ with $BC = CB = \Delta$. The uniqueness follows by familiar algebra from the existence. Write $C = B^{-1}$, and call C the *inverse* of B. Check that $(e^A)^{-1} = e^{-A}$, because A and $-A$ commute.

If f is a continuous function from $[0, \infty)$ to $\mathscr{A}$, and $0 \leq a < b < \infty$, then $\int_a^b f(t)\, dt$ is defined as the limit of the usual Riemann sums. The old arguments show that f is uniformly continuous on $[a, b]$, so the limit exists and depends linearly on f. Moreover,

$$\left\| \int_a^b f(t)\, dt \right\| \leq \int_a^b \| f(t) \|\, dt.$$

For $A \in \mathscr{A}$,

$$\int_a^b Af(t)\, dt = A \int_a^b f(t)\, dt \quad \text{and} \quad \int_a^b f(t)A\, dt = \left[\int_a^b f(t)\, dt \right] A.$$

For $c > 0$,

$$\int_a^b f(t + c)\, dt = \int_{a+c}^{b+c} f(t)\, dt.$$

For $a < b < c$,

$$\int_a^b f(t)\, dt + \int_b^c f(t)\, dt = \int_a^c f(t)\, dt.$$

Finally, if g is a continuous, real-valued function on $[0, t]$, then "by Fubini,"

$$\int_0^t \int_0^u g(u)f(v)\, dv\, du = \int_0^t \int_v^t g(u)f(v)\, du\, dv.$$

Let $\{P(t) : 0 \leq t < \infty\}$ be a *uniform semigroup*, namely,

$$P(t) \in \mathscr{A}, \quad P(t + s) = P(t)P(s), \quad P(0) = \Delta;$$

and

$$t \to P(t) \text{ is continuous}$$

(23) Theorem. $t^{-1}[P(t) - \Delta] \to Q$ *in* $\mathscr{A}$ *as* $t \to 0$, *and* $P(t) = e^{Qt}$.

PROOF.   Let $A(t) = \int_0^t P(u)\, du$. Then $t^{-1}A(t) \to \Delta$ as $t \to 0$, so $A(s)$ is invertible for some small positive s. Let

$$Q = [P(s) - \Delta]A(s)^{-1}.$$

Now

$$P(u)A(s) = \int_0^s P(u)P(v)\, dv$$

$$= \int_0^s P(u + v)\, dv$$

$$= \int_u^{u+s} P(v)\, dv$$

$$= A(u + s) - A(u).$$

So

$$[P(u) - \Delta]A(s) = A(u + s) - A(u) - A(s)$$
$$= [P(s) - \Delta]A(u)$$
$$= A(u)[P(s) - \Delta].$$

Multiply on the right by $A(s)^{-1}$:

$$P(u) = \Delta + A(u)Q.$$

By induction,

$$P(t) = \Delta + Qt + \cdots + \frac{Q^n t^n}{n!} + R_n,$$

where

$$R_n = \frac{1}{n!}\left[\int_0^t (t - u)^n P(u)\, du\right] Q^{n+1}.$$

Indeed, substituting $\Delta + A(u)Q$ for $P(u)$ in the formula for R_n shows

$$R_n = \frac{Q^{n+1} t^{n+1}}{(n + 1)!} + R_{n+1}.$$

Fix $T > 0$ and let $M = \max \{\|P(t)\| : 0 \leqq t \leqq T\}$. Then for $0 \leqq t \leqq T$,

$$\|R_n\| \leqq M \frac{\|Q\|^{n+1} T^{n+1}}{(n + 1)!} \to 0. \qquad \qquad \bigstar$$

For more information, see (Dunford and Schwartz, 1958, VIII.1).

(24) Remark. If $Q \in \mathscr{A}$, then $t \to e^{Qt}$ is a uniform semigroup. What (23) says is that all uniform semigroups are of this exponential type.

(25) Remark. Let $Q \in \mathscr{A}$ and $P(t) = e^{Qt}$. Let $h > 0$. Then

$$P(t + h) - P(t) = (e^{Qh} - \Delta)P(t) = P(t)(e^{Qh} - \Delta)$$
$$P(t) - P(t - h) = (e^{Qh} - \Delta)P(t - h) = P(t - h)(e^{Qh} - \Delta).$$

By looking at the power series,

$$\lim_{h \to 0} \frac{1}{h}(e^{Qh} - \Delta) = Q.$$

Thus, $P'(t) = QP(t) = P(t)Q$.

Let f be a function from $[0, \infty)$ to $\mathscr{A}$. Say f is *differentiable* iff

$$\frac{d}{dt} f(t) = f'(t) = \lim_{\varepsilon \to 0} \varepsilon^{-1}[f(t + \varepsilon) - f(t)]$$

exists for all $t \geqq 0$. If f and g are differentiable, so is fg; namely, $(fg)' = f'g + fg'$. If f is differentiable and $A \in \mathscr{A}$, then (Af) and $(A + f)$ are differentiable: $(Af)' = Af'$ and $(A + f)' = f'$. If f is differentiable and $f' \equiv 0$, then f is constant. Indeed, replace f by $-f(0) + f$ to get $f(0) = 0$. Then

$$f(s) = f(s) - f(t) + f(t),$$

so

$$\|f(s)\| - \|f(t)\| \leq \|f(s) - f(t)\|.$$

In particular, $t \to \|f(t)\|$ is real-valued, vanishes at 0, and has vanishing derivative. So $\|f(t)\| = 0$ for all t.

Fix $Q \in \mathscr{A}$. If $P(t) = e^{Qt}$, remark (25) shows

$$P'(t) = Qe^{Qt} = e^{Qt}Q.$$

Conversely, if P is differentiable, $P(0) = \Delta$, and either

$$P'(t) = QP(t) \quad \text{for all } t \geq 0$$

or

$$P'(t) = P(t)Q \quad \text{for all } t \geq 0,$$

then $P(t) = e^{Qt}$. Indeed, with the first condition,

$$\frac{d}{dt} [e^{-Qt}P(t)] = e^{-Qt} P'(t) - e^{-Qt}QP(t)$$

$$= e^{-Qt}[P'(t) - QP(t)]$$

$$= 0.$$

So, $e^{-Qt}P(t)$ is constant; put $t = 0$ to see the constant is Δ. Thus, $P(t) = e^{Qt}$. For the second condition, work on $P(t)e^{-Qt}$. In particular, P is a semigroup. This result can be summarized as

(26) Theorem. *Let $Q \in \mathscr{A}$. Then $P(t) = e^{Qt}$ is the unique solution of either the forward system*

$$P(0) = \Delta \quad \text{and} \quad P'(t) = P(t)Q$$

or the backward system

$$P(0) = \Delta \quad \text{and} \quad P'(t) = QP(t).$$

I learned the argument from Harry Reuter.

4. UNIFORM SUBSTOCHASTIC SEMIGROUPS

Suppose the standard substochastic semigroup $P(\cdot)$ on the finite or countably infinite set I is *uniform*, that is,

$$\lim_{t \to 0} P(t, i, i) = 1 \quad \text{uniformly in } i.$$

This condition is automatically satisfied if I is finite, a case treated in (Doob, 1953, VI.I). If A is a matrix on I, define $\|A\|$ as $\sup_i \Sigma_j |A(i,j)|$. The set $\mathscr{A}$ of matrices A with $\|A\| < \infty$ is a Banach algebra with norm $\|\cdot\|$, and identity Δ, where $\Delta(i,j)$ is 1 or 0 according as $i = j$ or $i \neq j$. And $\|A\| \leq 1$ if A is substochastic. If $0 \leq s < t < \infty$ and $h = t - s$, then

$$(27) \qquad \|P(t) - P(s)\| = \|(P(h) - \Delta)P(s)\|$$
$$\leq \|P(h) - \Delta\|$$
$$\leq 2 \sup_i [1 - P(h, i, i)].$$

So P is uniform in the sense of Section 3. As (23) implies, P has derivative $Q \in \mathscr{A}$ at 0, and $P(t) = e^{Qt}$. Clearly, $Q(i,j) = P'(0, i, j)$, where ' is the calculus derivative.

What matrices can arise? Plainly,

$$(28a) \qquad\qquad Q(i, i) \leq 0 \quad \text{for all } i$$

$$(28b) \qquad\qquad Q(i, j) \geq 0 \quad \text{for all } i \neq j.$$

Inequality (28a) implies $-Q(i, i) \leq \|Q\|$; so $Q \in \mathscr{A}$ implies

$$(28c) \qquad\qquad \inf_i Q(i, i) > -\infty.$$

By rearranging $\Sigma_j P(t, i, j) \leq 1$,

$$\frac{P(t, i, i) - i}{t} + \Sigma_{j \neq i} \frac{P(t, i, j)}{t} \leq 0.$$

By Fatou:

$$(28d) \qquad\qquad \Sigma_j Q(i, j) \leq 0.$$

(29) Theorem. *Let P be a uniform substochastic semigroup on I. Then $Q = P'(0)$ exists and satisfies condition (28). Conversely, let Q be a matrix on I which satisfies (28). Then there exists a unique uniform substochastic semigroup P with $P'(0) = Q$, namely*

$$P(t) = e^{Qt}.$$

Finally, for any $t > 0$:

$$P(t) \quad \text{is stochastic iff} \quad \Sigma_j Q(i, j) = 0 \quad \text{for all } i.$$

NOTE. Suppose Q satisfies (28) and P is a standard substochastic semigroup on I with $P'(0) = Q$. Then P is uniform by (13).

PROOF The first assertion has already been argued. For "conversely," $P(t) = e^{Qt}$ is a uniform $\mathscr{A}$-valued semigroup with $P'(0) = Q$ by (24–25). The uniqueness of P is part of (23). When is P stochastic or substochastic?

Choose a positive, finite q with

$$Q(i, i) \geqq -q \quad \text{for all } i.$$

Let

$$\hat{Q} = \frac{1}{q} Q + \Delta,$$

where Δ is the identity matrix. Then $\hat{Q}$ is always substochastic. And $\hat{Q}$ is stochastic iff $\Sigma_j Q(i, j) = 0$ for all i. But $Qt = -qt\Delta + qt\hat{Q}$, so

$$e^{Qt} = e^{-qt} e^{qt\hat{Q}}$$

$$= e^{-qt} \Sigma_{n=0}^{\infty} \frac{(qt)^n}{n!} \hat{Q}^n$$

is the Poisson average of powers of $\hat{Q}$, all of which are substochastic. Thus, e^{Qt} is always substochastic. Let $t > 0$. Then $\hat{Q}$ appears with positive weight, so e^{Qt} is stochastic iff $\hat{Q}$ is: that is, iff $\Sigma_j Q(i, j) = 0$ for all i.   ★

This proof was suggested by John Kingman.

5. THE EXPONENTIAL DISTRIBUTION

Let τ be a random variable on the probability triple $(\Omega, \mathscr{F}, \mathscr{P})$.

DEFINITION. *Let* $0 < q < \infty$. *Say* τ *has* exponential distribution with parameter q, *abbreviated as* τ *is* $e(q)$, *iff* $\mathscr{P}\{\tau > t\} = e^{-qt}$ *for all* $t \geqq 0$. *Then* $\mathscr{P}\{\tau = t\} = 0$ *and* $\mathscr{P}\{\tau \geqq t\} = e^{-qt}$ *for* $t \geqq 0$.

To state (30), suppose T is $e(q)$ on $(\Omega, \mathscr{F}, \mathscr{P})$. Let Σ be a sub-σ-field of $\mathscr{F}$, independent of T. Let S be Σ-measurable; allow $\mathscr{P}\{S = \infty\} > 0$. Let $A \in \Sigma$. Let u and t be nonnegative and finite.

(30) Lemma.

$$\mathscr{P}\{A \text{ and } S \leqq t \leqq t + u < S + T\} = e^{-qu} \mathscr{P}\{A \text{ and } S \leqq t < S + T\}.$$

PROOF.   Here is a computation.

$$\mathscr{P}\{A \text{ and } S \leqq t \leqq t + u < S + T\} = \int_{\{A \text{ and } S \leqq t\}} e^{-q(t+u-S)} \, d\mathscr{P}$$

$$= e^{-qu} \int_{\{A \text{ and } S \leqq t\}} e^{-q(t-S)} \, d\mathscr{P}$$

$$= e^{-qu} \mathscr{P}\{A \text{ and } S \leqq t < S + T\}.$$

The first and last equalities come from Fubini (10.21). I will set the first one up for you. Use $(\Omega, \mathscr{F}, \mathscr{P})$ for the basic triple. Let $X_1(\omega) = \omega$ and let $(\Omega_1, \mathscr{F}_1) = (\Omega, \Sigma)$. Let $X_2 = T$; let $(\Omega_2, \mathscr{F}_2)$ be the Borel half-line $[0, \infty)$. Let

$$f(\omega, y) = 1 \quad \text{if} \quad \omega \in A \quad \text{and} \quad S(\omega) \leq t \leq t + u < S(\omega) + y$$
$$= 0 \quad \text{otherwise.}$$

Then f is $\mathscr{F}_1 \times \mathscr{F}_2$-measurable. And $f(\omega, \cdot) \equiv 0$ unless $\omega \in A$ and $S(\omega) \leq t$. If $\omega \in A$ and $S(\omega) \leq t$, then

$$\int f[\omega, T(\omega')] \, \mathscr{P}(d\omega') = e^{-q[t+u-S(\omega)]}.$$

Let $\hat{\mathscr{P}}$ be $\mathscr{P}$ retracted to Σ. So

$$\mathscr{P}\{A \text{ and } S \leq t \leq t + u < S + T\} = \int f[X_1(\omega), X_2(\omega)] \, \mathscr{P}(d\omega)$$

$$= \iint f[\omega, T(\omega')] \, \mathscr{P}(d\omega') \, \hat{\mathscr{P}}(d\omega)$$

$$= \int_{\{A \text{ and } S \leq t\}} e^{-q(t+u-S)} \, d\mathscr{P}. \qquad \bigstar$$

(31) Lemma.   *If τ is $e(q)$ and $\lambda > -q$, the expectation of $e^{-\lambda\tau}$ is $q/(q + \lambda)$. In particular, the expectation of τ^n is $n!/q^n$, and the variance of τ is $1/q^2$.*

PROOF.   Integrate and expand. $\bigstar$

Let τ and σ be independent random variables on $(\Omega, \mathscr{F}, \mathscr{P})$.

(32) Lemma.   **(a)** *If τ has a density bounded above by B, so does $\tau + \sigma$.*

   (b) *If τ has a continuous density bounded above by B, so does $\tau + \sigma$.*

PROOF.   Claim (a) is easy. Use (a) and dominated convergence for (b). $\bigstar$

NOTE.   If τ is $e(q)$, its density is bounded above by q, but is discontinuous at 0, as a function on the real line.

To avoid exceptional cases, I will sometimes write $e(0)$ for the distribution concentrated at ∞, and $e(\infty)$ for the distribution concentrated at 0. Then $e(q)$ has mean $1/q$, where 0 and ∞ are reciprocals. For lemmas (33) and (34), let τ_m be independent $e(q_m) \infty$, for nonnegative integer m, on $(\Omega, \mathscr{F}, \mathscr{P})$, with $0 \leq q_m < \infty$. Let $\infty + x = \infty$ for $x > -\infty$. Write E for $\mathscr{P}$-expectation.

(33) Lemma.   *Let M be a finite or infinite subset of the nonnegative integers. Then $\Sigma_{m \in M} \tau_m < \infty$ a.e. if $\Sigma_{m \in M} 1/q_m < \infty$, and $\Sigma_{m \in M} \tau_m = \infty$ a.e. if $\Sigma_{m \in M} 1/q_m = \infty$.*

PROOF. Abbreviate $S = \Sigma_{m \in M} \tau_m$. As (31) implies, $E(S) = \Sigma_{m \in M} 1/q_m$, proving the first assertion. For the second, let $e^{-\infty} = 0$, so $0 \le e^{-x} \le 1$ is continuous and decreasing on $[0, \infty]$. By (31) and monotone convergence,

$$E(e^{-S}) = \Pi_{m \in M} q_m/(q_m + 1) = 1/\Pi_{m \in M}(1 + q_m^{-1}) = 0$$

when $\Sigma_{m \in M} 1/q_m = \infty$. Then $e^{-S} = 0$ a.e., forcing $S = \infty$ a.e.   ★

(34) Lemma. *Let M be a finite or infinite subset of the nonnegative integers containing at least 0 and 1. For $m \in M$, suppose $0 < q_m < \infty$, and suppose that $\Sigma_{m \in M} 1/q_m < \infty$. Then $\Sigma_{m \in M} \tau_m$ has a continuous density vanishing on $(-\infty, 0]$ and bounded above by q_0.*

PROOF. Lemma (32b) reduces M to $\{0, 1\}$. Then, the density at $t \ge 0$ is

$$q_0 q_1 \int_0^t e^{-q_0(t-s)} e^{-q_1 s}\, ds,$$

while the density at $t < 0$ is 0. This function of t is continuous. For $t \ge 0$, the integrand is at most $e^{-q_1 s}$, and $\int_0^\infty q_1 e^{-q_1 s}\, ds = 1$. So the density is at most q_0.   ★

6. THE STEP FUNCTION CASE

In this section, I will construct the general Markov chain whose sample functions are initially step functions, and are constant after the first bad discontinuity. The generality follows from (7.33). This construction will be used in Section 7 and in Chapters 6–8, so I really want you to go through it, even though the first reading may be difficult. I hope you will eventually feel that the argument is really simple.

Let I be a finite or countably infinite set. Give I the discrete topology. Let $0 < a \le \infty$. Let f be a function from $[0, a)$ to I. Then f is an *I-valued right continuous step function* on $[0, a)$ provided

$$f(t) = \lim \{f(s): s \downarrow t\} \qquad \text{for } 0 \le t < a$$
$$f(t-) = \lim \{f(s): s \uparrow t\} \text{ exists in } I \quad \text{for } 0 < t < a.$$

Let Γ be a substochastic matrix on I, with

$$\Gamma(i, i) = 0 \quad \text{for all } i.$$

Let q be a function from I to $[0, \infty)$.

Informal statement of (39)

Construct a process X as follows: The sample functions of X are I-valued right-continuous step functions, at least initially. The length of each visit to j is exponential with parameter $q(j)$, independent of everything else. The sequence of states visited by X is a discrete time Markov chain, with stationary transitions Γ and starting state $i \in I$. This may define the X sample function only on a finite time interval. Indeed, if $\Sigma_k \Gamma(j, k) < 1$, the jump process may disappear on leaving j. And even if X succeeds in visiting a full infinite sequence of states, the sum of the lengths of the visits may be finite. In either case, continue the X sample function over the rest of $[0, \infty)$ by making it equal to $\partial \notin I$. Then X is Markov with stationary transitions R, where R is a standard stochastic semigroup on $I \cup \{\partial\}$, with ∂ absorbing. Moreover,

$$R'(0, j, j) = -q(j) \qquad \text{for } j \text{ in } I$$
$$R'(0, j, k) = q(j)\Gamma(j, k) \quad \text{for } j \neq k \text{ in } I.$$

Formalities resume

Fix $\partial \notin I$. Extend Γ to $\hat{I} = I \cup \{\partial\}$ by setting

$$\hat{\Gamma}(i, j) = \Gamma(i, j) \qquad\qquad \text{for } i, j \in I$$
$$\hat{\Gamma}(i, \partial) = 1 - \Sigma_{j \in I} \Gamma(i, j) \quad \text{for } i \in I$$
$$\hat{\Gamma}(\partial, i) = 0 \qquad\qquad\quad \text{for } i \in I$$
$$\hat{\Gamma}(\partial, \partial) = 1.$$

Thus, $\hat{\Gamma}$ is stochastic on $\hat{I}$. Extend q to $\hat{I}$ by setting $q(\partial) = 0$. Define a matrix Q on $\hat{I}$ as follows:

$$Q(i, i) = -q(i) \qquad \text{for } i \in \hat{I}$$
$$Q(i, j) = q(i)\hat{\Gamma}(i, j) \quad \text{for } i \neq j \text{ in } \hat{I}.$$

Introduce the set $\mathscr{X}$ of pairs $x = (\omega, w)$, where ω is a sequence of elements of $\hat{I}$, and w is a sequence of elements of $(0, \infty]$.

Let

$$\xi_n(\omega, w) = \omega(n) \in \hat{I} \quad \text{and} \quad \tau_n(\omega, w) = w(n) \in (0, \infty]$$

for $n = 0, 1, \ldots$. Give $\mathscr{X}$ the product σ-field, namely the smallest σ-field over which $\xi_0, \xi_1, \ldots$ and $\tau_0, \tau_1, \ldots$ are measurable. Of course, $\mathscr{X}$ is Borel.

INFORMAL NOTE. The process X begins by visiting $\xi_0, \xi_1, \ldots$ with holding times $\tau_0, \tau_1, \ldots$.

For each $i \in \hat{I}$, let π_i be the unique probability on $\mathscr{X}$ for which, semi-formally:

(a) $\xi_0, \xi_1, \ldots$ is a discrete time Markov chain with stationary stochastic transitions $\hat{\Gamma}$ on $\hat{I}$ and starting state i;

(b) given $\xi_0, \xi_1, \ldots$, the random variables $\tau_0, \tau_1, \ldots$ are conditionally independent and exponentially distributed, with parameters $q(\xi_0)$, $q(\xi_1), \ldots$.

INFORMAL NOTE. π_i makes X a Markov chain with starting state i and generator Q.

More rigorously: introduce W, the space of sequences w of elements of $(0, \infty]$. Give W the product σ-field. For each function r from $\{0, 1, \ldots\}$ to $[0, \infty)$, let η_r be the probability on W making the coordinates independent and exponentially distributed, the nth coordinate having parameter $r(n)$. For $\omega \in \hat{I}^\infty$, the set of $\hat{I}$-sequences, let $q(\omega)$ be the function on $\{0, 1, \ldots\}$ whose value at n is $q(\omega(n))$. Think of $\xi = (\xi_0, \xi_1, \ldots)$ as projecting $\mathscr{X}$ onto $\hat{I}^\infty$, and $\tau = (\tau_0, \tau_1, \ldots)$ as projecting $\mathscr{X}$ onto W. Then π_i, a probability on $\mathscr{X}$ is uniquely defined by the two requirements:

(a) $\pi_i \xi^{-1} = \hat{\Gamma}_i$;

(b) $\eta_q(\omega)$ is a regular conditional π_i-distribution for τ given $\xi = \omega$.

By (10.43), this amounts to defining

$$(35) \qquad \pi_i\{A\} = \int_\Omega \eta_{q(\omega)}\{A(\omega)\}\, \hat{\Gamma}_i(d\omega),$$

where:

(a) $\hat{\Gamma}_i$ was defined in Section 1.3 as the probability on $\hat{I}^\infty$ making the coordinate process a $\hat{\Gamma}$-chain starting from i;

(b) A is a product measurable subset of $\mathscr{X}$;

(c) $A(\omega) = \{w : w \in W \text{ and } (\omega, w) \in A\}$ is the ω-section of A.

There is a more elementary characterization of π_i, which is also useful: π_i is the unique probability on $\mathscr{X}$ for which

$$(36) \qquad \pi_i\{\xi_0 = i_0, \ldots, \xi_n = i_n \text{ and } \tau_0 > t_0, \ldots, \tau_n > t_n\} = pe^{-t},$$

where

$$t = q(i_0)t_0 + \cdots + q(i_n)t_n$$

and p is the product

$$\hat{\Gamma}(i_0, i_1) \cdots \hat{\Gamma}(i_{n-1}, i_n);$$

this must hold for all n and all $\hat{I}$-sequences $i_0 = i, i_1, \ldots, i_n$, and all non-negative numbers $t_0, \ldots, t_n$. By convention, an empty product is 1. Use (10.16) for the uniqueness of π_i. For an easy construction of π_i, let $\zeta_0, \zeta_1, \ldots$ be Markov with transitions $\hat{\Gamma}$ starting from i, on some triple. Independent of

ζ, let $T_0, T_1, \ldots$ be independent and exponential with parameter 1. Let $\theta_n = T_n/q(\zeta_n)$. Then π_i is the distribution of (ζ, θ).

If π_i satisfies the first characterization (35), it satisfies the second (36) by this computation. Let A be the set of $\omega \in \hat{I}^\infty$ with $\omega(m) = i_m$ for $m = 0, \ldots, n$. Let B be the set of $w \in W$ such that $w(m) > t_m$ for $m = 0, \ldots, n$. Then

$$\pi_i(A \times B) = \int_A \eta_{q(\omega)}(B) \, \hat{\Gamma}_i(d\omega) = pe^{-t},$$

because $\hat{\Gamma}_i(A) = p$ and $\eta_{q(\omega)}(B) = e^{-t}$ for all $\omega \in A$. Since both characterizations pick out a unique probability, they are equivalent.

From either characterization,

(37) $\qquad \pi_i\{\xi_0 = i_0, \ldots, \xi_n = i_n \text{ and } (\tau_0, \ldots, \tau_n) \in B\} = pq,$

where: B is an arbitrary Borel subset of R^{n+1}; and q is the probability that

$$(U_0, \ldots, U_n) \in B,$$

$U_0, \ldots, U_n$ being independent exponential random variables, with parameters $q(i_0), \ldots, q(i_n)$, respectively; and p is defined as for (36).

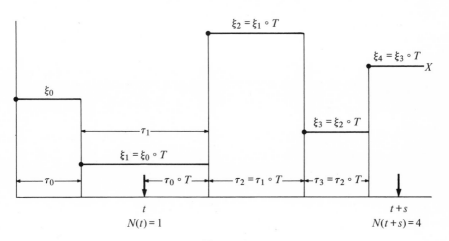

Figure 1.

Let $\sigma = \Sigma_{n=0}^\infty \tau_n$, so $0 < \sigma \leqq \infty$. For $x \in \mathcal{X}$ and $t < \sigma(x)$, let $N(t, x)$ be the least m with

$$\tau_0(x) + \cdots + \tau_m(x) > t.$$

For $t \geqq \sigma(x)$, let $N(t, x) = \infty$. Define a process X on $\mathcal{X}$ as in Figures 1 and 2:

$$X(t, x) = \xi_m(x) \quad \text{when } t < \sigma(x) \text{ and } N(t, x) = m;$$
$$= \partial \qquad \text{when } t \geqq \sigma(x).$$

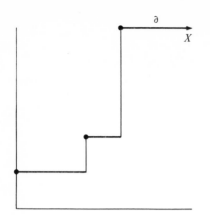
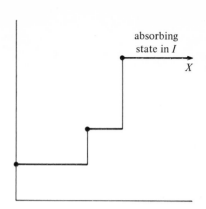

$d < \infty$ so $\sigma = \infty$

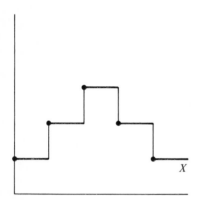
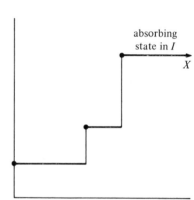

$d = \infty$ and $\sigma = \infty$

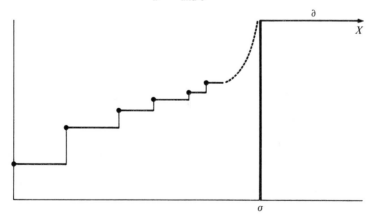

$d = \infty$ and $\sigma < \infty$

Figure 2.

158

Check that X is jointly measurable and $\hat{I}$-valued. *Joint measurability* mean relative to the product of the Borel σ-field on $[0, \infty)$ and the σ-field on $\mathscr{X}$.

FACT.   If $X(\cdot, x)$ is I-valued everywhere, then $\sigma(x) = \infty$, and $X(\cdot, x)$ is a right continuous I-valued step function.

DEFINITION.   *Let d be the least n if any with $\xi_n = \partial$, and $d = \infty$ if none. Call $j \in \hat{I}$ absorbing iff $q(j) = 0$. Let $\mathscr{X}_0$ be the subset of $\mathscr{X}$ where for all n:*

$$\tau_n = \infty \quad \text{iff} \quad \xi_n \text{ is absorbing};$$

$$\hat{\Gamma}(\xi_n, \xi_{n+1}) > 0.$$

Discussion

Use (37) to check $\pi_i(\mathscr{X}_0) = 1$. Confine x to $\mathscr{X}_0$. Then $\xi_n = \partial$ implies $\xi_{n+1} = \partial$, because $\hat{\Gamma}(\partial, \partial) = 1$. And $\xi_n \neq \partial$ implies $\xi_{n+1} \neq \xi_n$, because $\Gamma(j, j) = 0$. Look at Figure 2 while you think about the next lot of assertions. If $d < \infty$, then $\sigma = \infty$, and the X sample function is a step function, which usually terminates with a ∂; but if ξ visits an absorbing state in I before the first visit to ∂, then the sample function of X is an I-valued step function, with a finite number of steps. If $d = \infty$ and $\sigma = \infty$, then the X sample function is an I-valued step function, which typically has infinite number of steps; but if ξ visits an absorbing state in I, there are only a finite number. If $d = \infty$ and $\sigma < \infty$, then ξ cannot visit an absorbing state; the X sample function is an I-valued step function on $[0, \sigma)$, has a bad discontinuity at σ, and is ∂ after σ. In any case, each X sample function is continuous from the right.

NOTE.   If ξ_n is absorbing, so $\tau_n = \infty$, then the values of ξ_{n+m} and τ_{n+m} for $m > 0$ have no effect on X. Consequently, Γ could be normalized so $\Gamma(j, k) = 0$ for absorbing $j \in I$. Suppose this done. Keep x confined to $\mathscr{X}_0$. Then ξ_n is the nth state visited by X, with holding time τ_n, for $0 \leq n < d$. If $d = \infty$, then ξ visits no absorbing states, and X performs an infinite sequence of visits but may still ooze into ∂. If $d < \infty$ and ξ_{d-1} is absorbing, then ξ_{d-1} is the last state visited by X. If $d < \infty$ and ξ_{d-1} is not absorbing, then $\xi_d = \partial$ is the last state visited by X. If the row sums of Γ are 1 except at the absorbing states, then $d < \infty$ implies that X terminates with a visit of infinite length at $\xi_{d-1} \in I$.

WARNING.   Outside $\mathscr{X}_0$, it is likely that $\xi_1 = \xi_0$. So τ_0 and ξ_1 cannot be recovered from X. If x is confined to $\mathscr{X}_0$, however, then τ_0 is measurable on X; if x is confined to $\{\mathscr{X}_0$ and $\tau_0 < \infty\}$, then ξ_1 is measurable on X. If

$\Gamma(i, i) > 0$, contrary to assumption, then $\xi_1 = \xi_0$ would have positive π_i-probability. So ξ_1 would not be the state X visits on leaving ξ_0. The holding time parameter for i would be

$$q(i)[1 - \Gamma(i, i)].$$

The probability of jumping to j on leaving i would be

$$\Gamma(i, j)/[1 - \Gamma(i, i)].$$

And X would still be Markov with stationary transitions. Transition (i, i) would have derivative

$$-q(i)[1 - \Gamma(i, i)].$$

Transition (i, j) would still have derivative

$$q(i)\Gamma(i, j).$$

The argument is about the same, although it's harder to compute the derivative.

Theorems

(38) Theorem.   *Let $i \in I$. Then X has I-valued sample functions, which are right continuous step functions on $[0, \infty)$, with π_i-probability 1 iff*

$$\Sigma_n \{1/q(\xi_n): 0 \leqq n < d\} = \infty$$

with π_i-probability 1. More precisely, let θ be the least t if any with $X(t) = \partial$, and $\theta = \infty$ if none. Then $A = \{\theta = \infty\}$ differs by a π_i-null set from the set B where

$$\Sigma_n \{1/q(\xi_n): 0 \leqq n < d\} = \infty.$$

SHORT PROOF.   Condition on ξ and use (33).   ★

LONG PROOF.   The problem is to show $\pi_i\{A \Delta B\} = 0$; as usual $A \Delta B = (A \backslash B) \cup (B \backslash A)$. For $C \subset \mathscr{X} = \Omega \times W$ and $\omega \in \Omega$, let $C(\omega)$ be the ω-section of C, namely the set of $w \in W$ with $(\omega, w) \in C$. By (35),

$$\pi_i\{A \Delta B\} = \int \eta_{q(\omega)}\{A(\omega) \Delta B(\omega)\} \, \hat{\Gamma} \, (d\omega).$$

Temporarily, let $\mathbf{d}(\omega)$ be the least n if any with $\omega(n) = \partial$, and $\mathbf{d}(\omega) = \infty$ if none. Let $\mathbf{B}$ be the set of $\omega \in \hat{I}^\infty$ such that

$$\Sigma_n \{1/q(\omega(n)): 0 \leqq n < \mathbf{d}(\omega)\} = \infty.$$

Then $B = \mathbf{B} \times W$, so $B(\omega) = \varnothing$ for $\omega \notin \mathbf{B}$ and $B(\omega) = W$ for $\omega \in \mathbf{B}$. And

$$\pi_i\{A \Delta B\} = \int_{\omega \notin \mathbf{B}} \eta_{q(\omega)}\{A(\omega)\} \hat{\Gamma}_i(d\omega) + \int_{\omega \in \mathbf{B}} \eta_{q(\omega)}\{W \backslash A(\omega)\} \hat{\Gamma}_i(d\omega).$$

Check

$$\theta = \Sigma_n \{\tau_n : 0 \leq n < d\}.$$

Temporarily, let τ_n be the coordinate process on W. Then $A(\omega)$ is the subset of W where

$$\Sigma_n \{\tau_n : 0 \leq n < \mathbf{d}(\omega)\} = \infty.$$

With respect to $\eta_{q(\omega)}$, the variables τ_n are independent and exponential with parameter $q(\omega(n))$. For $\omega \notin \mathbf{B}$,

$$\eta_{q(\omega)}\{A(\omega)\} = 0$$

by the first assertion in (33). For $\omega \in \mathbf{B}$,

$$\eta_{q(\omega)}\{W \backslash A(\omega)\} = 0$$

by the second assertion in (33).   ★

(39) Theorem. *With respect to π_i, the process X is Markov with stationary transitions, say R, and starting state i. Here R is a standard stochastic semigroup on $\hat{I}$, for which ∂ is absorbing. Moreover, $R'(0) = Q$.*

NOTE. The retract of R to I is stochastic iff for any $i \in I$, with π_i-probability 1, the sample functions of X are I-valued everywhere: you can prove this directly. You know that an I-valued sample function is automatically a right continuous step function.

PROOF. Let $0 \leq t < \infty$ and let $0 < s < \infty$. Let $\Sigma(t)$ be the smallest σ-field over which all the $X(u)$ are measurable, for $0 \leq u \leq t$. The main thing to prove is the Markov property:

(40) $\pi_i\{A \text{ and } X(t + s) = k\} = \pi_i\{A\} \cdot \pi_j\{X(s) = k\},$

where

$$A \subset \{X(0) = i \text{ and } X(t) = j\} \quad \text{and} \quad A \in \Sigma(t) \quad \text{and} \quad i, j, k \text{ are in } \hat{I}.$$

First, I will argue the easy case $j = \partial$. Then $X(t + s) = \partial$, at least on $\{\mathscr{X}_0 \text{ and } X(t) = j\}$, while $\pi_\partial\{X(s) = \partial\} = 1$. If $k \neq \partial$, both sides of (40) vanish. If $k = \partial$, both sides of (40) are $\pi_i(A)$. This settles the case $j = \partial$.

From now on, assume $j \in I$. Abbreviate

$$D_m = \{X(0) = i \text{ and } X(t) = j \text{ and } N(t) = m\}.$$

Let $\mathscr{A}_m$ be the σ-field of subsets of D_m of the form $D_m \cap A^*$ with $A^* \in \Sigma(t)$. For $m = 0, 1, \ldots, \infty$, the sets D_m are pairwise disjoint, and their union is

$\{X(0) = i$ and $X(t) = j\}$. So it is enough to prove (40) for $A \in \mathscr{A}_m$: both sides are countably additive in A. If $i = \partial$ or $m = \infty$, both sides of (40) vanish. So fix $i \in I$ and $m < \infty$ for the rest. By definition, $X(t) = \xi_m$ on $\{N(t) = m\}$, so $\xi_m = j$ on D_m.

Call a set A *special* iff A is the event that

$$\xi_0 = i_0, \ldots, \xi_m = i_m \text{ and } \tau_0 > t_0, \ldots, \tau_{m-1} > t_{m-1} \text{ and}$$

$$\tau_0 + \cdots + \tau_{m-1} \leqq t < \tau_0 + \cdots + \tau_{m-1} + \tau_m,$$

where $i_0 = i, \ldots, i_m = j$ are in $\hat{I}$, and $t_0, \ldots, t_{m-1}$ are nonnegative numbers. So $A \subset D_m$. Let $\mathscr{D}_m$ be the σ-field in D_m generated by the special A. In other terms: $\mathscr{D}_m$ is the σ-field in D_m generated by $\xi_0, \ldots, \xi_m$ and $\tau_0, \ldots, \tau_{m-1}$, all retracted to D_m. I say $\mathscr{A}_m \subset \mathscr{D}_m$. Indeed, let $0 \leqq u \leqq t$ and let $h \in I$. Then

$$\{D_m \text{ and } X(u) = h\} = \bigcup_{n=0}^{m} D_{mn},$$

where

$$D_{mn} = D_m \cap \{\xi_n = h \text{ and } \tau_0 + \cdots + \tau_{n-1} \leqq u < \tau_0 + \cdots + \tau_{n-1} + \tau_n\}.$$

You get $D_{mn} \in \mathscr{D}_m$ for $n < m$. If $h \neq j$, then $D_{mm} = \varnothing$. If $h = j$, then

$$D_{mm} = D_m \cap \{\tau_0 + \cdots + \tau_{m-1} \leqq u\}.$$

Either way, $D_{mm} \in \mathscr{D}_m$. This proves $\mathscr{A}_m \subset \mathscr{D}_m$. Two different special A are disjoint or nested; so the class of special A, with the null set added, is closed under intersection. The union over $i_1, \ldots, i_{m-1}$ of the special A with $t_0 = \cdots = t_{m-1} = 0$ is D_m. And both sides of (40) are countably additive in A. If I manage to get (40) for the special A, then (10.16) will extend (40) to $\mathscr{D}_m$, which is more than enough.

Both sides of (40) vanish if one or more of $i_1, \ldots, i_{m-1}$ are equal to ∂, so fix them all in I. For the next part of the proof, I will construct a measurable mapping T of $\{N(t) = m\}$ into $\mathscr{X}$, such that

(41) $X(t + s) = X(s) \circ T \quad \text{on} \quad \{N(t) = m\}$

and

(42) $\pi_i\{A \text{ and } T^{-1}B\} = \pi_i\{A\} \cdot \pi_j\{B\}$

for all special A and all measurable subsets B of $\{X(0) = j\}$. Let

$$B = \{X(0) = j \text{ and } X(s) = k\}.$$

Using (41), check

$$T^{-1}B = \{N(t) = m \text{ and } X(t) = j \text{ and } X(t + s) = k\}.$$

Remember $A \subset \{N(t) = m$ and $X(t) = j\}$. So

$$A \cap T^{-1}B = \{A \text{ and } X(t + s) = k\}.$$

Clearly,

$$\pi_j\{B\} = \pi_j\{X(s) = k\}.$$

So (40) follows from (41–42).

INFORMAL NOTE. T is obtained by shifting the X sample function to the left by t, and then relabelling the jumps and holding times.

Define a formal T by the requirement that for $x \in \mathscr{X}$ with $N(t, x) = m$:

(43)
$$\begin{aligned}
\xi_n(Tx) &= \xi_{m+n}(x) \quad \text{for } n = 0, 1, \dots; \\
\tau_n(Tx) &= \tau_{m+n}(x) \quad \text{for } n = 1, 2, \dots; \\
\tau_0(Tx) &= \tau_0(x) + \cdots + \tau_m(x) - t.
\end{aligned}$$

Property (41) is fairly clear: look at Figure 1, and think this way. Fix $x \in \mathscr{X}$ with $t < \sigma(x)$ and $N(t, x) = m$. Fix $s \geq 0$. Suppose $t + s < \sigma(x)$. Then $N(t + s, x) \geq m$; say $N(t + s, x) = m + n$. In particular,

$$X(t + s, x) = \xi_{m+n}(x).$$

Let $\sigma_0(x) = 0$ and $\sigma_m(x) = \tau_0(x) + \cdots + \tau_{m-1}$ for $m = 1, 2, \dots$. Then $N(t + s, x) = m + n$ means

$$\sigma_{m+1}(x) + \tau_{m+1}(x) + \cdots + \tau_{m+n-1}(x)$$
$$\leq t + s < \sigma_{m+1}(x) + \tau_{m+1}(x) + \cdots + \tau_{m+n}(x);$$

that is,

$$\sigma_{m+1}(x) - t + \tau_{m+1}(x) + \cdots + \tau_{m+n-1}(x)$$
$$\leq s < \sigma_{m+1}(x) - t + \tau_{m+1}(x) + \cdots + \tau_{m+n}(x).$$

But $\sigma_{m+1}(x) - t = \tau_0(Tx)$ and $\tau_{m+\nu}(x) = \tau_\nu(Tx)$ for $\nu \geq 1$. Therefore, $N(t + s, x) = m + n$ means

$$\sigma_n(Tx) \leq s < \sigma_{n+1}(Tx),$$

that is,

$$N(s, Tx) = n.$$

So,

$$X(s, Tx) = \xi_n(Tx) = \xi_{m+n}(x) = X(t + s, x),$$

which disposes of (41) under the assumption $t + s < \sigma(x)$. The case $\sigma(x) \leq t + s$ is similar.

To handle (42), consider a special B of the form

$$B = \{\xi_0 = j_0, \ldots, \xi_n = j_n \text{ and } \tau_0 > u_0, \ldots, \tau_n > u_n\},$$

where $j_0 = j, \ldots, j_n$ are in $\hat{I}$ and $u_0, \ldots, u_n$ are nonnegative numbers. Remember $i_m = j_0 = j$. Now $A \cap T^{-1}B$ is the set where

$$\xi_0 = i_0, \ldots, \xi_m = i_m, \quad \xi_{m+1} = j_1, \ldots, \xi_{m+n} = j_n \quad \text{and}$$

$$\tau_0 > t_0, \ldots, \tau_{m-1} > t_{m-1}, \quad \tau_{m+1} > u_1, \ldots, \tau_{m+n} > u_n \quad \text{and}$$

$$\tau_0 + \cdots + \tau_{m-1} \leqq t \leqq t + u_0 < \tau_0 + \cdots + \tau_{m-1} + \tau_m.$$

As (37) shows, $\pi_i(A \cap T^{-1}B) = abce^{-u}$, where:

$$u = q(j_1)u_1 + \cdots + q(j_n)u_n;$$
$$a = \Pi_{v=0}^{m-1} \Gamma(i_v, i_{v+1});$$
$$b = \Pi_{v=0}^{n-1} \hat{\Gamma}(j_v, j_{v+1});$$

while c is the probability that

$$U_0 > t_0, \ldots, U_{m-1} > t_{m-1} \quad \text{and}$$

$$U_0 + \cdots + U_{m-1} \leqq t \leqq t + u_0 < U_0 + \cdots + U_{m-1} + U_m,$$

for independent exponential random variables $U_0, \ldots, U_m$, having parameters $q(i_0), \ldots, q(i_m)$. Now (30) implies that $c = de^{-v}$, where $v = q(j)u_0$ and d is the probability that

$$U_0 > t_0, \ldots, U_{m-1} > t_{m-1} \quad \text{and}$$

$$U_0 + \cdots + U_{m-1} \leqq t < U_0 + \cdots + U_{m-1} + U_m.$$

That is, $\pi_i(A \cap T^{-1}B) = abde^{-v-u}$. But $\pi_i(A) = ad$ and $\pi_j(B) = be^{-v-u}$, by (37). This completes the proof of (42) for one special A and all special B. Clearly, (42) holds for $B = \varnothing$ and $B = \{X(0) = j\}$; call these sets special also. Two different special B are disjoint or nested, and the class of special B's is closed under intersection and generates the full σ-field on $\{X(0) = j\}$. Both sides of (42) are countably additive in B, so (10.16) makes (42) hold for all measurable B. This completes the proof of (40).

Let $R(t, i, j) = \pi_i\{X(t) = j\}$. Use (40), and (4) with $\hat{I}$ for I to see: R is a stochastic semigroup on $\hat{I}$; while X is Markov with stationary transitions R and starting state i relative to π_i. I still have to show that R is standard, and $R'(0) = Q$. The ∂ row is easy. Fix $i \in I$. I will do the i row. I say

(44) $$\pi_i\{\tau_0 + \tau_1 \leqq t\} = o(t) \quad \text{as} \quad t \to 0.$$

Suppose i is not absorbing; the other case is trivial. Let U_i and U_j be independent and exponentially distributed, with parameters $q(i)$ and $q(j)$.

By (37),

$$\pi_i\{\tau_0 + \tau_1 \leq t \text{ and } \xi_1 = j\} = \hat{\Gamma}(i,j)\,\text{Prob}\{U_i + U_j \leq t\};$$

so

$$\pi_i\{\tau_0 + \tau_1 \leq t\} = \Sigma_j\,\hat{\Gamma}(i,j)\,\text{Prob}\{U_i + U_j \leq t\}.$$

The contribution to this sum from $j = \partial$ or absorbing j in I is clearly 0. When $q(j) > 0$, lemma (34) shows that

$$t^{-1}\,\text{Prob}\{U_i + U_j \leq t\}$$

is at most $q(i)$ and tends to 0 as $t \to 0$. Dominated convergence finishes (44). Confine x to the set $\{\xi_0 = i \text{ and } \xi_1 \neq i\}$, which has π_i-probability 1 because $\Gamma(i,i) = 0$. Suppose for a moment that $\tau_0 + \tau_1 > t$: then $X(t) = i$ iff $\tau_0 > t$; and $X(t) = j \neq i$ iff $\xi_1 = j$ and $\tau_0 \leq t$. As $t \to 0$:

$$\begin{aligned}
R(t,i,i) &= \pi_i\{\tau_0 > t\} + o(t) && \text{by (44)}\\
&= \text{Prob}\{U_i > t\} + o(t) && \text{by (37)}\\
&= e^{-q(i)t} + o(t);
\end{aligned}$$

and for $i \neq j$,

$$\begin{aligned}
R(t,i,j) &= \pi_i\{\xi_1 = j \text{ and } \tau_0 \leq t\} + o(t) && \text{by (44)}\\
&= \hat{\Gamma}(i,j)\,\text{Prob}\{U_i \leq t\} + o(t) && \text{by (37)}\\
&= \hat{\Gamma}(i,j)\,[1 - e^{-q(i)t}] + o(t).
\end{aligned}$$

The rest is calculus.   ★★

7. THE UNIFORM CASE

Fix a finite or countably infinite set I. Fix a uniform stochastic semigroup on I, as in Section 4. As (29) implies, $Q = P'(0)$ exists; the entries are uniformly bounded; the diagonal elements are nonpositive; the off-diagonal elements are nonnegative; and the row sums are zero. The first problem in this section is to construct a Markov chain with stationary transitions P, all of whose sample functions are *step functions*. Give I the discrete topology. A function f from $[0, \infty)$ to I is a *right continuous step function* iff $f(t) = \lim_{s\downarrow t} f(s)$ for all $t \geq 0$ and $f(t-) = \lim_{s\uparrow t} f(s) \in I$ exists for all $t > 0$. The discontinuity set of f is automatically finite on finite intervals, and may be enumerated as $\sigma_1(f) < \sigma_2(f) < \cdots$. If f has infinitely many discontinuities, then $\sigma_n(f) \to \infty$ as $n \to \infty$. If f has only n discontinuities, it is convenient to set $\sigma_{n+1}(f) = \sigma_{n+2}(f) = \cdots = \infty$. In any case, it is convenient to set $\sigma_0(f) = 0$. Then f is a constant, say $\xi_n(f) \in I$, on $[\sigma_n(f), \sigma_{n+1}(f))$. If f has only n discontinuities, leave $\xi_{n+1}(f)$, $\xi_{n+2}(f)$, ... undefined. Informally, $\xi_0(f)$, $\xi_1(f)$, ... are the successive *jumps* in f, or the successive states f visits; f visits $\xi_n(f)$ on $[\sigma_n(f), \sigma_{n+1}(f))$ with *holding time* $\sigma_{n+1}(f) - \sigma_n(f) = \tau_n(f)$. See Figure 3.

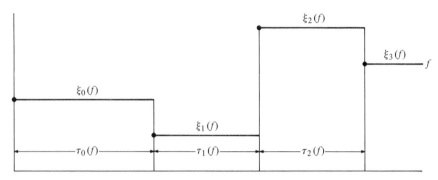

Figure 3.

Let S be the set of right continuous step functions from $[0, \infty)$ to I. Let $X(t, f) = f(t)$ for $t \geq 0$ and $f \in S$, and endow S with the smallest σ-field Σ over which all $X(t)$ are measurable. I claim that ξ_n and σ_n are Σ-measurable. The case $n = 0$ is easy: $\xi_0(f) = f(0)$ and $\sigma_0(f) = 0$. Suppose inductively that σ_n is Σ-measurable. Confine f to $\{\sigma_n < \infty\}$. Then

$$\xi_n(f) = \lim_{\varepsilon \downarrow 0} f[\sigma_n(f) + \varepsilon];$$

so $\xi_n(f) = j$ iff for all $m = 1, 2, \ldots$ there is a rational r with

$$\sigma_n(f) < r < \sigma_n(f) + \frac{1}{m} \quad \text{and} \quad f(r) = j.$$

So ξ_n is Σ-measurable. If $t > 0$, then

$$\sigma_{n+1}(f) - \sigma_n(f) < t$$

iff there are rational r, s with $0 < r < s < r + t$ and

$$r < \sigma_n(f) < s \quad \text{and} \quad f(s) \neq \xi_n(f).$$

So $\sigma_{n+1} = \sigma_n + \sigma_{n+1} - \sigma_n$ is Σ-measurable. Next, I claim that ξ_n and σ_n span Σ. For $f(t) = j$ iff $\xi_n(f) = j$ and $\sigma_n(f) \leq t < \sigma_{n+1}(f)$ for some $n = 0, 1, \ldots$.

Let $q(i) = -Q(i, i) \geq 0$. Define the substochastic *jump matrix* Γ on I as follows:

$$\Gamma(i, j) = Q(i, j)/q(i) \quad \text{for } i \neq j \quad \text{and} \quad q(i) > 0$$
$$= 0 \qquad\qquad \text{for } i = j \quad \text{or} \quad q(i) = 0.$$

Because $\{\xi_n, \sigma_n\}$ span Σ, there is at most one probability P_i on Σ for which:

(a) $\xi_0, \xi_1, \ldots$ is a discrete time Markov chain with stationary transitions Γ and starting state i;

(b) given $\xi_0, \xi_1, \ldots$, the random variables $\tau_n = \sigma_{n+1} - \sigma_n$ are conditionally independent and exponential with parameter $q(\xi_n)$, for $n = 0, 1, \ldots$.

NOTE. If Γ is substochastic, then ξ_n may be undefined with positive probability. By (29), the jth row sum of Γ is 1 or 0, according as $q(j) > 0$ or $q(j) = 0$.

The condition on P_i can be restated as follows, using (15.16): for any nonnegative integer n, and sequence $i_0 = i, i_1, \ldots, i_n$ in I, and nonnegative real numbers $t_0, \ldots, t_n$:

$$P_i\{\xi_m = i_m \text{ and } \tau_m > t_m \text{ for } m = 0, \ldots, n\} = pe^{-t},$$

where

$$p = \Pi_{m=0}^{n-1} \Gamma(i_m, i_{m+1})$$

$$t = \Sigma_{m=0}^{n} q(i_m)t_m.$$

By convention, an empty product is 1.

(45) Theorem. *The probability P_i exists. With respect to P_i, the process $\{X(t): 0 \leq t < \infty\}$ is a Markov chain with stationary transitions P and starting state i.*

FIRST PROOF. Use the setup of Section 6, with the present Γ and q. Check that the two Q's coincide on I. Fix $i \in I$. With respect to π_i, the process X on $\mathcal{X}$ is Markov with stationary standard and stochastic transitions R on $\hat{I}$, by (39); and ∂ is absorbing for R. So R is a standard substochastic semigroup when retracted to I. And $R'(0) = Q$ on I by (39). Furthermore,

$$R(t, i, i) \geq e^{-q(i)t},$$

either from the construction or from (12). So R is uniform. Now $R = P$ on I, by (29). To summarize: with respect to π_i, the process X on $\mathcal{X}$ is Markov with stationary transitions P and starting state i.

As before, let $\mathcal{X}_0$ be the subset of $\mathcal{X}$ where for all n:

$$\tau_n = \infty \quad \text{iff} \quad q(\xi_n) = 0;$$

$$\hat{\Gamma}(\xi_n, \xi_{n+1}) > 0.$$

Confine x to $\{\mathcal{X}_0 \text{ and } \xi_0 \in I\}$. Then d is one plus the least n if any with ξ_n absorbing, and $d = \infty$ if none. Indeed, $\hat{\Gamma}(j, \partial)$ is 0 or 1, according as $q(j) > 0$

or $q(j) = 0$. So $d < \infty$ makes $\tau_{d-1} = \infty$. Let

$$q = \sup_j q(j) < \infty.$$

Then

$$1/q(\xi_n) \geq 1/q,$$

so

$$\Sigma_n \{1/q(\xi_n) : 0 \leq n < d\} = \infty,$$

whether $d < \infty$ or $d = \infty$. Let $\mathscr{X}_1$ be the set of $x \in \mathscr{X}_0$ such that $X(\cdot, x)$ is I-valued everywhere. Remember

$$\pi_i\{\mathscr{X}_0 \text{ and } \xi_0 \in I\} = 1.$$

Now (38) makes $\pi_i(\mathscr{X}_1) = 1$.

If $x \in \mathscr{X}_1$, you know that $X(\cdot, x)$ is a right continuous I-valued step function on $[0, \infty)$. Visualize X as the mapping from $\mathscr{X}_1$ to S, which sends $x \in \mathscr{X}_1$ to $X(\cdot, x) \in S$. Check that X is measurable. Let $P_i = \pi_i X^{-1}$. Then P_i is a probability on Σ, because $\pi_i(\mathscr{X}_1) = 1$. For $0 \leq n < d$, check that ξ_n on $\mathscr{X}_1$ is the composition of ξ_n on S with X on $\mathscr{X}_1$, and τ_n on $\mathscr{X}_1$ is the composition of τ_n on S with X on $\mathscr{X}_1$; while ξ_n or τ_n on S applied to $X(\cdot, x)$ is undefined for $n \geq d(x)$. Indeed, $\xi_n \neq \partial$ implies $\xi_{m+1} \neq \xi_m$ and $\tau_m < \infty$ for $m < n$, while $\xi_n = \partial$ implies $\tau_m = \infty$ for some $m < n$, on $\mathscr{X}_1$. Consequently, the P_i-distribution of $\{\xi_n, \tau_n\}$ on S coincides with the π_i-distribution of

$$\{\xi_n, \tau_n : 0 \leq n < d\} \quad \text{on } \mathscr{X}.$$

Namely, $\{\xi_n\}$ is Markov with stationary transitions Γ and starting state i. Given ξ, the holding times τ_n are conditionally independent and exponentially distributed, the parameter for τ_n being $q(\xi_n)$. So, I constructed the right P_i. The P_i-distribution of the coordinate process X on S coincides with the π_i-distribution of the process X on $\mathscr{X}$: both are Markov with transitions P and starting state i.   ★

SECOND PROOF.   Use (38) and (7.33).   ★

What does (45) imply about an abstract Markov chain with transitions P? Roughly, there is a standard modification, all of whose sample functions are right continuous step functions. Then the jump process is a Γ-chain. Given the jumps, the holding times are conditionally independent and exponentially distributed, the parameter for visits to j being $q(j)$.

More exactly, let $(\Omega, \mathscr{F}, \mathscr{P})$ be an abstract probability triple. Let $\{Y(t) : 0 \leq t < \infty\}$ be an $\bar{I}$-valued process on $(\Omega, \mathscr{F})$. With respect to $\mathscr{P}$, suppose Y is a Markov chain with stationary transitions P. Remember that P is uniform, by assumption. For simplicity, suppose $\mathscr{P}\{Y(0) = i\} = 1$. Let Ω_0 be the set of $\omega \in \Omega$ such that $Y(\cdot, \omega)$ retracted to the rationals agrees with some $f \in S$ retracted to the rationals; of course, f depends on ω.

(46) Proposition. $\Omega_0 \in \mathscr{F}$ *and* $\mathscr{P}(\Omega_0) = 1$.

PROOF. Consider the set of functions ψ from the nonnegative rationals R to I, with the product σ-field. The set F of ψ which agree with some $f = f(\psi) \in S$ retracted to the rationals is measurable by this argument. Let $\theta_0(\psi) = 0$. Let

$$\zeta_n(\psi) = \lim \{\psi(r) : r \in R \text{ and } r \downarrow \theta_n(\psi)\}$$

$$\theta_{n+1}(\psi) = \sup \{r : r \in R \text{ and } \psi(s) = \zeta_n(\psi) \text{ for } s \in R \text{ with } \theta_n(\psi) < s < r\}.$$

Then F is the set of ψ such that: either

$$\theta_0(\psi) < \theta_1(\psi) < \theta_2(\psi) < \cdots < \infty$$

all exist, and $\theta_n(\psi) \to \infty$ as $n \to \infty$, and

$$\zeta_0(\psi), \zeta_1(\psi), \ldots$$

all exist, and

$$\zeta_n(\psi) = \psi[\theta_n(\psi)] \quad \text{whenever } \theta_n(\psi) \in R;$$

or for some n,

$$\theta_0(\psi) < \theta_1(\psi) < \cdots < \theta_n(\psi) < \infty \quad \text{and} \quad \theta_{n+1}(\psi) = \infty$$

all exist, and

$$\zeta_0(\psi), \zeta_1(\psi), \ldots, \zeta_n(\psi)$$

all exist and

$$\zeta_m(\psi) = \psi[\theta_m(\psi)] \quad \text{whenever } m = 0, \ldots, n \text{ and } \theta_m(\psi) \in R.$$

By discarding a null set, suppose $Y(r) \in I$ for all $r \in R$. Let Y_R be Y with time domain retracted to the rationals. Then Y_R is measurable, and Ω_0 is $\{Y_R \in F\}$. Let X be the coordinate process on S, and let X_R be X with time domain retracted to the rationals. The $\mathscr{P}$-distribution of Y_R coincides with the P_i-distribution of X_R. But $\{X_R \in F\}$ is all of S, and therefore has P_i-probability 1. So

$$\mathscr{P}\{\Omega_0\} = \mathscr{P}\{Y_R \in F\} = P_i\{X_R \in F\} = 1. \qquad \bigstar$$

NOTATION. Suppose $Y(r, \omega) \in I$ for all $r \in R$. The value of Y_R at ω is the function $r \to Y(r, \omega)$ from R to I. Similarly for X_R.

For $\omega \in \Omega_0$, let $Y^*(t, \omega)$ be the limit of $Y(r, \omega)$ as rational r decreases to t. For $\omega \notin \Omega_0$, let $Y^*(\cdot, \omega) = i$.

(47) Proposition. $Y^*(\cdot, \omega) \in S$ and $\mathscr{P}\{Y(t) = Y^*(t)\} = 1$ for each $t \geqq 0$.

PROOF. The first claim is easy. For the second, fix $t \geqq 0$. Let $Rt = R \cup \{t\}$ be the set of nonnegative rationals R, and t. Consider the set of functions ψ from Rt to I, with the product σ-field. Let G be the set of ψ with

$$\psi(t) = \lim \psi(r) \quad \text{as} \quad r \in R \text{ decreases to } t.$$

Then G is measurable. Confine ω to the subset of Ω_0 where $Y(t) \in I$. This subset has $\mathscr{P}$-probability 1. Let Y_{Rt} be Y with time domain retracted to Rt. Then Y_{Rt} is measurable, and

$$\{Y(t) = Y^*(t)\} = \{Y_{Rt} \in G\}.$$

Let X be the coordinate process on S, and let X_{Rt} be X with time domain retracted to Rt. The $\mathscr{P}$-distribution of Y_{Rt} coincides with the P_i-distribution of X_{Rt}. But $\{X_{Rt} \in G\}$ is all of S, and therefore has P_i-probability 1. So

$$\mathscr{P}\{Y(t) = Y^*(t)\} = \mathscr{P}\{Y_{Rt} \in G\} = P_i\{X_{Rt} \in G\} = 1. \qquad \bigstar$$

NOTATION.   Suppose $\omega \in \Omega_0$ and $Y(t, \omega) \in I$. The value of Y_{Rt} at ω is the function $s \to Y(s, \omega)$ from Rt to I. Similarly for X_{Rt}.

In the terminology of (Doob, 1953), the process Y^* is a standard modification of Y. For (48), suppose all the sample functions of Y are in S; if not, replace Y by Y^*. The mapping M which sends ω to $Y(\cdot, \omega)$ is a measurable mapping from Ω to S. Let $Y(\cdot, \omega)$ visit the states $\hat{\xi}_0(\omega), \hat{\xi}_1(\omega), \ldots$ with holding times $\hat{\tau}_0(\omega), \hat{\tau}_1(\omega), \ldots$. That is,

$$\hat{\xi}_n(\omega) = \xi_n(M\omega) \quad \text{and} \quad \hat{\tau}_n(\omega) = \tau_n(M\omega).$$

(48) Proposition.   *With respect to $\mathscr{P}$: the process $\hat{\xi}_0, \hat{\xi}_1, \ldots$ is a Markov chain with stationary transitions Γ and starting state i; given $\hat{\xi}_0, \hat{\xi}_1, \ldots$, the random variables $\hat{\tau}_0, \hat{\tau}_1, \ldots$ are conditionally independent and exponentially distributed, with parameters $q(\hat{\xi}_0), q(\hat{\xi}_1), \ldots$. That is, for $i_0 = i, i_1, \ldots, i_n$ in I and nonnegative $t_0, \ldots, t_n$,*

$$\mathscr{P}\{\hat{\xi}_0 = i_0, \ldots, \hat{\xi}_n = i_n \text{ and } \hat{\tau}_0 > t_0, \ldots, \hat{\tau}_n > t_n\} = pe^{-t},$$

where p is the product

$$\Gamma(i_0, i_1) \cdots \Gamma(i_{n-1}, i_n)$$

and

$$t = q(i_0)t_0 + \cdots, q(i_n)t_n.$$

PROOF.   With respect to $\mathscr{P}M^{-1}$, the coordinate process X on S is Markov with stationary transitions P and starting state i. There is at most one such probability, so $\mathscr{P}M^{-1} = P_i$. And the $\mathscr{P}$-distribution of $\{\hat{\xi}_n, \hat{\tau}_n\}$ coincides with the P_i-distribution of $\{\xi_n, \tau_n\}$. $\qquad \bigstar$

NOTE.   If Γ is substochastic, then $\hat{\xi}_n$ may be undefined with positive probability. By convention, an empty product is 1.

There is a useful way to restate (48). Define the probability π_i on $\mathscr{X}$ as in Section 6, using the present Γ and q. Suppose for a moment that $q(j) > 0$ for all j. Let $\hat{\Omega}$ be the subset of Ω where $\hat{\xi}_n$ is defined and $\hat{\tau}_n < \infty$ for all n. Then $\mathscr{P}(\hat{\Omega}) = 1$. Let $\hat{M}$ map $\hat{\Omega}$ into $\mathscr{X}$:

$$\xi_n(\hat{M}\omega) = \hat{\xi}_n(\omega) \quad \text{and} \quad \tau_n(\hat{M}\omega) = \hat{\tau}_n(\omega).$$

Then

(49) $$\mathscr{P}\hat{M}^{-1} = \pi_i.$$

Now drop the assumption that $q(j) > 0$. Let

$$\hat{d} = \inf\{n : \hat{\xi}_n \text{ is undefined}\} \quad \text{on } \hat{\Omega}.$$

Remember

$$d = \inf\{n : \xi_n = \partial\} \quad \text{on } \mathscr{X}.$$

Then

(50) the $\mathscr{P}$-distribution of $\{(\hat{\xi}_n, \hat{\tau}_n) : 0 \leqq n < \hat{d}\}$ coincides with the π_i-distribution of $\{(\xi_n, \tau_n) : 0 \leqq n < d\}$.

6

EXAMPLES FOR THE STABLE CASE

1. INTRODUCTION

I will prove some general theorems on stable chains in Chapter 7. In this chapter, I construct examples; Section 6 is about Markov times and I think you'll find it interesting. This chapter is quite difficult, despite my good intentions; and it isn't really used in the rest of the book, so you may want to postpone a careful reading.

Each example provides a countable state space I, and a reasonably concrete stochastic process $\{X(t):0 \leq t < \infty\}$, which is Markov with stationary transitions P. Here P is a standard stochastic semigroup on I. In most of the examples, the sample functions are not step functions, and P is not uniform. The generator $Q = P'(0)$ of P will be computed explicitly, but P will be given only in terms of the process. Each state $i \in I$ will be stable: that is,

$$q(i) = -Q(i, i) < \infty.$$

Each example lists

(a) *Description*

(b) *State space I*

(c) *Holding time q*

(d) *Generator Q*

(e) *Formal construction.*

Only the ordering of the states is specified. The holding time for each visit to i is exponential with parameter $q(i)$, independent of all other parts of the

I want to thank Howard Taylor and Charles Yarbrough for reading drafts of this chapter.

construction. On a first reading of the chapter, you should skim Sections 2 and 4, and ignore the formal constructions, which are based on theorems (27) and (108). The examples will give you some idea of the possible pathology. You can study theorems (27, 108) and the formal constructions later, when you want a proof that the processes described in the examples really are Markov with the properties I claim. The only thing in this chapter used elsewhere in the book is example (132): it is part of the proof of (7.51). This chapter only begins to expose the corruption. More is reported in Chapter 8, in Chapter 3 of *ACM*, and in Sections 2.12–13 of *B & D*.

The class of processes that can be constructed by the present theorems (27, 108) is small. One main restriction is that all states are stable. So as not to contradict later theorems, the sample functions will be constant in *I* over maximal half-open intervals $[a, b)$, called the intervals of constancy, which cover Lebesgue almost all of $[0, \infty)$. On the exceptional set, the sample function will be infinite. Within the class of stable processes, the processes of this chapter have a very special feature. Loosely stated, the intervals of constancy can be indexed in order by part of a fixed set *C* in such a way that as *c* increases through *C*, the *c*th state visited by the process—namely, the value of the process on the *c*th interval of constancy—evolves in a Markovian way. At first sight, this may not look like an assumption. Sections 8.5–8.6 and *B & D*, Sec. 2.13 show that it is a serious one. Incidentally, the set of intervals of constancy is countable and linearly ordered, so these two properties are forced on *C*.

In the first class of processes I want to construct, the order type of the intervals of constancy depends on the starting state. These processes have a countable, linearly ordered state space *I*. Starting from $i \in I$, they move through all $j \geq i$, in order. I will make this precise in Section 2. The examples of Section 3 fall into this class. In the second class of processes, the order type of the intervals of constancy, until you reach the end of time, is fixed. I will describe this class in Section 4. The examples of Sections 5 and 7 fall into the second class. Naturally, there is one theorem that covers both constructions, and other things as well. But it seems too complicated for the examples I really want.

A general method for constructing stable chains, in the spirit of this chapter, is not yet known.

2. THE FIRST CONSTRUCTION

The parameters

(1)   Let *I* be a countable set, linearly ordered by $<$, without a last element. For $i \in I$, let

$$I_i = \{j: j \in I \text{ and } j \geq i\}.$$

The set I will be the state space. Starting from $i \in I$, my process will move through I_i in order.

(2)   Let q be a function from I to $(0, \infty)$.

The holding time in j will be exponential with parameter $q(j)$, independent of everything else.

Suppose

(3) $\Sigma_j \{1/q(j) : j \in I_i\} = \infty$ for all $i \in I$

and

(4) $\Sigma_j \{1/q(j) : j \in I_i \text{ and } j \leqq k\} < \infty$ for all $i \in I$ and $k \in I_i$.

Condition (3) guarantees that my process is defined on all of $[0, \infty)$. Condition (4) guarantees that it moves through all of I_i, when it starts from i.

The construction

Let i range over I.

(5)   Let W_i be the set of functions w from I_i to $(0, \infty)$. For $j \in I_i$ and $w \in W_i$, let
$$\tau(j, w) = w(j).$$

Strictly speaking, τ should have a subscript i. Informally speaking, $\tau(j)$ will be the holding time in j.

(6)   Give W_i the smallest σ-field making all the $\tau(j)$ measurable.

(7)   For $j \in I_i$ and $w \in W_i$, let
$$\lambda(j) = \Sigma_k \{\tau(k) : k \in I_i \text{ and } k < j\}$$
$$\rho(j) = \lambda(j) + \tau(j) = \Sigma_k \{\tau(k) : k \in I_i \text{ and } k \leqq j\}.$$

(8)   Give I the discrete topology, and let $\bar{I} = I \cup \{\varphi\}$ be its one-point compactification.

(9)   Define a process X_i on W_i:
$$X_i(t) = j \quad \text{if } \lambda(j) \leqq t < \rho(j) \quad \text{for some } j \in I_i$$
$$\quad\quad = \varphi \quad \text{if } \lambda(j) \leqq t < \rho(j) \quad \text{for no } j \in I_i.$$

EXPLANATION.   The process should visit j on an interval of length $\tau(j)$. The process should spend total time 0 in the fictitious state φ. So I put the left endpoint of the j-interval at the sum of the lengths of the preceding intervals, namely $\lambda(j)$. To prevent various difficulties with the sample functions, I will have to cut W_i down to W_i^* in (12).

You should check (10–14).

(10a) $X_i(t) = i$ iff $0 \leq t < \tau(i)$.

If i has an immediate successor j, then

(10b) $X_i(t) = j$ iff $\tau(i) \leq t < \tau(i) + \tau(j)$.

If $i < k < j$, then

(10c) $\{X_i(t) = j\} \subset \{\tau(i) + \tau(k) \leq t\}$.

(11) X_i is jointly measurable.

That is, $(t, w) \to X_i(t, w)$ is product measurable on $[0, \infty) \times W_i$, where $[0, \infty)$ has the Borel σ-field, and W_i has the σ-field (6).

(12) Let W_i^* be the set of $w \in W_i$ such that

$$\Sigma_j \{w(j) : j \in I_i\} = \infty, \quad \text{but}$$
$$\lambda(j, w) < \infty \quad \text{for all } j \in I_i.$$

(13) Let π_i be the probability on W_i which makes the $\tau(j)$ independent and exponentially distributed, the parameter for $\tau(j)$ being $q(j)$.

Use conditions (3–4) and (5.33):

(14) $\pi_i\{W_i^*\} = 1$.

Lemmas

Let i range over I, and j over I_i. Let $w \in W_i^*$, as defined in (12).

(15) $X_i(t, w) = j$ iff $\lambda(j, w) \leq t < \rho(j, w)$. This interval has length $\tau(j, w)$.

(16) $j \to [\lambda(j, w), \rho(j, w))$ is 1-1 and order preserving on I_i. The union of $[\lambda(j, w), \rho(j, w))$ as j ranges over I_i is precisely $\{t : X_i(t, w) \in I_i\}$.

(17) $X_i(\cdot, w)$ is regular in the sense of (7.2).

(18) **Lemma.**   Lebesgue $\{t : X(t, w) = \varphi\} = 0$, for $w \in W_i^*$.

PROOF.   Relations (15–16) show

$$\text{Lebesgue } \{t : t \leq \lambda(j, w) \text{ and } X_i(t, w) \in I\} = \lambda(j, w).$$

So

$$\text{Lebesgue } \{t : t \leq \lambda(j, w) \text{ and } X_i(t, w) = \varphi\} = 0.$$

But $\lambda(j, w)$ increases to ∞ as j increases through I_i, by definition (12) of W_i^*.   ★

(19) Lemma. $\pi_i\{X_i(t) \in I\} = 1.$

PROOF.   Define a new process Y on W_i:

$$Y(s, w) = X_i[w(i) + s, w].$$

As (11) shows, Y is jointly measurable: it's the composition of X_i with the mapping $(s, w) \to (w(i) + s, w)$. As (18) shows,

(20)   Lebesgue $\{s: Y(s, w) = \varphi\} = 0$ for each $w \in W_i^*$.

Let

$$E = \{s : 0 \leq s < \infty \text{ and } \pi_i[Y(s) = \varphi] > 0\}.$$

Use Fubini on (14, 20). The set E is Borel, and

(21)   Lebesgue $E = 0.$

Check

(22) $\{X_i(t) = \varphi\} = \{\tau(i) \leq t \text{ and } Y[t - \tau(i)] = \varphi\}.$

I say Y is measurable on $\{\tau(j):j > i\}$. For $Y(s) > i$; and $Y(s) = k > i$ iff

$$\Sigma_j \{\tau(j):i < j < k\} \leq s < \Sigma_j \{\tau(j):i < j \leq k\}.$$

Review (13) to see that Y is π_i-independent of $\tau(i)$, and $\tau(i)$ is exponential with parameter $q(i)$. Fubini up (22) and use (21):

$$\pi_i\{X_i(t) = \varphi\} = q(i) \int_0^t \pi_i\{Y(t-s) = \varphi\} \cdot e^{-q(i)s}\, ds = 0. \qquad \bigstar$$

(23)   Let I^s be the set of $i \in I$ which have an immediate successor, $s(i)$.

(24)   Let Q be this matrix on I:

$$Q(i, i) = -q(i) \quad \text{for all } i \in I$$
$$Q(i, j) = 0 \qquad \text{if } j \neq i \text{ or } s(i)$$
$$Q(i, s(i)) = q(i) \qquad \text{for } i \in I^s.$$

(25)   Let $P(t, i, j) = \pi_i\{X_i(t) = j\}.$

(26) Lemma. $P'(0, i, j) = Q(i, j)$; *the definitions are* (23–25).

PROOF.   This is easy for $j < i$. Suppose $j = i$. Then $\{X_i(t) = i\} = \{\tau(i) > t\}$ by (10). So

$$P(t, i, i) = \pi_i\{\tau(i) > t\} = e^{-q(i)t} \quad \text{by (13).}$$

Suppose $i \in I^s$ and $j = s(i)$. By (10),

$$\{X_i(t) = j\} = \{\tau(i) \leq t < \tau(i) + \tau(j)\}.$$

So
$$P(t, i, j) = \pi_i\{\tau(i) \leq t < \tau(i) + \tau(j)\}$$
$$= \pi_i\{\tau(i) \leq t\} - \pi_i\{\tau(i) + \tau(j) \leq t\}.$$

But
$$\pi_i\{\tau(i) \leq t\} = 1 - e^{-q(i)t} \quad \text{by (13)}$$
$$\pi_i\{\tau(i) + \tau(j) \leq t\} = o(t) \quad \text{by (13) and (5.34)}.$$

Suppose $j > i$ but $j \neq s(i)$. Then there is a state k with $i < k < j$. By (10),
$$\{X_i(t) = j\} \subset \{\tau(i) + \tau(k) \leq t\},$$

so
$$P(t, i, j) = o(t) \quad \text{by (13) and (5.34)}. \qquad \bigstar$$

The theorem

(27) **Theorem.** *Suppose* (1–4). *Define the process* X_i *on the probability triple* (W_i, π_i) *by* (5–9) *and* (13). *Define* Q *and* P *by* (23–25).

  (a) *P is a standard stochastic semigroup on* I, *with generator* Q.

  (b) X_i *is Markov with stationary transitions* P *and starting state* i.

NOTE. X_i has properties (15–18) on W_i^*, which has π_i-probability 1 by (14).

PROOF.   Let $i \leq j \leq k$ be in I, and let s, t be nonnegative.

(28)   Let $\mathscr{F}(t)$ be the σ-field in W_i spanned by $X(u)$ for $0 \leq u \leq t$.

Let $A \in \mathscr{F}(t)$ with $A \subset \{X_i(t) = j\}$. I will argue

(29) $\pi_i\{A \text{ and } X_i(t + s) = k\} = \pi_i\{A\} \cdot P(s, j, k)$.

Take (29) on faith for a minute. Use lemma (5.4) on (15), (16), (19), (29) to make P a stochastic semigroup on I, and X_i a P-chain starting from i. This proves (b), and (26) completes the proof of (a).
  To start on (29),

(30)   let T map $\{X_i(t) = j\}$ into W_j as follows:
$$(Tw)(j) = \lambda(j, w) + \tau(j, w) - t;$$
$$(Tw)(h) = \tau(h, w) \quad \text{for } h > j.$$

From definition (9),

(31) $X_i(t + s) = X_j(s) \circ T \quad \text{on } \{X_i(t) = j\};$

so

(32) $T^{-1}\{X_j(s) = k\} = \{X_i(t) = j \text{ and } X_i(t + s) = k\}$.

Next,

(33)   let $B = \{w: w \in W_j$ and $w(j_m) > u_m$ for $m = 0, \ldots, n\}$, where
$j_0 = j < j_1 < \cdots < j_n$ are in I, and $u_0, u_1, \ldots, u_n$ are nonnegative numbers.

I claim

(34) $$\pi_i\{A \cap T^{-1}B\} = \pi_i\{A\} \cdot \pi_j\{B\}$$
for all $A \in \mathscr{F}(t)$ with $A \subset \{X_i(t) = j\}$
and all B of the form (33).

Take (34) on faith for a minute. The equality holds for $B = \varnothing$ or W_j.
This enlarged set of B's is closed under intersection and generates the full
σ-field on W_j. So (10.16) makes the equality for all measurable subsets B
of W_j. Put $B = \{X_j(s) = k\}$ and use (32) to get (29).

I will now prove (34). Fix $A \in \mathscr{F}(t)$ with $A \subset \{X_i(t) = j\}$. Remember
$\rho(j) = \lambda(j) + \tau(j)$ from (7). Review (9).

(35)   Abbreviate $C = \{W_i$ and $\lambda(j) \leq t < \lambda(j) + \tau(j)\} = \{X_i(t) = j\}$.

(36)   Let Σ be the σ-field in W_i generated by $\tau(h)$ with $i \leq h < j$.

I claim

(37)   There is an $\hat{A} \in \Sigma$ with $A = \hat{A} \cap C$.

Let $0 \leq u \leq t$. Check $i \leq X(u) \leq j$ on C, by (15–16). Let $i \leq h < j$. Then
$$\{X_i(u) = h \text{ and } X_i(t) = j\} = \hat{A} \cap C, \text{ where}$$
$$\hat{A} = \{\lambda(h) \leq u < \rho(h)\} \in \Sigma.$$

Similarly,
$$\{X_i(u) = X_i(t) = j\} = \hat{A} \cap C, \text{ where}$$
$$\hat{A} = \{\lambda(j) \leq u\} \in \Sigma.$$

This proves (37).

NOTE.   If $\hat{A} \in \Sigma$, then $\hat{A} \cap C \in \mathscr{F}(t)$.
Fix B of the form (33). Let

(38) $C_0 = \{W_i$ and $\lambda(j) \leq t \leq t + u_0 < \lambda(j) + \tau(j)\}$

(39) $B_1 = \{W_i$ and $\tau(j_m) > u_m$ for $m = 1, \ldots, n\}$.

Use the $\hat{A}$ of (37) and definition (30) to get

(40) $A \cap T^{-1}B = \hat{A} \cap C_0 \cap B_1.$

So

(41) $$\pi_i\{A \cap T^{-1}B\} = \pi_i\{\hat{A} \cap C_0\} \cdot \pi_i\{B_1\},$$

because $\hat{A}$ and C_0 are measurable on $\tau(h)$ with $h \leq j$; while B_1 is measurable on $\tau(h)$ with $h > j$, by (33, 39); and definition (13) makes these two lots π_i-independent.

Check that $\hat{A}$ and $\lambda(j)$ are Σ-measurable. The definitions are (36–37) and (7). Definition (13) makes $\tau(j)$ independent of Σ and exponential with parameter $q(j)$, relative to π_i. Abbreviate $u = u_0$, and remember $j = j_0$. Use (5.30) to conclude

(42) $$\pi_i\{\hat{A} \cap C_0\} = \pi_i\{\hat{A} \cap C\} \cdot e^{-q(j)u};$$

where C was defined in (35).

Use definitions (13, 39, 33) to check

(43) $$e^{-q(j)u}\, \pi_i\{B_1\} = \pi_j\{B\}.$$

Combine (41–43) and (37) to settle (34).   ★

3. EXAMPLES ON THE FIRST CONSTRUCTION

(44) **Example.** (a) *Description.* The Poisson process with parameter λ moves through $0, 1, \ldots$ in order. The holding times are independent and exponentially distributed, with parameter λ. See Figure 1.

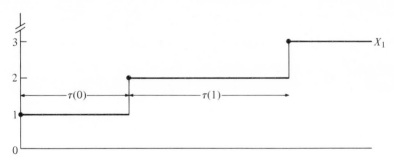

Figure 1.

(b) *State space.* $I = \{0, 1, \ldots\}$, with the usual order.

(c) *Holding times.* $q(i) = \lambda$ for all i.

(d) *Generator.* $Q(i, j) = 0$ unless $j = i$ or $i + 1$; and $Q(i, i) = -\lambda$; and $Q(i, i + 1) = \lambda$.

(e) *Formal construction.*   Use (27).   ★

(45) Example. (a) *Description.* The semigroup is not uniform, but almost all sample functions are right continuous step functions. The states are the integers *I*. Starting from $i \in I$, the process moves successively through i, $i + 1, \ldots$. See Figure 1.

(b) *State space.* *I* is the integers, with the usual order.

(c) *Holding times.* $q(i)$ is arbitrary subject to: $0 < q(i) < \infty$; and $q(i) \to \infty$ as $i \to \infty$; and $\Sigma_{i=1}^{\infty} 1/q(i) = \infty$.

(d) *Generator.* $Q(i, j) = 0$ unless $j = i$ or $i + 1$; and $Q(i, i) = -q(i)$; and $Q(i, i + 1) = q(i)$.

(e) *Formal construction.* Use (27). The semigroup can't be uniform, by (5.29). If $w \in W_i^*$, as defined in (12), then $X(\cdot, w)$ is a right continuous step function, by (15–16). And $\pi_i\{W_i^*\} = 1$ by (14).   ★

(46) Example. (a) *Description.* The process moves through the non-negative rationals, in order. The set of infinities is dense in itself, closed from the right, open from the left, and has the power of the continuum. All off-diagonal elements in the generator vanish, and the generator need not determine the semigroup.

(b) *State space.* *I* is the set of nonnegative rationals, in the usual order.

(c) *Holding times.* q is arbitrary, subject to (2–4).

(d) *Generator.* $Q(i, j) = 0$ unless $j = i$; and $Q(i, i) = -q(i)$.

(e) *Formal construction.* Use (27). Fix $w \in W_i^*$, as defined in (12); remember (14). As (15–16) show,

$$\{t : X_i(t, w) \in I\}$$

is a countable union of maximal intervals $[a, b)$ which are ordered like the rationals: between any two, there is a third. So

$$S_\varphi(w) = \{t : X_i(t,w) = \varphi\}$$

is dense in itself; if $t_n \in S_\varphi(w)$ decreases to t, then $t \in S_\varphi(w)$; if $t \in S_\varphi(w)$, there are $s > t$ but arbitrarily close with $s \in S_\varphi(w)$. And $S_\varphi(w)$ has power c.

Suppose $i \neq j$ but $q(i) = q(j)$. Interchanging i and j does not affect Q, but does change P: so Q cannot determine P.   ★

HINT. There are only countably many left endpoints of intervals of finitude. Temporarily add them to $S_\varphi(w)$, and make this larger set have power c.

Example (46) is due to Lévy.

(47) Example. (a) *Description.* Let C be a closed subset of $[0, \infty)$, which contains 0 but has no interior. For almost all sample functions: the set of infinities is homeomorphic to C; and the sample function is continuous at its infinities. The process moves through its states, in order.

(b) *State space.* Let $\mathscr{I}$ be the set of maximal subintervals of $(-\infty, \infty)$ complementary to C. Let I consist of all pairs (α, m), where $\alpha \in \mathscr{I}$ and m is an integer. Let $(\alpha, m) < (\beta, n)$ iff

$$\alpha \text{ is to the left of } \beta \quad \text{or} \quad \alpha = \beta \text{ and } m < n.$$

(c) *Holding times.* q is arbitrary subject to (2–4).

(d) *Generator.* Let $i = (\alpha, n)$. Then $Q(i, j) = 0$ unless $t = (\alpha, n)$ or $(\alpha, n + 1)$. And

$$Q[i, (\alpha, n + 1)] = -Q(i, i) = q(i).$$

(e) *Formal construction.* Use (27). Let ϕ be the complementary interval whose right endpoint is 0. Let $i = (\phi, 0)$ and $w \in W_i^*$, as defined in (12). Remember (14) that $\pi_i\{W_i^*\} = 1$. Remember I_i from (1). For $\alpha \in \mathscr{I}$,

let $\lambda(\alpha, w)$ be the sum of $\tau(\beta, m)$ over all $(\beta, m) \in I_i$ with β to the left of α;

let $\rho(\alpha, w)$ be the sum of $\tau(\beta, m)$ over all $(\beta, m) \in I_i$ with β not to the right of α.

By (15–16), the set

$$\{t : X(t, w) \in I\}$$

is a countable union of maximal intervals $(\lambda(\alpha, w), \rho(\alpha, w))$ for $\alpha \in \mathscr{I}$. Furthermore,

$$\alpha \to (\lambda(\alpha, w), \rho(\alpha, w))$$

is 1-1 and order-preserving. On interval ϕ, the process runs in order through $\{(\phi, n) : n = 0, 1, \ldots\}$. On interval $\alpha > \phi$, the process runs in order through $\{(\alpha, n) : n = \ldots, -1, 0, 1, \ldots\}$. So

$$S_\varphi(w) = \{w : X_i(t, w) = \varphi\}$$

is closed and homeomorphic to C; the function $X_i(\cdot, w)$ is continuous on $S_\varphi(w)$.

★

RELEVANT FACT. Let S and T be two closed subsets of $[0, \infty)$, without interior. Then S is homeomorphic to T iff the set of complementary intervals of S is order-isomorphic to the set of complementary intervals of T.

4. THE SECOND CONSTRUCTION

Informal description

Let I be a countably infinite state space. Let C be a countably infinite set, linearly ordered by $<$, with first element 0. The intervals of constancy will be indexed by some initial segment of C. Let $\xi(c)$ be the value of the process on interval c. I want $\xi(c)$ to be Markovian. What does this mean? Let Γ_i be the distribution of ξ when the starting state is i. This explains condition (63). Fix a present index $d \in C$. Let

$$C_d = \{c : c \in C \text{ and } c \geq d\}.$$

The past is

$$\{\xi(c) : c \in C \text{ and } c \leq d\}.$$

The future is

$$\{\xi(c) : c \in C_d\}.$$

Given the past and $\xi(d) = j$, the conditional Γ_i-distribution of the future should be the Γ_j-distribution of the whole jump process $\{\xi(c)\}$. As far as Γ_j is concerned, c runs over all of C. The index c in the future only runs over C_d. So I have to wish these index sets order-isomorphic. More explicitly, there should be a strictly increasing map

$$M_d = M(d, \cdot)$$

of C_d onto C. Suppose $a < b < c$ are in C. There are now two ways to compute the position of c relative to b. The direct method gives $M(b, c)$. The indirect method maps first by $M(a, \cdot)$, getting

$$0 = M(a, a) < M(a, b) < M(a, c);$$

and then computes the position of $M(a, c)$ relative to $M(a, b)$. I want the two methods to agree:

$$M(b, c) = M[M(a, b), M(a, c)].$$

You should now accept condition (51) on the order structure of C and (66) on the Markov property.

DIGRESSION.   Let $c = M_a^{-1}b$, so $c \in C_a$ and $b = M_a c$. I claim

$$M_c = M_b M_a.$$

Both sides have domain C_c. Take $d \in C_c$:

$$M_c d = M(M_a c, M_a d) = M_b M_a d.$$

So C is a semigroup with identity 0, where

$$a + b = M_a^{-1} b.$$

And $b > a$ iff $-a + b = M b \in C$. So you're really facing the nonnegative part of a countably infinite, linearly ordered, non-commutative group.   ★

Here is the final slogan: Given the visiting process, the holding times are conditionally independent and exponential, with parameter depending only on what the process is currently holding onto. Let q specify these parameters: namely, q is a nonnegative function on I. Given ξ, the length $\tau(c)$ of the cth interval of constancy is conditionally exponential with parameter $q[\xi(c)]$, and these lengths are conditionally independent as c varies over C.

How could you construct such an object? It is easy to generate a process $\{[\xi(c), \tau(c)]: c \in C\}$ of states and holding times which has the right properties. You might as well use the coordinate process on the set of functions from C to $I \times (0, \infty]$. This is done in (75). The process should be $\xi(c)$ on the cth interval of constancy, which has length $\tau(c)$. But where do you put this interval? The sample function should spend Lebesgue almost no time outside intervals of constancy. This suggests making the left endpoint $\lambda(c)$ of the cth interval equal to the sum of the lengths of the previous intervals:

$$\lambda(c) = \Sigma_d \{\tau(d): d \in C \text{ and } d < c\}.$$

The right endpoint of the cth interval should be

$$\rho(c) = \lambda(c) + \tau(c),$$

so the length is $\tau(c)$. And

$$X(t) = \xi(c) \quad \text{for} \quad \lambda(c) \leq t < \rho(c),$$

so X is $\xi(c)$ in the cth interval, which is half-open. This is done in (72).

How much sample functions have I got? If $\lambda(c) < \infty$, then the time domain of the sample function extends to $\rho(c)$ at least. To cover the line, I want

$$\sup_c \{\rho(c): \lambda(c) < \infty\} = \infty.$$

To get it, assume (65). The intervals of constancy now cover Lebesgue almost all of $[0, \infty)$. On the exceptional null set, put the process in its fictitious state φ, namely the point in the one-point compactification of discrete I.

WARNING.   Give C the order topology, and complete it. Suppose x is in this completion, and is a limit point of C from both sides. If you don't take precautions, you get stuck with

$$\sigma = \Sigma \{\tau(c): c \leq x\} < \infty,$$

but

$$\Sigma \{\tau(c): x < c < c^*\} = \infty \quad \text{for all } c^* > x.$$

Then you can't continue the sample function past σ.

I want $[\lambda(c), \rho(c))$ to be the cth interval of constancy. Part of this is free:

$$c \to [\lambda(c), \rho(c)) \quad \text{for} \quad c \in C$$

is order-preserving and 1-1. But why is $[\lambda(c), \rho(c))$ a maximal interval of constancy? If there happens to be a least element $1 > 0$ in C, I assume (64). This encourages $[\lambda(c), \rho(c))$ to be maximal. The argument is in (94).

The parameters and the space Ω

(48) Let I be a countably infinite set. Give I the discrete topology, and let $\bar{I} = I \cup \{\varphi\}$ be its one-point compactification.

(49) Let q be a function from I to $[0, \infty)$.

Do not assume $q(i) > 0$.

(50) Let C be a countably infinite set, linearly ordered by $<$, with a first element 0. For $d \in C$, let

$$C_d = \{c : c \in C \text{ and } c \geqq d\}.$$

(51) For each $d \in C$, assume that there is a 1-1 order preserving map $M_d = M(d, \cdot)$ of C_d onto C, such that:

M_0 is the identity;

$M[M(a, b), M(a, c)] = M(b, c)$ for $a < b < c$ in C.

(52) Say C is *discrete* iff there is a least $c > 0$. If C is discrete, each c has an immediate successor, call it $s(c)$: use (51). Let $1 = s(0)$. Otherwise, say C is *indiscrete*: then C is order-isomorphic to the nonnegative rationals.

(53) Illustration. Let $C = \{0, 1, 2, \ldots\}$ with the usual order. Define $M(d, n) = n - d$ for $n = d, d + 1, \ldots$.

(54) Illustration. Let C consist of all pairs (m, n) of nonnegative integers. Let $(m, n) < (m', n')$ iff

$$m < m', \quad \text{or}$$
$$m = m' \quad \text{and} \quad n < n'.$$

Let $d = (m, n) \leqq (m', n')$. Define

$$M_d(m', n') = (0, n' - n) \quad \text{for} \quad m' = m$$
$$= (m' - m, n') \quad \text{for} \quad m' > m.$$

You check (51).

(55) Illustration. Let C consist of all pairs (m, n) of integers, such that: $n \geqq 0$ when $m = 0$. Let $(m, n) < (m', n')$ iff

$$m < m', \quad \text{or}$$
$$m = m' \quad \text{and} \quad n' < n.$$

Let $d = (m, n) \leq (m', n')$. Define

$$M_d(m', n') = (0, n' - n) \quad \text{for} \quad m' = m$$
$$= (m' - m, n') \quad \text{for} \quad m' > m.$$

You check (51).

(56) Illustration. Let C be the nonnegative rationals, with the usual order. Let $M_d(r) = r - d$ for $r \geq d$.

(57)   Let Ω be the set of all functions ω from C to I. Let

$$\xi(c, \omega) = \omega(c) \quad \text{for} \quad c \in C \quad \text{and} \quad \omega \in \Omega.$$

Give Ω the smallest σ-field over which all $\xi(c)$ are measurable.

(58)   For each $i \in I$, let Γ_i be a probability on Ω.

There is a condition (65) on q and Γ. To state it, and for later use, make the following definitions.

(59)   For $c \in C$ and $\omega \in \Omega$ let,

$$\lambda^*(c, \omega) = \Sigma_d \{1/q[\omega(d)]: d \in C \text{ and } d < c\}$$
$$\rho^*(c, \omega) = \Sigma_d \{1/q[\omega(d)]: d \in C \text{ and } d \leq c\}.$$

(60). Observation. Let $c \in C$ and $\omega \in \Omega$ with $\lambda^*(c, \omega) < \infty$. For each $i \in I$, there are only finitely many indices $d < c$ with $\omega(d) = i$.

(61)   Let Ω^* be the set of all $\omega \in \Omega$ such that

$$\sup_c \{\rho^*(c, \omega): c \in C \text{ and } \lambda^*(c, \omega) < \infty\} = \infty.$$

(62) Observation. $\omega \in \Omega^*$ iff

$$\Sigma_c \{1/q[\omega(c)]: c \in C \text{ and } \lambda^*(c, \omega) < \infty\} = \infty.$$

The conditions

For each $i \in I$, suppose:

(63) $\Gamma_i\{\xi_0 = i\} = 1$;

(64) $\Gamma_i\{\xi(1) \neq i\} = 1$ if C is discrete, as defined in (52);

(65) $\Gamma_i\{\Omega^*\} = 1$,   where Ω^* was defined in (59,61).

The next assumption is the Markov property. Suppose

(66) $\Gamma_i\{\xi(c_m) = i_m$ for $m = 0, \ldots, n\} = \Pi_{m=0}^{n-1} \gamma[M(c_m, c_{m+1}), i_m, i_{m+1}]$

where

$$c_0 = 0 < c_1 < \cdots < c_n \quad \text{are in } C, \text{ and}$$

$$i_0 = i, i_1, \ldots, i_n \qquad\qquad \text{are in } I, \text{ and}$$

(67)
$$\gamma(c, j, k) = \Gamma_j\{\xi(c) = k\}.$$

Here is a digression. What γ can appear in (67)? Clearly,

$$\gamma(c, j, k) \geqq 0$$

$$\Sigma_k\, \gamma(c, j, k) = 1$$

$$\gamma(0, i, i) = 1$$

$$\gamma(1, i, i) = 0 \quad \text{if } C \text{ is discrete}$$

$$\gamma(b, i, k) = \Sigma_j\, \gamma(a, i, j) \cdot \gamma(M_a b, j, k) \quad \text{for} \quad a < b.$$

Conversely, suppose γ satisfies these conditions. Then you can define Γ by (66) and Kolmogorov: use (51) to help the consistency. Properties (63–64) are immediate. Property (65) remains an assumption.

The construction

(68)   Let W be the set of all functions w from C to $(0, \infty]$. Let

$$\tau(c, w) = w(c) \quad \text{for} \quad c \in C \quad \text{and} \quad w \in W.$$

Give W the smallest σ-field over which all $\tau(c)$ are measurable.

(69)   For $c \in C$ and $w \in W$, let

$$\lambda(c, w) = \Sigma_d\, \{\tau(d, w) : d \in C \text{ and } d < c\}$$
$$\rho(c, w) = \lambda(c, w) + \tau(c, w) = \Sigma_d\, \{\tau(d, w) : d \in C \text{ and } d \leqq c\}.$$

(70)   Let $\mathscr{X} = \Omega \times W$, with the product σ-field.

(71)   Let $x = (\omega, w)$ with $\omega \in \Omega$ and $w \in W$. Let $c \in C$. Define

$$\xi(c, x) = \xi(c, \omega) \quad \text{and} \quad \tau(c, x) = \tau(c, w)$$
$$\lambda(c, x) = \lambda(c, w) \quad \text{and} \quad \rho(c, x) = \rho(c, w).$$

(72)   Define a process X on $\mathscr{X}$:

$$\begin{aligned}
X(t, x) &= \xi(c, x) \quad &\text{if} \quad \lambda(c, x) \leqq t < \rho(c, x) \quad &\text{for some} \quad c \in C \\
&= \varphi \quad &\text{if} \quad \lambda(c, x) \leqq t < \rho(c, x) \quad &\text{for no} \quad c \in C.
\end{aligned}$$

(73) For each $\omega \in \Omega$, let $\eta_{q(\omega)}$ be the probability on W which makes the $\tau(c)$ independent and exponentially distributed, the parameter for $\tau(c)$ being $q[\omega(c)]$.

(74) For $A \subset \mathscr{X}$ and $\omega \in \Omega$, let $A(\omega)$ be the ω-section of A:

$$A(\omega) = \{w : w \in W \text{ and } (\omega, w) \in A\}.$$

(75) For each $i \in I$, let π_i be the following probability on $\mathscr{X}$:

$$\pi_i\{A\} = \int_\Omega \eta_{q(\omega)}\{A(\omega)\} \, \Gamma_i(d\omega);$$

the prior definitions are (57–58) and (68–70) and (73–74).
    To state (76), let

$$c_0 = 0 < c_1 < \cdots < c_n \quad \text{be in } C$$
$$i_0 = i, i_1, \ldots, i_n \qquad \text{be in } I;$$

and let

$$U_0, U_1, \ldots, U_n$$

be independent and exponentially distributed, with respective parameters $q(i_0), q(i_1), \ldots, q(i_n)$. Let B be a Borel subset of Euclidean $(n+1)$-space.

(76) $\pi_i\{\xi(c_m) = i_m \text{ for } m = 0, \ldots, n \text{ and } (\tau(c_0), \tau(c_1), \ldots, \tau(c_n)) \in B\}$
      $= \Gamma_i\{\xi(c_m) = i_m \text{ for } m = 0, \ldots, n\} \cdot \text{Prob} \{(U_0, U_1, \ldots, U_n) \in B\}.$

Use (76) and (63):

(77) $\pi_i\{\xi(0) = i\} = 1 \quad \text{and} \quad \pi_i\{\tau(0) > t\} = e^{-q(i)t}.$

Properties of X

(78) X is jointly measurable.
(79) $X(t) = \xi(0) \quad \text{for} \quad 0 \leq t < \tau(0).$
(80) $X(0) = \xi(0) \quad \text{by (79)}.$
(81) $\pi_i\{X(0) = i\} = 1 \qquad \text{by (80, 77)}.$
(82) If C is discrete in the sense of (52), then

$$X(t) = \xi(1) \quad \text{for} \quad \tau(0) \leq t < \tau(0) + \tau(1).$$

(83) Review (68–69). Let

$$C_f[w] = \{c : c \in C \text{ and } \lambda(c, w) < \infty\}$$
$$\rho_f(w) = \Sigma_c \{\tau(c, w) : c \in C_f[w]\}$$
$$C_f[\omega, w] = C_f[w] \quad \text{and} \quad \rho_f(\omega, w) = \rho_f(w).$$

The f is for finite. The square bracket is to prevent confusion with sections. Fix $x = (\omega, w) \in \mathcal{X}$. Review (70–72).

(84)   The map $c \to [\lambda(c, x), \rho(c, x))$ is 1-1 and strictly increasing on $C_f[x]$.

(85) $X(t, x) = \xi(c, x)$ for $\lambda(c, x) \leq t < \rho(c, x)$, an interval of length $\tau(c, x)$.

(86)   The union of $[\lambda(c, x), \rho(c, x))$ as c varies over $C_f[x]$ is

$$\{t : t < \rho_f(x) \quad \text{and} \quad X(t, x) \in I\}.$$

(87)   Lebesgue $\{t : t < \rho_f(x) \quad \text{and} \quad X(t, x) = \varphi\} = 0.$

(88) $X(t, x) = \varphi \quad \text{for} \quad t \geq \rho_f(x).$

(89) WARNING. $X(\cdot, x)$ need not be regular in the sense of (7.2). And $[\lambda(c, x), \rho(c, x))$ need not be a maximal interval of constancy. See (94).

The set $\mathcal{X}_1$

(90)   Review (52, 57). If C is discrete, let

$$\Omega^s = \{\omega : \omega \in \Omega \text{ and } \omega[s(c)] \neq \omega(c) \text{ for all } c \in C\}.$$

If C is indiscrete, let $\Omega^s = \Omega$.

(91) Lemma. $\Gamma_i\{\Omega^s\} = 1$, where Ω^s is defined in (90).

PROOF.   Use (64, 66).   ★

(92)   Review (59, 61) and (68–70) and (83) and (90). Let $\mathcal{X}_1$ be the set of $x = (\omega, w) \in \mathcal{X}$ such that:

$$\omega \in \Omega^* \cap \Omega^s, \text{ and}$$

$$\lambda(c, w) < \infty \quad \text{iff} \quad \lambda^*(c, \omega) < \infty \quad \text{for all } c \in C, \quad \text{and } \rho_f(w) = \infty.$$

(93) Lemma. $\pi_i\{\mathcal{X}_1\} = 1$, where $\mathcal{X}_1$ is defined in (92).

PROOF.† For $c \in C$ and $\omega \in \Omega$, let

$$W[c, \omega] = \{w : w \in W \text{ and } \lambda(c, w) < \infty\} \quad \text{when} \quad \lambda^*(c, \omega) < \infty$$
$$= \{w : w \in W \text{ and } \lambda(c, w) = \infty\} \quad \text{when} \quad \lambda^*(c, \omega) = \infty.$$

For $\omega \in \Omega$, let

$$W_\omega = \bigcap_c \{W[c, \omega] : c \in C\}.$$

Let

$$W_\infty = \{w : w \in W \text{ and } \rho_f(w) = \infty\}.$$

Let

$$\Omega_1 = \Omega^* \cap \Omega^s.$$

† There's less here than meets the eye; but keep track of the notation.

Then $\mathscr{X}_1$ is the set of pairs (ω, w) with $\omega \in \Omega_1$ and $w \in W_\omega \cap W_\infty$. By (75),

$$\pi_i\{\mathscr{X}_1\} = \int_{\Omega_1} \eta_{q(\omega)}\{W_\omega \cap W_\infty\} \, \Gamma_i(d\omega).$$

But $\Gamma_i\{\Omega_1\} = 1$ by (65) and (91). Fix $\omega \in \Omega$. By (73) and (5.33),

$$\eta_{q(\omega)}\{W[c, \omega]\} = 1 \quad \text{for each } c;$$

so

$$\eta_{q(\omega)}\{W_\omega\} = 1.$$

Review (59, 61). Fix $\omega \in \Omega^*$. Let

$$C_f^*[\omega] = \{c : \lambda^*(c, \omega) < \infty\}$$

$$\sigma_f(\omega, w) = \Sigma_c \{w(c) : c \in C_f^*[\omega]\}.$$

By (62, 73) and (5.33),

$$\eta_{q(\omega)}\{w : w \in W \text{ and } \sigma_f(\omega, w) = \infty\} = 1.$$

Review (83). If $w \in W_\omega$, then $C_f^*[\omega] = C_f[w]$, so

$$\sigma_f(\omega, w) = \rho_f(\omega, w) = \rho_f(w).$$

Consequently,

$$W_\omega \cap W_\infty = W_\omega \cap \{w : w \in W \text{ and } \sigma_f(\omega, w) = \infty\}$$

has $\eta_{q(\omega)}$-probability 1.   ★

(94) Proposition.   (a) *If $x \in \mathscr{X}_1$, then $X(\cdot, x)$ is regular in the sense of* (7.2).

(b)   *If $x \in \mathscr{X}_1$ and $c \in C$ and $\lambda(c, x) < \infty$, then $[\lambda(c, x), \rho(c, x))$ is a maximal interval of constancy in $X(\cdot, x)$.*

PROOF.   *Claim (a).* I will argue that $X(\cdot, x)$ is continuous from the right at t. The existence of a limit from the left is similar. If

$$t \in \bigcup_c [\lambda(c, x), \rho(c, x)),$$

it's easy. Otherwise, $X(t, x) = \varphi$. Let $t_n \downarrow t$, with $X(t_n, x) = j_n \in I$. I have to make $j_n \to \varphi$. By (84–86), you can find a nonincreasing sequence $c_n \in C$ with

$$\lambda(c_n, x) \leqq t_n < \rho(c_n, x) \quad \text{and} \quad \xi(c_n, x) = j_n.$$

As definition (92) implies, $\lambda^*(c_n, x) < \infty$. Use (60) and definition (71).

Claim (b). Suppose C is discrete in the sense of (52); the argument for indiscrete C is easier. Suppose $\rho(c, x) < \infty$. Then

$$\rho(c, x) = \lambda(s(c), x).$$

But $\xi(s(c), x) \neq \xi(c, x)$, because $x \in \Omega^s \times W$, as defined in (90). So $X(\cdot, x)$

changes at $\rho(c, x)$. If $c = s(d)$ for some d, the same argument makes $X(\cdot, x)$ change at $\lambda(c, x)$. If $c > 0$ and $c = s(d)$ for no d, then c is a limit point of C from the left. Use (60) to find $c(x) < c$, such that

$$\xi(d, x) \neq \xi(c, x) \quad \text{for} \quad c(x) < d < c.$$

So

$$X(t, x) \neq \xi(c, x) \quad \text{for} \quad \rho(c(x), x) \leqq t < \lambda(c, x).$$

This forces $X(\cdot, x)$ to change at $\lambda(c, x)$.   ★

(95) Lemma. $\pi_i \{X(t) \in I\} = 1$.

   PROOF.   If $q(i) = 0$, use (77) and (79). Suppose $q(i) > 0$. By (76),

(96) $\tau(0)$ is exponential with parameter $q(i)$, and is independent of $\{\xi(c), \tau(c): c > 0\}$, relative to π_i.

Let

$$Y(s, x) = X[\tau(0, x) + s, x].$$

So Y is jointly measurable by (78). If $x \in \mathcal{X}_1$, then $\rho_j(x) = \infty$ by definitions (83, 92), so

$$\text{Lebesgue } \{s: Y(s, x) = \varphi\} = 0$$

by (87). Temporarily, let

$$\lambda_0(c) = \Sigma_d \{\tau(d): d \in C \text{ and } 0 < d < c\}.$$

Then

$$Y(s) = \xi(c) \quad \text{if} \quad \lambda_0(c) \leqq s < \lambda_0(c) + \tau(c) \quad \text{for some } c \in C$$
$$= \varphi \qquad \text{if} \quad \lambda_0(c) \leqq s < \lambda_0(c) + \tau(c) \quad \text{for no } c \in C.$$

Therefore, Y is measurable on $\{\xi(c), \tau(c): c > 0\}$. Now (96) makes $\tau(0)$ exponential with parameter $q(i)$, independent of Y, relative to π_i. Complete the argument as in (19).   ★

The σ-fields $\mathcal{A}$ and $\mathcal{B}$

   Fix i and j in I, and $d \in C$.

(97)   Let $\mathcal{A}$ be the σ-field in Ω spanned by $\xi(c)$ with $c \leqq d$.

(98)   Let

$$0 = c_0 < c_1 < \cdots < c_n \quad \text{be in } C$$
$$e_m = M_d^{-1} c_m \quad \text{for } m = 0, \ldots, n$$
$$j_0 = j, j_1, \ldots, j_n \quad \text{be in } I$$
$$B_1 = \{\omega: \omega \in \Omega \text{ and } \omega(c_m) = j_m \text{ for } m = 0, \ldots, n\}$$
$$\hat{B}_1 = \{\omega: \omega \in \Omega \text{ and } \omega(e_m) = j_m \text{ for } m = 0, \ldots, n\}.$$

The prior definitions are (51) of M and (57) of Ω.

(99) Lemma. *Let h be a nonnegative, $\mathscr{A}$-measurable function on Ω, as defined in (97). Define B_1 and $\hat{B}_1$ as in (98). Then*

$$\int_{\hat{B}_1} h \, dF_i = \left[\int_{\{\xi(d)=j\}} h \, d\Gamma_i \right] \cdot \Gamma_j\{B_1\}.$$

PROOF.   This restates the Markov property (66): time d is the present, h is in the past, $\hat{B}_1$ is in the future, and B_1 is $\hat{B}_1$ shifted to start at time 0. Formally, let

$$d_0 = 0 < d_1 < \cdots < d_N = d \quad \text{be in } C$$

$$i_0 = i, i_1, \ldots, i_N = j \quad \text{be in } I$$

$$D = \{\omega : \omega \in \Omega \text{ and } \omega(d_m) = i_m \text{ for } m = 0, \ldots, N\}.$$

Now

$$e_0 = M_d^{-1} c_0 = M_d^{-1} 0 = d = d_N$$

$$d_0 < d_1 < \cdots < d_N < e_1 < \cdots < e_n$$

$$j_0 = j = i_N.$$

By (66–67),

$$\Gamma_i\{D \text{ and } \hat{B}_1\} = pq,$$

where

$$p = \Pi_{m=0}^{N-1} \gamma[M(d_m, d_{m+1}), i_m, i_{m+1}] = \Gamma_i\{D\}$$

and

$$q = \Pi_{m=0}^{n-1} \gamma[M(e_m, e_{m+1}), j_m, j_{m+1}] = \Gamma_j\{B_1\}:$$

because (51) makes

$$M(e_m, e_{m+1}) = M(M_d e_m, M_d e_{m+1}) = M(c_m, c_{m+1}).$$

So (99) holds for $h = 1_D$. By (10.16), the result holds for $h = 1_A$ with $A \in \mathscr{A}$. Now extend.   ★

(100)   Let $\mathscr{B}$ be the σ-field in W spanned by $\tau(c)$ with $c \leqq d$.

(101) Lemma.   *Review (73). If $B \in \mathscr{B}$, as defined in (100), then $\omega \to \eta_{q(\omega)}\{B\}$ is $\mathscr{A}$-measurable, as defined in (97)*

PROOF.   This is easy when

$$B = \{w : w \in W \text{ and } w(c_m) < t_m \text{ for } m = 0, \ldots, n\}$$

with

$$c_0 < c_1 < \cdots < c_n \leqq d \quad \text{in } C.$$

Now extend.   ★

The generator

(102)   Let $P(t, i, j) = \pi_i\{X(t) = j\}$.

(103)   Define a matrix Q on I as follows:

$$Q(i, i) = -q(i);$$

when C is discrete in the sense of (52),

$$Q(i, s(i)) = q(i)$$

$$Q(i, j) = 0 \quad \text{for } j \neq i, s(i);$$

when C is indiscrete,

$$Q(i, j) = 0 \quad \text{for } j \neq i.$$

(104) Lemma $P'(0) = Q$, *as defined in* (102–103).

PROOF   When C is discrete, you can use the corresponding argument in (5.39). The results you need are (64, 76, 79–82).

Suppose C is indiscrete. Fix i and j in I, with $j = i$ allowed. The case $q(i) = 0$ is easy, so assume $q(i) > 0$. Fix $\infty \notin C$, and pretend $\infty > c$ for all $c \in C$. Review definition (59, 61) of Ω^*. Define a measurable mapping K from Ω^* to $C \cup \{\infty\}$ as follows. If $\lambda^*(c, \omega) < \infty$ and $\omega(c) = j$ for some $c > 0$, there is a least such c by (60); and $K(\omega)$ is this least c. Otherwise, $K(\omega) = \infty$. Count C off as $\{c_1, c_2, \ldots\}$. Define a measurable mapping L from Ω^* to C as follows:

$$L(\omega) \text{ is the } c_n \text{ with least } n \text{ satisfying } 0 < c_n < K(\omega).$$

Because C is indiscrete, L is properly defined. Define a measurable mapping ζ from Ω^* to I:

$$\zeta(\omega) = \omega[L(\omega)].$$

For each $k \in I$, let U_i and U_k be independent, exponential random variables, with parameters $q(i)$ and $q(k)$. If $\omega \in \Omega^*$ and $\xi(0, \omega) = i$ and $\zeta(\omega) = k$, then (73) shows:

(105)   the $\eta_{q(\omega)}$-distribution of $\tau(0)$ and $\tau[L(\omega)]$ coincides with the distribution of U_i and U_k.

I claim:

(106) $\pi_i\{\tau(0) \leqq t \text{ and } X(t) = j\} = o(t) \quad \text{as } t \to 0.$

To argue (106), abbreviate

$$A_t = \{\tau(0) \leqq t \text{ and } X(t) = j\}.$$

By (65) and (75),

$$(107) \qquad \pi_i\{A_t\} = \int_{\Omega^*} \eta_{q(\omega)}\{A_t(\omega)\}\, \Gamma_i(d\omega).$$

Fix $\omega \in \{\Omega^*$ and $\xi(0) = i$ and $\zeta = k\}$. Abbreviate

$$E[t, \omega] = \{W \text{ and } \tau(0) + \tau[L(\omega)] \leq t\}.$$

Define W_ω as in (93), and remember $\eta_{q(\omega)}\{W_\omega\} = 1$ by (5.33). I claim that $W_\omega \cap A_t(\omega) \subset E[t, \omega]$. Indeed, fix an $x = (\omega, w)$ with $w \in W_\omega \cap A_t(\omega)$. Then $X(t, x) = j$. So (72) there is a $c \in C$ with

$$\lambda(c, w) \leq t < \rho(c, w) \quad \text{and} \quad \xi(c, \omega) = j.$$

As (92) implies, $\lambda^*(c, \omega) < \infty$. So $K(\omega) \in C$, and

$$0 < L(\omega) < K(\omega) \leq c.$$

Now (84) shows

$$\tau(0, w) + \tau[L(\omega), w] \leq \rho[L(\omega), w] < \lambda(c, w) \leq t,$$

proving $W_\omega \cap A_t(\omega) \subset E[t, \omega]$. Conclude

$$\eta_{q(\omega)}\{A_t(\omega)\} \leq \eta_{q(\omega)}\{E[t, \omega]\} = \text{Prob } \{U_i + U_k \leq t\}$$

by (105). Combine this with (107):

$$\pi_i\{A_t\} \leq \Sigma_k\, \pi_i\{\zeta = k\} \cdot \text{Prob } \{U_i + U_k \leq t\}.$$

But (5.34) makes

$$q(i) \geq t^{-1} \text{Prob } \{U_i + U_k \leq t\} \to 0 \quad \text{as } t \to 0.$$

Now dominated convergence settles (106).

If $j \neq i$, then (79) makes

$$\{\xi(0) = i \text{ and } X(t) = j\} \subset \{\tau(0) \leq t\};$$

so (77) and (106) prove

$$\pi_i\{X(t) = j\} = o(t) \quad \text{as } t \to 0.$$

This proves $P'(0, i, j) = 0 = Q(i, j)$ for $i \neq j$.

I will now compute $P'(0, i, i)$. Check that $\{\xi(0) = i \text{ and } X(t) = j\}$ equals

$$\{\xi(0) = i \text{ and } \tau(0) > t\} \cup \{\xi(0) = i \text{ and } \tau(0) \leq t \text{ and } X(t) = i\}.$$

Use (77) and (106):

$$\pi_i\{X(t) = i\} = e^{-q(i)t} + o(t) \quad \text{as } t \to 0. \qquad \bigstar$$

The theorem

(108) Theorem. *Suppose* (48–51) *and* (63–66). *Define the probability triple* $(\mathcal{X}, \pi_i)$ *by* (70, 75). *Define the process* X *on* $\mathcal{X}$ *by* (72). *Define P and Q by* (102–103). *Then*

(a) *P is a standard stochastic semigroup on I, with generator Q.*

(b) *X is Markov with stationary transitions P and starting state i, relative to* π_i.

NOTE. The construction has properties (78–88) and (93–94).

PROOF. To start with, fix i and j in I, fix $t \geq 0$, and fix $d \in C$.

(109)   Let $D = \{W$ and $\lambda(d) \leq t < \lambda(d) + \tau(d)\}$; the definitions are (68–69).

(110)   Define a mapping T_1 of Ω into Ω:

$$(T_1\omega)(c) = \omega(M_d^{-1}c) \quad \text{for } c \in C;$$

the prior definitions are (51, 57).

(111)   Defining a mapping T_2 of D into W:

$$(T_2w)(0) = \lambda(d, w) + \tau(d, w) - t;$$

$$(T_2w)(c) = w(M_d^{-1}c) \quad \text{for } c \in C \quad \text{with } c > 0;$$

the prior definitions are (51), (68–69), and (109).

(112)   Define a mapping T of $\Omega \times D$ into $\mathcal{X}$:

$$T(\omega, w) = (T_1\omega, T_2w).$$

You have to argue

(113) $X(t + s) = X(s) \circ T$ for all s on $\Omega \times D$.

This is a straightforward and boring project, using (84–86) and (88).

(114)   Define a subset A of $\mathcal{X}$ as follows.

$A = A_1 \times (A_2 \cap D)$, where:

D was defined in (109);

$A_1 = \{\omega : \omega \in \Omega$ and $\omega(d_m) = i_m$ for $m = 0, \ldots, N\}$;

$A_2 = \{w : w \in W$ and $w(d_m) > t_m$ for $m = 0, \ldots, N - 1\}$;

$d_0 = 0 < d_1 < \cdots < d_N = d$ are in C;

$i_0 = i, i_1, \ldots, i_N = j$ are in I;

$t_0, t_1, \ldots, t_{N-1}$ are nonnegative numbers.

NOTE. $m < N$ in A_2.

(115) Define a subset B of $\mathscr{X}$ as follows.

$$B = B_1 \times B_2, \quad \text{where:}$$
$$B_1 = \{\omega : \omega \in \Omega \text{ and } \omega(c_m) = j_m \text{ for } m = 0, \ldots, n\};$$
$$B_2 = \{w : w \in W \text{ and } w(c_m) > u_m \text{ for } m = 0, \ldots, n\};$$
$$c_0 = 0 < c_1 < \cdots < c_n \quad \text{are in } C;$$
$$j_0 = j, j_1, \ldots, j_n \quad \text{are in } I;$$
$$u_0, u_1, \ldots, u_n \quad \text{are nonnegative numbers.}$$

I claim:

(116) $$\pi_i(A \cap T^{-1}B) = \pi_i(A) \cdot \pi_j(B).$$

To start on (116), make the following definition.

(117) With the notation of (115), let

$$e_m = M_d^{-1} c_m \quad \text{for } m = 0, \ldots, n,$$

the M coming from (51); so

$$d_0 = 0 < d_1 < \cdots < d_N = d = e_0 < e_1 < \cdots < e_n,$$

the d_m coming from (114); let

$$\hat{B}_1 = \{\omega : \omega \in \Omega \text{ and } \omega(e_m) = j_m \text{ for } m = 0, \ldots, n\};$$
$$\hat{B}_2 = \{w : w \in W \text{ and } w(e_m) > u_m \text{ for } m = 1, \ldots, n\};$$
$$\hat{D} = \{w : w \in D \text{ and } t + u_0 < \lambda(d, w) + \tau(d, w)\},$$

where D was defined in (109).

NOTE. $m > 0$ in $\hat{B}_2$.

Remember $i_N = j = j_0$ and $d_N = d = e_0$. Confirm

(118) $$A \cap T^{-1}B = (A_1 \cap \hat{B}_1) \times (A_2 \cap \hat{D} \cap \hat{B}_2).$$

By (75),

(119) $$\pi_i(A \cap T^{-1}B) = \int_{A_1 \cap \hat{B}_1} \eta_{q(\omega)}(A_2 \cap \hat{D} \cap \hat{B}_2) \, \Gamma_i(d\omega).$$

But A_2 and $\hat{D}$ are measurable on the σ-field $\mathscr{B}$ of (100). Use definition (73):

(120) $$\eta_{q(\omega)}(A_2 \cap \hat{D} \cap \hat{B}_2) = \eta_{q(\omega)}(A_2 \cap \hat{D}) \cdot e^{-v}, \quad \text{where}$$
$$v = q(j_1)u_1 + \cdots + q(j_n)u_n \quad \text{and } \omega \in \hat{B}_1.$$

Let Σ be the σ-field in W spanned by $\tau(c)$ with $c < d$. Then A_2 and $\lambda(d)$ are Σ-measurable: definitions (114) and (69). Abbreviate $u = u_0$. Remember

$i_N = j = j_0$ and $d_N = d = e_0$. Use (73) and (5.30):

(121) $\eta_{q(\omega)}(A_2 \cap \hat{D}) = \eta_{q(\omega)}(A_2 \cap D) \cdot e^{-q(j)u}$ when $\omega(d) = j$;

the set D comes from (109).

   Combine (119–121):

(122) $\pi_i(A \cap T^{-1}B) = e^{-s} \int_{A_1 \cap \hat{B}_1} \eta_{q(\omega)}(A_2 \cap D) \, \Gamma_i(d\omega)$,   where

$$s = q(j_0)u_0 + q(j_1)u_1 + \cdots + q(j_n)u_n.$$

But $A_2 \cap D \in \mathscr{B}$; the definitions are (114, 109) and (100). So (101) makes

$$\omega \to \eta_{q(\omega)}(A_2 \cap D)$$

$\mathscr{A}$-measurable, as defined in (97). Check $A_1 \in \mathscr{A}$ and $A_1 \subset \{\xi(d) = j\}$ from definition (114). By (99),

(123) $\int_{A_1 \cap \hat{B}_1} \eta_{q(\omega)}(A_2 \cap D) \, \Gamma_i(d\omega) = \left[\int_{A_1} \eta_{q(\omega)}(A_2 \cap D) \, \Gamma_i(d\omega) \right] \cdot \Gamma_j(B_1).$

By (75) and (114),

(124) $\pi_i(A) = \int_{A_1} \eta_{q(\omega)}(A_2 \cap D) \, \Gamma_i(d\omega).$

By (76) and (115),

(125) $\pi_j(B) = e^{-s} \, \Gamma_j(B_1)$,   where s comes from (122).

Combine (122–125) to get (116).

   The class of sets B of the form (115), with variable n, c_m, j_m, and u_m, is closed under intersection, modulo the null set, and generates the full σ-field on $\{\mathscr{X}$ and $\xi(0) = j\}$. Each B is a subset of $\{\mathscr{X}$ and $\xi(0) = j\}$; and this set is of the form (115), with $n = 0$ and $u_0 = 0$. Both sides of (116) are countably additive in B. By (10.16), equality holds in (116) for any A of the form (114), and any measurable subset B of $\{\mathscr{X}$ and $\xi(0) = j\}$.

   I now have to vary A.

(126)   Let $A(d) = \{\xi(0) = i$ and $\xi(d) = j$ and $\lambda(d) \leq t < \rho(d)\} \subset \mathscr{X}$; let $\mathscr{A}(d)$ be the σ-field of subsets of $A(d)$ generated by sets of the form (114).

The class of sets A of the form (114), with variable N, d_m, i_m, and t_m is closed under intersection, modulo the null set. Each A is a subset of $A(d)$, and $A(d)$ is of the form (114), with $N = 1$ and $t_0 = 0$. Both sides of (116) are countably additive in A. By (10.16), equality stands in (116) for all $A \in \mathscr{A}(d)$ and all measurable subsets B of $\{\mathscr{X}$ and $\xi(0) = j\}$. Put

$$B = \{\xi(0) = j \text{ and } X(s) = k\}.$$

Use (113):

$$T^{-1}B = \{\Omega \times D \text{ and } \xi(d) = j \text{ and } X(t+s) = k\}.$$

From (126),

(127) $$A(d) \subset \{\Omega \times D \text{ and } \xi(d) = j\}.$$

So extended (116) makes

(128) $\pi_i\{A \text{ and } X(t+s) = k\} = \pi_i\{A\} \cdot \pi_j\{X(s) = k\}$ for all $A \in \mathscr{A}(d)$.

How big is $\mathscr{A}(d)$?

(129) Let $\mathscr{F}(t)$ be the σ-field in $\mathscr{X}$ spanned by $X(u)$ for $0 \leq u \leq t$.

I claim

(130) If $A \in \mathscr{F}(t)$ as defined in (129), then $A(d) \cap A \in \mathscr{A}(d)$, as defined in (126).

NOTE. I do not claim $A(d) \in \mathscr{F}(t)$.

To argue (130), let $0 \leq u \leq t$ and let $h \in I$. Then

$$\{A(d) \text{ and } X(u) = h\} = \bigcup_c \{G(c) : c \in C \text{ and } c \leq d\}, \quad \text{where}$$

$$G(c) = \{A(d) \text{ and } \xi(c) = h \text{ and } \lambda(c) \leq u < \rho(c)\}.$$

If $c < d$, then $G(c) \in \mathscr{A}(d)$ by definitions (126) and (69). If $c = d$, then $\rho(c) = \rho(d) > u$ is free on $A(d)$, so $G(c)$ is still in $\mathscr{A}(d)$. This proves (130). The sets $A(d)$ of (126) are disjoint as d varies over C, and their union is

$$\{X(0) = i \text{ and } X(t) = j\}.$$

Use (128–130):

(131) $\pi_i\{A \text{ and } X(t+s) = k\} = \pi_i\{A\} \cdot \pi_j\{X(s) = k\}$

for all $A \in \mathscr{F}(t)$ with $A \subset \{X(0) = i \text{ and } X(t) = j\}$.

Now use lemma (5.4). Condition (5.3a) holds by (81). Condition (5.3b) holds by (95). Condition (5.3c) holds by (131). This and (104) prove (108). ★★★

5. EXAMPLES ON THE SECOND CONSTRUCTION

The first example will be useful in proving (7.51).

(132) Example. (a) Description. Let I be a countably infinite set. Let Q be a matrix on I, such that

$$q(i) = -Q(i, i) \geq 0$$
$$Q(i, j) \geq 0 \quad \text{for } i \neq j$$
$$\Sigma_j Q(i, j) = 0.$$

Let

$$\Gamma(i,j) = Q(i,j)/q(i) \quad \text{for } i \neq j \quad \text{and} \quad q(i) > 0$$
$$= 0 \quad \text{elsewhere.}$$

So

$$\Gamma(i, i) = 0$$
$$\Gamma(i,j) = 0 \quad \text{when } q(i) = 0$$
$$\Sigma_j \, \Gamma(i,j) = 1 \quad \text{when } q(i) > 0.$$

Let p be a probability on I. Starting from i, the process jumps according to Γ, and the holding times are filled in according to q. If the process hits an absorbing state j, that is $q(j) = 0$, the visit to j has infinite length and the sample function is completely defined. Otherwise, the sample function makes an infinite number of visits. However, the time θ to perform these visits may be finite. If so, start the process over again at a state chosen from p, independent of the past sample function. Repeat this at any future exceptional times. See Figure 2. If θ is finite with positive probability, then there is a 1-1 correspondence between p and the transitions P^p.

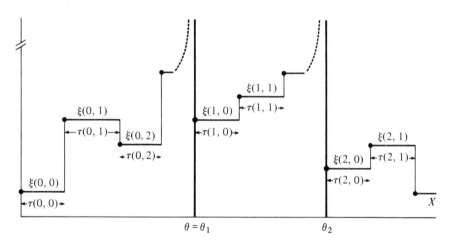

$$\theta = \theta_1 \qquad\qquad \theta_2$$

Figure 2.

(b) *State space.* Fix $a \neq b$ outside I. The state space (48) is $\hat{I} = I \cup \{a, b\}$.

(c) *Holding times.* Extend q to vanish at a and b. So q is defined on $\hat{I}$.

(d) *Generator.* Extend Q so $Q(i, c) = Q(c, i) = 0$ for $i \in \hat{I}$ and $c = a$ or b.

(e) *Formal construction.* Define C, $<$, and M as in (54). Extend Γ to a matrix on $\hat{I}$ as follows:

$\Gamma(i, a) = 0$ or 1 according as $q(i) > 0$ or $q(i) = 0$,   for $i \in I$;

$\Gamma(i, b) = 0$ for $i \in I$;

$\Gamma(c, i) = 0$ for $c = a$ or b and $i \in I$;

$\Gamma(a, b) = \Gamma(b, a) = 1$

$\Gamma(a, a) = \Gamma(b, b) = 0$.

I need a and b to get (64). The extended matrix is stochastic on $\hat{I}$. Define the probability Γ_i' of (57–58) by the requirements that it makes:

$\{\xi(m, n): n = 0, 1, \ldots\}$ independent Markov chains with stationary transitions Γ for $m = 0, 1, \ldots$;

$\xi(0, 0) = i$ almost surely;

$\xi(m, 0)$ have distribution p, for $m > 0$.

You should check (48–51) and (63–64) and (66). I will check (65) for $i \in I$; you do $i = a$ or b. Relative to Γ_i^p, the variables $1/q[\xi(m, 0)]$ are independent and identically distributed for $m = 1, 2, \ldots$. They are positive. So

(133) $\qquad \Sigma_{c \in C}\, 1/q[\xi(c)] \geq \Sigma_{m=1}^{\infty}\, 1/q[\xi(m, 0)] = \infty$

with Γ_i^p-probability 1. Fix one ω satisfying (133). I say $\omega \in \Omega^*$, as defined in (59, 61). If $\lambda^*(c, \omega) < \infty$ for all c, this follows from (62). If $\lambda^*(c, \omega) = \infty$ for some c, then there is a least such c, call it $c(\omega)$: because C is well-ordered. So,

$$\lambda^*(c, \omega) < \infty \quad \text{iff} \quad c < c(\omega);$$

and

$$\Sigma_c\, \{1/q[\omega(c)]: c \in C \text{ and } \lambda^*(c, \omega) < \infty\} = \Sigma_c\, \{1/q[\omega(c)]: c \in C \text{ and } c < c(\omega)\}$$
$$= \lambda^*(c(\omega), \omega)$$
$$= \infty.$$

So (62) works again. Theorem (108) completes the formal construction. ★

Write

$$\pi_i^p = \Gamma_i^p \times \eta,$$

as defined in (75), to show the dependence on p. I would now like to isolate the properties of the construction that will be useful in (7.51). Fix $i \in I$. Use (76):

(134) $\quad \xi(1, 0)$ is independent of $\{\xi(0, n), \tau(0, n): n = 0, 1, \ldots\}$ and has distribution p, relative to π_i^p.

Let

$$\sigma_n = \tau(0, 0) + \cdots + \tau(0, n - 1) \quad \text{for } n = 1, 2, \ldots$$
$$\theta = \Sigma_{n=0}^{\infty}\, \tau(0, n) = \lim_n \sigma_n.$$

Use (134):

(135) $\quad \pi_i^p\{\theta < \infty \text{ and } \xi(1, 0) = j\} = \pi_i^p\{\theta < \infty\} \cdot p(j) \quad \text{for } j \in I$.

To state (136), let $\mathcal{X}_0$ be the subset of $\mathcal{X}_1$, as defined in (92), with:

$$\xi(m, 0) \in I \quad \text{for } m = 0, 1, \ldots;$$
$$\Gamma[\xi(m, n), \xi(m, n + 1)] > 0 \quad \text{for } (m, n) \in C.$$

(136) Lemma. (a) $\pi_i^p\{\mathcal{X}_0\} = 1$ *   for $i \in I$.*

   (b) *If $x \in \mathcal{X}_0$, then $X(t, x) \in I$ for all $t \geq 0$.*

   (c) *If $x \in \mathcal{X}_0$, then $X(\cdot, x)$ is regular in the sense of (7.2).*

PROOF.   *Claim (a).*   Use (93) and (76).

   Claim (b).   Let $x \in \mathcal{X}_0$. Suppose $\xi(\cdot, \cdot)(x)$ visits a or b. Let (m, n) be the first index with $\xi(m, n) = a$ or b. Get $n > 0$ and $q[\xi(m, n - 1)(x)] = 0$. So $\lambda^*(m, n)(x) = \infty$ by definition (59), forcing $\lambda(m, n)(x) = \infty$ by definitions (69, 92). This prevents $X(\cdot, x)$ from reaching a or b, by definition (72). You check $X(t, x) \neq \varphi$.

   Claim (c).   Use (94).   ★

(137) Lemma.   *Let $P^p(t, i, j) = \pi_i^p\{X(t) = j\}$ for i and j in I. Then P^p is a standard stochastic semigroup on I, with generator Q.*

PROOF.   Use (108) and (136).   ★

DISCUSSION.   Fix $x \in \mathcal{X}_0$, as defined for (136). Here is a description of $X(\cdot, x)$. Let

$$\theta_M(x) = \Sigma \{\tau(m, n)(x) : (m, n) \in C \text{ and } m < M\};$$

so $\theta_0(x) = 0$ and $\theta_1(x) = \theta(x)$. Suppose $\lambda(M, N)(x) < \infty$. Let m be one of $0, \ldots, M - 1$.

   $X(\cdot, x)$ is a step function on $[\theta_m(x), \theta_{m+1}(x))$, visiting $\xi(m, n)(x)$ with holding time $\tau(m, n)(x)$ for $n = 0, 1, \ldots$.
   $X(\cdot, x)$ is a step function on $[\theta_M(x), \rho(M, N)(x))$, visiting $\xi(M, n)(x)$ with holding time $\tau(M, n)(x)$ for $n = 0, \ldots, N$.

And

$$\lim_{n \to \infty} \xi(m, n)(x) = \varphi;$$

in fact,

$$\Sigma_{n=0}^{\infty} 1/q[\xi(m, n)(x)] < \infty.$$

Part of this I need. Keep $x \in X_0$, and check (138–140); the times θ and σ_n were defined after (134).

(138) $\sigma_n < \infty$ iff there are at least $n + 1$ intervals of constancy in X. If $\sigma_n(x) < \infty$, then $X(\cdot, x)$ begins by visiting

$$\xi(0, 0)(x), \xi(0, 1)(x), \ldots, \xi(0, n)(x)$$

on intervals of length

$$\tau(0, 0)(x), \tau(0, 1)(x), \ldots, \tau(0, n)(x).$$

(139) $\xi(1, 0) = \lim X(r)$ as rational r decreases to θ, on $\{\theta < \infty\}$.

Use (138–139).

(140) The sets $\{\theta < \infty\}$ and $\{\theta < \infty$ and $\xi(1, 0) = j\}$ are in the σ-field spanned by $\{X(r): r$ is rational$\}$, on $\mathscr{X}_0$.

Here is a more explicit proof of (140). For any real t, the event that $\theta \leqq t$ coincides with the event that for any finite subset J of I, there is a rational $r \leqq t$ with $X(r) \notin J$. The event that $\theta < \infty$ and $\xi(1, 0) = j$ coincides with the event that for any pair of rationals r and s,

either $\theta \notin (r, s)$

or there is a rational t with $r < \theta < t < s$ and $X(t) = j$.

WARNING. π_i^p has mysterious features not controlled by the semigroup of transition probabilities, like the beauty of the sample functions. However, the π_i^p-distribution of X retracted to rational times has no mysteries at all: it is completely controlled by the semigroup and the starting state i. Since Q is silent about p, it does not determine the semigroup.

(141) Note. To get the simplest case of (132), let I be the integers. Let $\Gamma(n, n + 1) = 1$ for all n. Let $0 < q(n) < \infty$ with

$$\Sigma_{n=-\infty}^{\infty} 1/q(n) < \infty.$$

(142) Example. (a) Description. For each sample function, there are exceptional times t such that for any $\varepsilon > 0$, on $(t - \varepsilon, t)$ and on $(t, t + \varepsilon)$, the function assumes infinitely many values in the integers I. Give I the discrete topology, and let $I \cup \{\varphi\}$ be the one-point compactification. There is no natural way to assign an I-value to the sample function at an exceptional t. But if the sample function is set equal to φ at exceptional t, continuity in $I \cup \{\varphi\}$ is secured there. Starting from i, the process moves successively through $i, i + 1, \ldots$. This specifies the process only on a finite interval, $[0, \theta)$. At θ and future exceptional times, restart the process with $i = -\infty$. See Figure 3.

(b) State space. I is the integers.

(c) Holding times. $0 < q(i) < \infty$; and $\Sigma_{i \in I} 1/q(i) < \infty$.

(d) Generator. $Q(i, j) = 0$ unless $j = i$ or $i + 1$; and $Q(i, i) = -q(i)$; and $Q(i, i + 1) = q(i)$.

(e) Formal construction. Define C, $<$, and M as in (55). Define the

Γ_i of (57–58) by the requirements:

$\xi(0, n) = i + n$ for all $n = 0, 1, \ldots$, almost surely;

$\xi(m, n) = n$ for all $m = 1, 2, \ldots$ and integer n, almost surely.

You should check (48–51) and (63–66). Then use (108).   ★

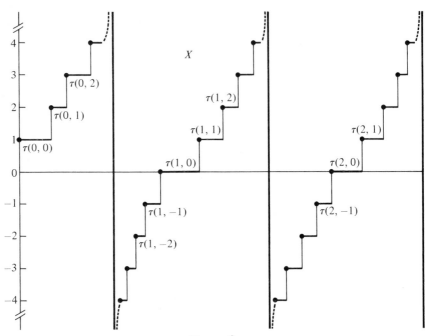

Figure 3.

(143)   The generators in (141) and (142) are equal. But the sample functions are very different.

(144) **Example** (a) *Description.* The process moves cyclically through the rationals in $[0, 1)$.

(b) *State space.* I is the set of rationals in $[0, 1)$.

(c) *Holding times.* q is arbitrary, subject to

$$q(i) > 0 \quad \text{for all } i, \quad \text{and} \quad \Sigma_{i \in I} \, 1/q(i) < \infty.$$

(d) *Generator.* $Q(i, i) = -q(i)$ for all i, and $Q(i, j) = 0$ for $i \neq j$.

(e) *Formal construction.* Define C, $<$, and M as in (56). Define the Γ_i of (57–58) by the requirement that

$\xi(c) \in I$ and $\xi(c) = i + c$ modulo 1   for all c, almost surely.

You should check (48–51) and (63–66). Then use (108).   ★

6. MARKOV TIMES

Example (160) and the results of this section may help you understand some of the technicalities in my formulation of strong Markov, Section 7.4. The results of this section will not be used in other chapters of the book. Let $(\mathcal{X}, \mathcal{F})$ be a measurable space. Let V be a compact metric set. Endow V with the Borel σ-field. For each $t \geq 0$, let $X(t)$ be a V-valued and $\mathcal{F}$-measurable function on $\mathcal{X}$.

(145a)   Let $\mathcal{F}(t)$ be the σ-field in $\mathcal{X}$ spanned by $X(s)$ for $0 \leq s \leq t$.

(145b)   Let $\mathcal{F}(t+) = \bigcap_{n=1}^{\infty} \mathcal{F}\left(t + \dfrac{1}{n}\right).$

(146a)   Say σ is a *strict Markov time* iff $0 \leq \sigma \leq \infty$ and
$$\{\sigma \leq t\} \in \mathcal{F}(t) \quad \text{for all } t \geq 0.$$

(146b)   Say τ is a *Markov time* iff $0 \leq \tau \leq \infty$ and
$$\{\tau < t\} \in \mathcal{F}(t) \quad \text{for all } t \geq 0.$$

Suppose ρ is measurable and $0 \leq \rho \leq \infty$.

(146c)   Let $\mathcal{F}(\rho)$ be the collection of $A \in \mathcal{F}$ such that
$$A \cap \{\rho \leq t\} \in \mathcal{F}(t) \quad \text{for all } t \geq 0.$$

(146d)   Let $\mathcal{F}(\rho+)$ be the collection of $A \in \mathcal{F}$ such that
$$A \cap \{\rho < t\} \in \mathcal{F}(t) \quad \text{for all } t \geq 0.$$

You can check (147).

(147a) $A \in \mathcal{F}(\rho+)$ iff $A \cap \{\rho \leq t\} \in \mathcal{F}(t+)$ for all $t \geq 0$.

(147b) $\mathcal{F}(\rho) \subset \mathcal{F}(\rho+)$.

(147c) ρ is Markov iff $\{\rho \leq t\} \in \mathcal{F}(t+)$ for all $t \geq 0$.

(147d) ρ is Markov iff $\mathcal{X} \in \mathcal{F}(\rho+)$: then $\mathcal{F}(\rho+)$ is a σ-field.

(147e) ρ is strict Markov iff $\mathcal{X} \in \mathcal{F}(\rho)$; then $\mathcal{F}(\rho)$ is a σ-field.

(147f)   Strict Markov times are Markov.

(148) Lemma. *Let σ and τ be two strict Markov times.*

    (a) $\{\sigma < \tau\} \in \mathcal{F}(\tau)$.

    (b) $\{\sigma < \tau\} \in \mathcal{F}(\sigma)$.

    (c) $\{\sigma \leq \tau\} \in \mathcal{F}(\sigma) \cap \mathcal{F}(\tau)$.

    (d) $\{\sigma \leq \tau\} \cap \mathcal{F}(\sigma) \subset \{\sigma \leq \tau\} \cap \mathcal{F}(\tau)$.

    (e) $\{\sigma = \tau\} \in \mathcal{F}(\sigma) \cap \mathcal{F}(\tau)$.

    (f) $\{\sigma = \tau\} \cap \mathcal{F}(\sigma) = \{\sigma = \tau\} \cap \mathcal{F}(\tau)$.

PROOF. Let r range over a countable dense subset of $[0, t]$, which contains t.

Claim (a). $\{\sigma < \tau\} \cap \{\tau \leq t\} = \bigcup_r \{\sigma < r < \tau \leq t\}$.

Claim (b). $\{\sigma < \tau\} \cap \{\sigma \leq t\} = \bigcup_r \{\sigma \leq r < \tau\}$.

Claim (c). $\{\sigma \leq \tau\} = \mathscr{X} \backslash \{\tau < \sigma\}$. Use (a) and (b).

Claim (d). Let $A \subset \{\sigma \leq \tau\}$ and $A \in \mathscr{F}(\sigma)$. Then

$$A \cap \{\tau \leq t\} = \bigcup_r \{A \text{ and } \sigma \leq r\} \cap \{\tau \leq r\}.$$

Claims (e) and (f) follow from (c) and (d), because

$$\{\sigma = \tau\} = \{\sigma \leq \tau\} \cap \{\tau \leq \sigma\}. \qquad\qquad \bigstar$$

(149) Lemma. *Let τ be a Markov time. For each n, let σ_n be a strict Markov time. For each x, suppose*

$$\tau(x) < \sigma_{n+1}(x) \leq \sigma_n(x) \quad \text{and} \quad \sigma_n(x) \downarrow \tau(x).$$

Then $\mathscr{F}(\sigma_n)$ is nonincreasing with n, and

$$\mathscr{F}(\tau+) = \bigcap_{n=1}^\infty \mathscr{F}(\sigma_n).$$

PROOF. To begin with, $\mathscr{F}(\tau+) \subset \mathscr{F}(\sigma_n)$: adapt the argument for (148). Next, $\mathscr{F}(\sigma_n)$ nonincreases by (148d). Finally, let $A \in \mathscr{F}(\sigma_n)$ for all n. I have to get $A \in \mathscr{F}(\tau+)$. Let $B_n = \{\sigma_n \leq t\}$. Then $B_n \uparrow \{\tau < t\}$. So

$$A \cap B_n \in \mathscr{F}(t) \quad \text{and} \quad A \cap B_n \uparrow A \cap \{\tau < t\}. \qquad\qquad \bigstar$$

(150) Lemma. *Let τ be a Markov time. Then $\tau + \dfrac{1}{n}$ is a strict Markov time, and $\mathscr{F}\left(\tau + \dfrac{1}{n}\right)$ is nonincreasing with n, and*

$$\mathscr{F}(\tau+) = \bigcap_{n=1}^\infty \mathscr{F}\left(\tau + \frac{1}{n}\right).$$

PROOF. Use (149). $\qquad\qquad\qquad\qquad\qquad\qquad\qquad\qquad\qquad\qquad \bigstar$

(151) Definition. *If Σ is a sub σ-field of $\mathscr{F}$, then $\Sigma(x)$ is the Σ-atom containing x: namely, the intersection of all Σ-sets containing x. Alternatively, $y \in \Sigma(x)$ iff x and y are both in or both out of any Σ-set. Say Σ is separable iff Σ is the smallest σ-field which includes a countable collection $\mathscr{C}$ of sets. Say Σ is saturated iff any $A \in \mathscr{F}$ which is a union, even uncountable, of Σ-atoms is in Σ.*

I will quote the main result (152) on saturation from Blackwell (1954) without proof.

(152) Suppose $(\mathscr{X}, \mathscr{F})$ is Borel. Any separable sub σ-field of $\mathscr{F}$ is saturated.

Borel is defined in Section 10.9. Atoms are discussed in Section 10.5.

(153) Lemma. *Let* $\Sigma_1 \supset \Sigma_2 \supset \cdots$ *be sub σ-fields of* $\mathscr{F}$. *Let*

$$\Sigma_\infty = \bigcap_{n=1}^\infty \Sigma_n \quad and \quad \alpha(x) = \bigcup_{n=1}^\infty \Sigma_n(x).$$

(a) $\Sigma_1(x) \subset \Sigma_2(x) \subset \dots$.

(b) $\Sigma_\infty(x) = \alpha(x)$.

(c) *If each Σ_n is saturated, then Σ_∞ is saturated.*

The definitions are in (151).

PROOF. *Claim (a)* is easy.

Claim (b). Let $y \in \alpha(x)$. Then there is an $n = n(x, y)$ with $y \in \Sigma_n(x)$. No Σ_n-set can separate x and y: so no Σ_∞-set can do it. This proves:

$$\alpha(x) \subset \Sigma_\infty(x).$$

Let $y \notin \alpha(x)$. Then $y \notin \Sigma_n(x)$ for all n. For all n, there is an $A_n \in \Sigma_n$ with

$$x \in A_n \quad and \quad y \notin A_n.$$

Let $A = \liminf A_n$. Then $x \in A$, and $y \notin A$. But $A \in \Sigma_\infty$, because the Σ_n nonincrease. This proves $y \notin \Sigma_\infty(x)$, so

$$\Sigma_\infty(x) \subset \alpha(x).$$

Claim (c). Suppose $A \in \mathscr{F}$ is a union of Σ_∞-atoms. Now (b) stops A from splitting Σ_n-atoms, so $A \in \Sigma_n$, forcing $A \in \Sigma_\infty$.   ★

(154) Lemma. (a) *Points x and y are in the same $\mathscr{F}(t)$-atom iff*

$$X(s, x) = X(s, y) \quad for\ 0 \leq s \leq t.$$

(b) *Points x and y are in the same $\mathscr{F}(t+)$-atom iff there is a positive* $\varepsilon = \varepsilon(x, y)$ *such that*

$$X(s, x) = X(s, y) \quad for\ 0 \leq s \leq t + \varepsilon.$$

PROOF. Check (a). Then use (153b) to get (b).   ★

Temporarily,

$$X(\infty, z) = 0.$$

(155) Lemma. *Let σ be a strict Markov time.*

(a) *Then x and y are in the same $\mathscr{F}(\sigma)$-atom iff $\sigma(x) = \sigma(y)$ and*

$$X(t, x) = X(t, y) \quad for\ 0 \leq t \leq \sigma(x).$$

(b) *If $\sigma(x) = u < \infty$, and*

$$X(t, x) = X(t, y) \quad for\ 0 \leq t \leq u,$$

then $\sigma(y) = u$.

Let τ be a Markov time.

(c) *Then x and y are in the same $\mathcal{F}(\tau+)$-atom iff $\tau(x) = \tau(y)$ and there is an $\varepsilon = \varepsilon(x, y) > 0$ such that*

$$X(t, x) = X(t, y) \quad for \ 0 \leq t \leq \tau(x) + \varepsilon.$$

(d) *If $\tau(x) = u < \infty$, and $\varepsilon > 0$, and*

$$X(t, x) = X(t, y) \quad for \ 0 \leq t \leq u + \varepsilon,$$

then $\tau(y) = u$.

PROOF. *Claim (a).* Suppose x and y are in the same $\mathcal{F}(\sigma)$-atom. Then $\sigma(x) = \sigma(y)$ because σ is $\mathcal{F}(\sigma)$-measurable, and

$$X(t, x) = X(t, y) \quad for \ 0 \leq t \leq \sigma(x)$$

because

$$\{X(t) = v \text{ and } \sigma \geq t\} \in \mathcal{F}(\sigma).$$

Conversely, suppose $\sigma(x) = \sigma(y) = u$ say, and

$$X(t, x) = X(t, y) \quad for \ 0 \leq t \leq u.$$

Let $A \in \mathcal{F}(\sigma)$. I have to show that x and y are both in or both out of A. But

$$A \cap \{\sigma \leq u\} \in \mathcal{F}(u).$$

As (154a) shows, x and y are in the same $\mathcal{F}(u)$-atom: so both are in or both are out of $A \cap \{\sigma \leq u\}$. Since both are in $\{\sigma \leq u\}$, it follows that both are in or both are out of A.

Claim (b). The set $\{\sigma = u\}$ is in $\mathcal{F}(u)$, and can't split $\mathcal{F}(u)$-atoms. But x and y are in the same $\mathcal{F}(u)$-atom, by (154a).

Claim (c). Use (153), (150), and (a).

Claim (d) is like (b). ★

For the rest of the section, suppose

(156) $\qquad\qquad X(\cdot, x)$ is right continuous for all $x \in \mathcal{X}$.

(157) Proposition. *Suppose* (156), *and suppose* $(\mathcal{X}, \mathcal{F})$ *is Borel.*

(a) $\mathcal{F}(t)$ *is separable, and saturated.*

(b) $\mathcal{F}(t+)$ *is saturated.*

For (c), let σ be a strict Markov time. Define a process Y as follows:

$$Y(t) = X(t) \quad for \ all \ t \quad on \ \{\sigma = \infty\};$$

$$Y(t) = X(t) \quad for \ t \leq \sigma \quad on \ \{\sigma < \infty\};$$

$$= X(\sigma) \quad for \ t \geq \sigma \quad on \ \{\sigma < \infty\},$$

(c) *Y generates $\mathcal{F}(\sigma)$; in particular, $\mathcal{F}(\sigma)$ is separable and saturated. For (d), let τ be a Markov time.*

(d) $\mathscr{F}(\tau+)$ *is saturated.*

Separable and *saturated* are defined in (151).

PROOF. *Claim* (*a*). Use (152) to saturate $\mathscr{F}(t)$.

Claim (*b*). Use (a) and (153c).

Claim (*c*). Let $\mathscr{Y}$ be the σ-field generated by Y. Check that $Y(t)$ is $\mathscr{F}(\sigma)$-measurable, so $\mathscr{Y} \subset \mathscr{F}(\sigma)$. Check that Y is right continuous, so $\mathscr{Y}$ is separable. Using (155a, b), check that $\mathscr{Y}$ has the same atoms as $\mathscr{F}(\sigma)$. Now use (152).

Claim (*d*). Use (c) and (153c) and (150).   ★

NOTE. Suppose (156), and suppose $(\mathscr{X}, \mathscr{F})$ is Borel. Let $0 \leqq \sigma \leqq \infty$ be measurable, and satisfy (155b). Then σ is strict Markov, by (157a). Let $0 \leqq \tau \leqq \infty$ be measurable, and satisfy (155d). Then τ is Markov by (147c) and (157b).

If $(\mathscr{X}, \mathscr{F})$ isn't Borel, this characterization of stopping times fails, as does (157a); analyticity isn't enough. I don't know about (152c).

EXAMPLE. Let A be a non-Borel subset of $[0, 1]$. Let $B = [0, 1]\backslash A$. Let $\mathscr{X}$ be the following subset of $[0, 1] \times \{0, 1\}$:

$$(A \times \{0\}) \cup (B \times \{1\}).$$

For $t \geqq 0$ and $(u, v) \in \mathscr{X}$, let

$$\begin{aligned} X(t, (u, v)) &= u & \text{when } 0 \leqq t \leqq 1 \\ &= u + v(t - 1) & \text{when } t \geqq 1. \end{aligned}$$

Let $\mathscr{F}$ be the smallest σ-field in $\mathscr{X}$ which makes each $X(t)$ measurable. So X is a process with real-valued, continuous sample functions. Let

$$\sigma(u, v) = \tfrac{1}{2}v.$$

I claim:

(a) $\mathscr{F}(1)$ is not saturated;

(b) σ has property (155b) and is $\mathscr{F}$-measurable, but is not strict Markov.

PROOF. Let $\mathscr{B}$ be the full Borel σ-field in $[0, 1] \times \{0, 1\}$. Let $\mathscr{C}$ be the σ-field in $[0, 1] \times \{0, 1\}$ of sets of the form $C \times \{0, 1\}$, where C is a Borel subset of $[0, 1]$. For $\Sigma = \mathscr{B}$ or $\mathscr{C}$, let $\tilde{\Sigma}$ be the σ-field in $\mathscr{X}$ of all sets $\mathscr{X} \cap S$ with $S \in \Sigma$; the atoms of $\tilde{\Sigma}$ are the singletons.

I say $\mathscr{F} = \tilde{\mathscr{B}}$. Indeed, $X(t)$ is $\tilde{\mathscr{B}}$-measurable, so $\mathscr{F} \subset \tilde{\mathscr{B}}$. Conversely,

$$\{(u, v) : (u, v) \in \mathscr{X} \text{ and } u \leqq a\} = \{X(0) \leqq a\} \in \mathscr{F}$$
$$\{(u, v) : (u, v) \in \mathscr{X} \text{ and } v \leqq b\} = \{X(2) - X(1) \leqq b\} \in \mathscr{F};$$

so $\tilde{\mathscr{B}} \subset \mathscr{F}$. Similarly, $\mathscr{F}(1) = \tilde{\mathscr{C}}$.

I say $A \times \{0\}$ is in $\tilde{\mathscr{B}}$ but not in $\tilde{\mathscr{C}}$. First,

$$A \times \{0\} = \mathscr{X} \cap ([0, 1] \times \{0\}).$$

Second, if $C \subset [0, 1]$ and

$$\mathscr{X} \cap (C \times \{0, 1\}) = A \times \{0\},$$

then $C = A$; so C is not Borel.

   Claim (a).   The set $A \times \{0\}$ is in $\mathscr{F}$ and is a union of $\mathscr{F}(1)$-atoms, but is not in $\mathscr{F}(1)$.

   Claim (b).   The time σ is $\mathscr{F}$-measurable. And σ is 0 or $\frac{1}{2}$, according as $X(0) \in A$ or $X(0) \in B$. But

$$\{\sigma = 0\} = A \times \{0\} \notin \mathscr{F}(1). \qquad\qquad \bigstar$$

   Proposition (157c) identifies a generating class for $\mathscr{F}(\sigma)$: sets of the first kind $\{X(t) = j \text{ and } \sigma \geq t\}$, and sets of the second kind $\{X(\sigma) = j\}$. I once thought that sets of the first kind were enough, but this is seldom true.

   EXAMPLE.   Let $(\mathscr{X}, \mathscr{F})$ be the cartesian product of Borel $(0, \infty)$ and $\{2, 3\}$. For $t \geq 0$ and $(u, v) \in \mathscr{X}$, let

$$X(t, (u, v)) = 1 \quad \text{when } 0 \leq t < u$$
$$= v \quad \text{when } t \geq u.$$

So X is a right-continuous process. Let

$$\sigma(u, v) = u \quad \text{and} \quad \xi(u, v) = v,$$

so σ is a strict Markov time. Let Σ be the σ-field generated by σ, and let $\mathscr{E}$ be the σ-field generated by the sets

$$\{X(t) = j \text{ and } \sigma \geq t\}.$$

Clearly, $\Sigma \subset \mathscr{E} \subset \mathscr{F}(\sigma)$. Let $\mathscr{P}$ be the probability on $\mathscr{F}$ such that:

   σ and ξ are independent;

   σ is exponential with parameter 1;

   ξ is 2 or 3 with probability $\frac{1}{2}$ each.

Parenthetically, $\mathscr{P}$ makes X a Markov chain. I claim:

   (a) ξ is constant on $\mathscr{E}$-atoms;

   (b) each $\mathscr{E}$-set differs by a $\mathscr{P}$-null set from a Σ-set;

   (c) ξ is $\mathscr{P}$-independent of $\mathscr{E}$;

   (d) $\mathscr{E}$ is inseparable.

   PROOF.   *Claim (a).*   Suppose $\xi(x) = j \neq \xi(y)$. If $\sigma(x) \neq \sigma(y)$, then x and y can even be separated by a Σ-set. So let $\sigma(x) = \sigma(y) = t$. Then x and y are separated by the $\mathscr{E}$-set $\{X(t) = j \text{ and } \sigma \geq t\}$.

Claim (*b*).   The basic $\mathscr{E}$-set $\{X(t) = j \text{ and } \sigma \geq t\}$ differs by a $\mathscr{P}$-null set from the set $\{X(t) = j \text{ and } \sigma > t\}$. This set is empty unless $j = 1$, in which case this set reduces to $\{\sigma > t\}$. Either way, this set is in Σ.

Claim (*c*).   Use (b).

Claim (*d*).   Use (a, c) and (152).   ★

For the rest of this section, let I be a countably infinite set. Let $V = I \cup \{\varphi\}$ be the one-point compactification of discrete I.

EXAMPLE.   Let $(\mathscr{X}, \mathscr{F})$ be the Borel space of sequences of ± 1. Let $s_n(x) = x(n)$ for $n = 1, 2, \ldots$ and $x \in \mathscr{X}$. For $t \geq 0$ and $x \in \mathscr{X}$, let

$$X(t, x) = 0 \qquad \text{when } t \geq 1$$
$$= s_n(x)n \quad \text{when } \frac{1}{n+1} \leq t < \frac{1}{n} \quad \text{and} \quad n = 1, 2, \ldots$$
$$= \varphi \qquad \text{when } t = 0.$$

So X is a right-continuous process. As everybody knows, $\mathscr{F}(0+)$ is inseparable.

PROOF.   Let $\mathscr{P}$ be the probability on $\mathscr{F}$ which makes the s_n independent and ± 1 with probability $\frac{1}{2}$ each. Let Σ be the tail σ-field in $\mathscr{X}$. Each Σ-atom is a countable set: x and y are in the same Σ-atom iff

$$s_n(x) = s_n(y) \quad \text{for all } n \geq n(x, y),$$

by (153b). So $\mathscr{P}$ assigns measure 0 to each atom of Σ. But $\mathscr{P}$ is 0-1 on Σ, by Kolmogorov. Now (10.17) forestalls the separability of Σ. You have to check $\Sigma = \mathscr{F}(0+)$.   ★

(158) Proposition.   *Suppose* (156). *Then*

(a) $\{X(t) \in I\} \cap \mathscr{F}(t) = \{X(t) \in I\} \cap \mathscr{F}(t+)$.

More generally, for strict Markov σ,

(b) $\{X(\sigma) \in I\} \cap \mathscr{F}(\sigma) = \{X(\sigma) \in I\} \cap \mathscr{F}(\sigma+)$.

PROOF.   *Claim* (*a*).   Let $A \subset \{X(t) \in I\}$ and $A \in \mathscr{F}(t+)$. I have to get $A \in \mathscr{F}(t)$. Let

$$B_n = \left\{ X(t) \in I \text{ and } X(s) = X(t) \text{ for } t \leq s \leq t + \frac{1}{n} \right\}.$$

Using (156), you can get $B_n \in \mathscr{F}\left(t + \frac{1}{n}\right)$ and $B_n \uparrow \{X(t) \in I\}$. Because $A \in \mathscr{F}\left(t + \frac{1}{n}\right)$, you can use the monotone class argument to find $A_n \in \mathscr{F}(t)$ with $A_n \subset \{X(t) \in I\}$, such that

$$A \cap B_n = A_n \cap B_n \quad \text{for all } n.$$

Check $A = \liminf A_n \in \mathscr{F}(t)$.

Claim (b).   Let $A \subset \{X(\sigma) \in I\}$ and $A \in \mathscr{F}(\sigma+)$. I need $A \in \mathscr{F}(\sigma)$. But

$$\{A \text{ and } \sigma \leqq t\} = \{A \text{ and } \sigma < t\} \cup \{A \text{ and } \sigma = t\}.$$

The first set on the right is in $\mathscr{F}(t)$ by definition. The second one is at first sight only in $\mathscr{F}(t+)$, but (a) gets it into $\mathscr{F}(t)$.   ★

(159) Proposition.   *Suppose* (156), *and suppose* $X(t, x) \in I$ *for all* $t \geqq 0$ *and all* $x \in \mathscr{X}$. *Then every Markov time is strict.*

PROOF.   Use (147c) and (158a).   ★

NOTE.   Suppose (156). Let τ be a Markov time. Suppose $\tau(x) = \infty$ or $X[\tau(x), x] \in I$, for all $x \in \mathscr{X}$. Then τ is strict, as in (158b). This sharpens (159).

NOTE.   Suppose $\mathscr{P}$ is a probability on $(\mathscr{X}, \mathscr{F})$, which makes X an I-valued Markov chain: so $\mathscr{P}\{X(t) \in I\} = 1$ for all t; and $\mathscr{F}(t+)$ is larger than $\mathscr{F}(t)$ only on a $\mathscr{P}$-null set, which depends on t. There is (181) a strict Markov σ with $\mathscr{F}(\sigma+)$ really larger than $\mathscr{F}(\sigma)$; and (183) a Markov τ which is really different from any strict Markov time.

7.   CROSSING THE INFINITIES

(160) Example.   **(a)** *Description.*   The states are the pairs of integers. Starting from (a, b), the process moves successively through (a, b), $(a, b + 1), \ldots$. This defines the process only a finite interval $[0, \theta)$. Let S be a stochastic matrix on the integers. At θ, choose a' from $S(a, \cdot)$, independent of the past sample function, and restart the construction from $(a', -\infty)$. See Figure 4.

   (b) *State space.* $I = \{(u, v) : u \text{ and } v \text{ are integers}\}$.

   (c) *Holding times.* $q(u, v) = r(v)$, where $0 < r(v) < \infty$, and

$$\Sigma_{v=-\infty}^{\infty} 1/r(v) < \infty.$$

   (d) *Generator.*

$$Q[(u, v), (u, v)] = -r(v);$$
$$Q[(u, v), (u, v + 1)] = r(v);$$
$$Q[(u, v), (u', v')] = 0 \quad \text{unless } u' = u \text{ and } v' = v \text{ or } v + 1.$$

   (e) *Formal construction.*   Define, C, $<$, and M as in (55). Let $i = (a, b) \in I$.

(161)   Let Ω_i be the set of $\omega \in \Omega$, as defined in (57), such that: $\omega(0, n) = (a, b + n)$ for $n = 0, 1, \ldots$; the first coordinate $\zeta(m, \omega)$ of $\omega(m, n)$ depends on m, but not on n; $\omega(m, n) = (\zeta(m, \omega), n)$ for positive m and integer n.

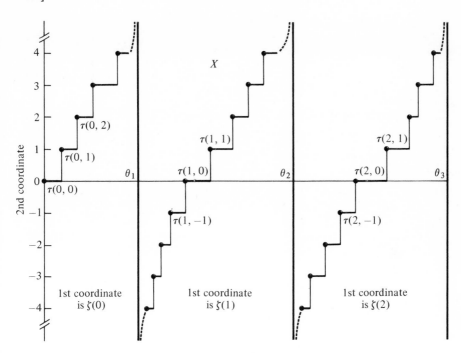

Figure 4.

(162)   Define the Γ_i of (58) by the requirements that $\Gamma_i(\Omega_i) = 1$, while $\{\zeta(m): m = 0, 1, \ldots\}$ is a discrete-time Markov chain with stationary transitions S and starting state a, relative to Γ_i.

You should check (48–51) and (63–66). Then use (108).   ★

Here are some of the features of (160).

(163) Q is silent about S.

(164)   Let $z = (0, 0) \in I$.

(165)   If $S(0, 0) = 1$,   then $P(t, z, (1, 0)) = 0$ for all $t > 0$.
        If $S(0, 1) > 0$,   then $P(t, z, (1, 0)) > 0$ for all $t > 0$.

Press (165) harder.

(166)   The classification of states as transient, null recurrent, or positive recurrent, and the partition into communicating classes, all depend on S.

So (163) prevents Q from determining these properties.

I would now like to consider the strong Markov properties of (160). It will help to clean the sample functions. Review (68–69). Let W_0 be the set of

$w \in W$ such that

$$\lambda(c, w) < \infty \quad \text{for all } c \in C,$$

and

$$\Sigma_{c \in C} \, w(c) = \infty.$$

Review (73, 161, 164). Check $\eta_{q(\omega)}(W_0) = 1$ for $\omega \in \Omega_z$. Let

$$\mathscr{X}_z = \Omega_z \times W_0, \quad \text{with the product } \sigma\text{-field.}$$

Review (75, 162). Check

(167) $\pi_z\{\mathscr{X}_z\} = 1, \quad \text{and} \quad \mathscr{X}_z \text{ is Borel.}$

Define $\zeta(m)$ on $\mathscr{X}_z$:

$$\zeta(m)(\omega, w) = \zeta(m, \omega).$$

Remember (71) that

$$\tau(m, n)(\omega, w) = w(m, n) \quad \text{and} \quad \lambda(m, n)(\omega, w) = \lambda(m, n)(w).$$

Retract these functions to $\mathscr{X}_z$. Use (76, 162):

(168)   The variables $\tau(m, n)$ with $(m, n) \in C$ and the stochastic process $\{\xi(m) : m = 0, 1, \ldots\}$ are all π_z-independent; $\tau(m, n)$ is exponential with parameter $r(n)$; and ζ is Markov with stationary transitions S and starting state 0.

Let $M = 1, 2, \ldots$.

(169a)   Let $\mathscr{A}(M)$ be the σ-field in $\mathscr{X}_z$ spanned by

$$\{\zeta(m) : m < M\} \text{ and } \{\tau(m, n) : (m, n) \in C \text{ and } m < M\}.$$

(169b)   Let $\mathscr{A}(M, N)$ be the σ-field in $\mathscr{X}_z$ spanned by

$$\{\zeta(m) : m \leq M\} \text{ and } \{\tau(m, n) : (m, n) \in C \text{ and } (m, n) < (M, N)\}.$$

(169c)   Let $\mathscr{A}(M, -\infty) = \bigcap_{N=1}^{\infty} \mathscr{A}(M, -N)$.

NOTE. $\zeta(M)$ is $\mathscr{A}(M, -\infty)$-measurable, but not $\mathscr{A}(M)$-measurable. Use (168):

(170) $S[\zeta(M-1), k]$ is a version of $\pi_z\{\zeta(M) = k \mid \mathscr{A}(M)\}$.

On $\mathscr{X}_z$, let

(171) $\theta_M = \Sigma \{\tau(m, n) : (m, n) \in C \text{ and } m < M\}.$

Check

(172) $0 = \theta_0 < \theta_1 < \theta_2 < \cdots \quad \text{and} \quad \theta_m \to \infty.$

Review (72). Retract $X(t)$ to $\mathscr{X}_z$. You should check

(173)   *Description of $X(\cdot, x)$ for $x \in \mathscr{X}_z$:*

(a) $X(t, x) = \varphi$ iff $t = \theta_m(x)$ for some $m = 1, 2, \ldots$;

(b) $X(\cdot, x)$ moves through $\{(0, n):n = 0, 1, \ldots\}$ in order on $[0, \theta_1(x))$, the holding time in $(0, n)$ being $\tau(0, n)(x)$;

(c) $X(\cdot, x)$ moves through $\{(\zeta(m, x), n):n = \ldots, -1, 0, 1, \ldots\}$ in order on $(\theta_m(x), \theta_{m+1}(x))$, the holding time in $(\zeta(m, x), n)$ being $\tau(m, n)(x)$, for $m = 1, 2, \ldots$.

Use $(\mathscr{X}_z, \pi_z)$ for the probability triple of Section 6. The mass is 1 and the σ-field is Borel by (167). Use the present X for the process of Section 6. Use (173a):

(174) $\theta_M(x)$ is the time of the Mth visit to φ by $X(\cdot, x)$.

I claim

(175) θ_M is strict Markov: definitions (171, 146a).

Indeed, let M be a positive integer and let t be a positive real number. The event $\theta_M \leqq t$ coincides with this event. There is a rational $r < t$ such that $\theta_{M-1} \leqq r$, and for any finite subset J of I there is a rational s with

$$r < s < t \quad \text{and} \quad X(s) \notin J.$$

I claim

(176) $\mathscr{F}(\theta_M) = \mathscr{A}(M)$: definitions (146c, 169a).

Indeed, $\mathscr{F}(\theta_M)$ is separable by (157c) and $\mathscr{A}(M)$ is separable by inspection. The atoms of $\mathscr{F}(\theta_M)$ and $\mathscr{A}(M)$ coincide by (155a) and (173). Now use (152).
  Review (68–71) and use (173):

(177) $\lambda(M, N)(x)$ is the least $t > \theta_M(x)$ such that the second coordinate of $X(t, x)$ is N.

So

(178) $\lambda(M, N)$ is a strict Markov time, and $\lambda(M, N) \downarrow \theta_M$ as $N \downarrow -\infty$.

As in (176),

(179) $\mathscr{F}[\lambda(M, N)] = \mathscr{A}(M, N)$: definitions (146c, 169b).

Use (149):

(180) $\mathscr{F}(\theta_M+) = \mathscr{A}(M, -\infty)$: definitions (146d, 169c).

(181) Proposition.   *θ_1 is strict Markov. If $S(0, \cdot)$ is nontrivial, then the π_z-measure algebra of $\mathscr{F}(\theta_1+)$ is strictly larger than the π_z-measure algebra of $\mathscr{F}(\theta_1)$.*

PROOF.   Use (175) for the first claim. For the second, $\zeta(1)$ is $\mathscr{F}(\theta_1+)$-measurable by (180). Use (170, 176) to see that $\zeta(1)$ is π_z-independent of $\mathscr{F}(\theta_1)$, and has π_z-distribution $S(0, \cdot)$.   ★

Let Y be the post-θ_2 process:

$$Y(t, x) = X[\theta_2(x) + t, x].$$

(182) Proposition. θ_2 *is a strict Markov time. The post-θ_2 process Y is identically φ at time 0. But Y is dependent on $\mathscr{F}(\theta_2)$, so on $\mathscr{F}(\theta_2+)$, provided*

$$S(0, 1) > 0 \quad and \quad S(0, 2) > 0 \quad and \quad S(1, 3) \neq S(2, 3).$$

PROOF.   Use (175) for the first assertion, and (173a) for the second. For the third, I say that $\zeta(2)$ is measurable on Y. Indeed, (173) makes $\zeta(2)$ the first coordinate of $Y(t)$ for all small positive t. Combine (170, 176):

$$\pi_z\{\zeta(2) = 3 \mid \mathscr{F}(\theta_2)\} = S(1, 3) \quad \text{on } \{\zeta(1) = 1\}$$

$$= S(2, 3) \quad \text{on } \{\zeta(1) = 2\}.$$

From (168),

$$\pi_z\{\zeta(1) = 1\} = S(0, 1) \quad \text{and} \quad \pi_z\{\zeta(1) = 2\} = S(0, 2). \qquad \bigstar$$

Let τ be the least θ_m if any with $\zeta(m) = 1$, and $\tau = \infty$ if none.

(183) Proposition. τ *is Markov in the sense of* (146b). *Let*

$$S(j, k) = \frac{1}{K + 1} \quad for \; j, k = 0, 1, \ldots, K.$$

Then $\pi_z\{\tau < \infty\} = 1$. But

$$\pi_z\{\sigma = \tau\} \leq \frac{1}{K + 1}$$

for any strict Markov time σ.

NOTE. $$\pi_z\{\theta_1 = \tau\} = \frac{1}{K + 1}.$$

PROOF.   Use (173) for the first assertion, and (168) for the second. For the third, let σ^* be a strict Markov time. Let $\theta_\infty = \infty$. Let σ be the least $\theta_m \geq \sigma^*$ for $m = 1, 2, \ldots, \infty$. Then σ is strict Markov,

$$\sigma = \theta_\nu \quad \text{for random } \nu = 1, 2, \ldots, \infty,$$

and

$$\pi_z\{\sigma^* = \tau\} \leq \pi_z\{\sigma = \tau\}.$$

Use (148e, 175):

$$\{\nu = m\} = \{\sigma = \theta_m\} \in \mathscr{F}(\theta_m) \quad \text{for } m = 1, 2, \ldots.$$

Use (170, 176):

$$\pi_z\{\nu = m \text{ and } \zeta(m) = 1\} = \pi_z\{\nu = m\} \frac{1}{K + 1}.$$

Abbreviate $J = \{1, 2, \ldots \}$. Then

$$\{\sigma = \tau\} \subset \bigcup_{m \in J} \{v = m \text{ and } \zeta(m) = 1\}.$$

So

$$\pi_z\{\sigma = \tau\} \leqq \Sigma_{m \in J} \; \pi_z\{v = m \text{ and } \zeta(m) = 1\}$$

$$= \Sigma_{m \in J} \; \pi_z\{v = m\} \frac{1}{K+1}$$

$$\leqq \frac{1}{K+1}.$$

★

7

THE STABLE CASE

1. INTRODUCTION

In this chapter, unless I say otherwise, let P be a standard stochastic semi-group on the countable set I, with all states stable:

$$q(i) = -P'(0, i, i) < \infty \quad \text{for all } i.$$

The first problem is to create a P-chain X, all of whose sample functions are regular: continuous from the right, with limits from the left at all times, when discrete I has been compactified by adjoining the point at infinity φ. This is done in Section 2. To see why compactification is a good idea, look at (6.142).

Let $Q(i, j) = P'(0, i, j)$, and let

$$\Gamma(i, j) = Q(i, j)/q(i) \quad \text{for } i \neq j \quad \text{and} \quad q(i) > 0$$
$$= 0 \qquad\qquad \text{elsewhere.}$$

Call Γ the *jump matrix*. Suppose the chain X starts from i with $q(i) > 0$. Let τ be the time of first leaving i. Let

$$Y(t) = X(\tau + t).$$

Call Y the *post-exit* process. Then:

 τ and Y are independent;

 τ is exponential with parameter $q(i)$;

 $Y(0) = j$ with probability $\Gamma(i, j)$;

 Y is Markov chain with stationary transitions P and regular sample functions;

 $Y(0) = \varphi$ is a distinct possibility.

This theorem is proved in Section 3.

I want to thank Howard Taylor and Victor Yohai for checking the final draft of this chapter.

Let τ be a general Markov time. Given $X(\tau) = j \in I$, the pre-τ sigma field and the post-τ process are conditionally independent. The post-τ process is a P-chain starting from j, all of whose sample functions are regular. This is proved in Section 4.

Let Q be a matrix on I, with

$$q(i) = -Q(i, i) \geqq 0 \quad \text{for all } i$$
$$Q(i, j) \geqq 0 \quad \text{for all } i \neq j$$
$$\Sigma_j \, Q(i, j) = 0 \quad \text{for all } i.$$

Then there is a minimal standard substochastic semigroup $\bar{P}$ with generator Q. If $\bar{P}$ isn't stochastic, there are a continuum of different standard stochastic semigroups P with generator Q. You can manufacture $\bar{P}$ as follows. Let

$$\Gamma(i, j) = Q(i, j)/q(i) \quad \text{for } i \neq j \quad \text{and} \quad q(i) > 0$$
$$= 0 \qquad\qquad \text{elsewhere.}$$

Start a chain jumping according to Γ, and waiting according to q. If the chain only covers part of the line according to this program, too bad for it. The transitions of this chain are $\bar{P}$. You pick up the other solutions by continuing the construction in different ways, as in (6.132). These results are proved in Section 5.

Here are the main results of Section 6. First,

$$t \to P(t, i, j)$$

is continuously differentiable. Second,

$$P'(t) = QP(t) \quad \text{for some } t > 0$$

iff there are jumps to φ on almost no sample functions, iff

$$\Sigma_j \, Q(i, j) = 0 \quad \text{for all } i.$$

Third,

$$P'(t) = P(t)Q \quad \text{for some } t > 0$$

iff there are jumps from φ on almost no sample functions. In fact,

$$f(t, i, j) = P'(t, i, j) - \Sigma_k P(t, i, k)Q(k, j)$$

is the renewal density for jumps from φ to j, in a chain starting from i.

2.   REGULAR SAMPLE FUNCTIONS

Throughout this chapter, except as noted in Section 5, let P be a standard stochastic semigroup on the finite or countably infinite set I. As (5.10) states, $0 \leqq q(i) = -P'(0, i, i)$ exists; throughout this chapter, except in (3–5) and

(18–20), make the

(1) Assumption. $q(i) < \infty$ for all i.

As (5.14) states, $P'(0) = Q$ exists and is finite, with $Q(i,j) \geq 0$ for $i \neq j$ and $\Sigma_j Q(i,j) \leq 0$. Give I the discrete topology and let $\bar{I} = I \cup \{\varphi\}$ be the one-point compactification of I for infinite I. The state φ is called *infinite* or *fictitious* or *adjoined* by contrast with *finite* or *real* states $i \in I$. Let $\bar{I} = I$ for finite I. Let f be a function from $[0, \infty)$ to $\bar{I}$.

(2) Definition. *Say f is* regular *iff*:

    (a) $\lim f(s) = f(t)$ *as s decreases to t, for all $t \geq 0$;*

    (b) $\lim f(s)$ *exists as s increases to t, for all $t > 0$.*

The main point of this section is to construct a Markov chain with stationary transitions P, starting from any i, such that all the sample functions are regular. This result is essentially due to Doob (1942) and Lévy (1951). For another treatment, see (Chung, 1960, II.5 and II.6).

Let R be the set of binary rationals in $[0, \infty)$, namely,

$$R = \{r : r = a2^{-b} \text{ for some nonnegative integers } a \text{ and } b\}.$$

Let Ω be the set of all functions from R to I. Endow I with the σ-field of all its subsets, and Ω with the product σ-field. Of course, Ω is Borel. Let $\{X(r) : r \in R\}$ be the coordinate process on Ω, namely,

$$X(r)(\omega) = \omega(r) \quad \text{for } \omega \in \Omega \text{ and } r \in R.$$

For each $i \in I$, let P_i be the probability on Ω for which $\{X(r) : r \in R\}$ is Markov with stationary transitions P and $X(0) = i$: for $0 = r_0 < r_1 < \cdots < r_n$ in R and $i_0 = i, i_1, \ldots, i_n$ in I,

(3) $P_i\{X(r_m) = i_m \text{ for } m = 0, \ldots, n\} = \prod_{m=0}^{n-1} P(r_{m+1} - r_m, i_m, i_{m+1})$.

By convention, an empty product is 1.

Let $A(i, s) = \{\omega : \omega(r) = i \text{ for } 0 \leq r \leq s\}$.

(4) Lemma. *$A(i, s)$ is measurable and $P_i\{A(i, s)\} = e^{-q(i)s}$, even without (1).*

PROOF. Suppose $s \in R$. Let n be so large that $N = 2^n s$ is a positive integer, and let

$$A(n, i, s) = \{\omega : \omega \in \Omega \text{ and } \omega(m/2^n) = i \text{ for } m = 0, \ldots, N\}.$$

Plainly,

(5) $P_i\{A(n, i, s)\} = [P(2^{-n}, i, i)]^N$.

As n increases, $A(n, i, s)$ decreases to $A(i, s)$, while the right side of (5) converges to $e^{-q(i)s}$. You move s.   ★

ASSUMPTION (1) IS NOW IN FORCE.

For $r \in R$ and $\varepsilon > 0$, let $G(r, \varepsilon)$ be the set of $\omega \in \Omega$ such that $\omega(s) = \omega(r)$ for all $s \in R$ with $|s - r| \leq \varepsilon$. Let

$$\Omega_g = \bigcap_{r \in R} \bigcup_{\varepsilon > 0} G(r, \varepsilon).$$

(6) Lemma. Ω_g is measurable, and $P_i\{\Omega_g\} = 1$.

PROOF. The set Ω_g is measurable, because ε can be confined to a sequence tending to 0. The main thing to prove is

$$\lim_{\varepsilon \to 0} P_i\{G(r, \varepsilon)\} = 1$$

for each $r \in R$. To avoid trivial complications, suppose $r > 0$. Fix a positive binary rational ε less than r. Using (4) and a primitive Markov property,

$P_i\{G(r, \varepsilon) \text{ and } X(r) = j\}$

$\quad = P_i\{X(r - \varepsilon) = j \quad \text{and} \quad X(r - \varepsilon + s) = j \text{ for } s \in R \text{ with } 0 \leq s \leq 2\varepsilon\}$

$\quad = P_i\{X(r - \varepsilon) = j\} \cdot P_j\{X(s) = j \text{ for } s \in R \text{ with } 0 \leq s \leq 2\varepsilon\}$

$\quad = P(r - \varepsilon, i, j) \cdot e^{-2q(j)\varepsilon}.$

Sum out j, and use Fatou; or note that $P(r - \varepsilon, i, \cdot) \to P(r, i, \cdot)$ in norm as $\varepsilon \to 0$.   ★

The variable U is *geometric with parameter* p iff U takes the value u with probability $(1 - p)p^u$ for $u = 0, 1, \dots$.

(7) Lemma. *Let* U_n *be geometric with parameter* $p_n < 1$. *Let* $p_n \to 1$. *Let* $0 \leq q \leq \infty$. *Let* $a_n > 0$ *and* $a_n \to 0$ *so that*

$$(1 - p_n)/a_n \to q.$$

Then the distribution of $a_n U_n$ *converges to exponential with parameter* q.

PROOF. Easy.   ★

For $\omega \in \Omega_g$ and $r \in R$, there is a maximal interval of $s \in R$ with r as interior point and $\omega(s) = \omega(r)$. Let u and v be the endpoints of this interval, which depend on ω. If $r > 0$, then $u < r < v$ and $\omega(s) = \omega(r)$ for all $s \in R$ with $u < s < v$, and this is false for smaller u or larger v. Either $u = 0$ or u is binary irrational; either $v = \infty$ or v is a binary irrational. The changes for $r = 0$ should be clear. If $\omega(r) = j$, the interval $(u, v) \cap R$ will be called a *j-interval* of ω. Let $\Omega_{j,s}$ be the set of $\omega \in \Omega_g$ such that only finitely many j-intervals of ω have nonempty intersection with $[0, s]$. Let

$$\Omega_v = \bigcap_{j \in I} \bigcap_{s > 0} \Omega_{j,s}.$$

(8) Lemma. Ω_v is measurable and $P_i\{\Omega_v\} = 1$.

PROOF. In the definition of Ω_v, the index s can be confined to R without changing the set. Consequently, it is enough to prove that $\Omega_{j,s}$ is measurable and $P_i\{\Omega_{j,s}\} = 1$. To avoid needless complications, suppose $s = 1$ and $q(j) > 0$. Let $A(N)$ be the set of $\omega \in \Omega_g$ such that N or more j-intervals of ω have nonempty intersection with $[0, 1]$. Then $A(N)$ decreases to $\Omega_g \backslash \Omega_{j,1}$ as N increases. Because Ω_g is measurable and $P_i\{\Omega_g\} = 1$ by (6), it is enough to prove that $A(N)$ is measurable and $P_i\{A(N)\} \to 0$ as $N \to \infty$. Let $Y(n, m) = X(m2^{-n})$, so $Y(n, 0), Y(n, 1), \ldots$ is a discrete time Markov chain with transitions $P(2^{-n})$ starting from i, with respect to P_i. For the moment, fix n. A *j-sequence* of ω is a maximal interval of times m for which $Y(n, m)(\omega) = j$. Let $C_1, C_2, \ldots$ be the cardinalities of the first, second, $\ldots$ j-sequence. Of course, the C's are only partially defined. Let $A(n, N)$ be the set of $\omega \in \Omega_g$ such that N or more j-sequences of ω have nonempty intersection with the interval $m = 0, \ldots, 2^n$. Plainly, $A(n, N)$ is measurable. On $A(n, N)$, there are N or more j-sequences, of which the first $N - 1$ are disjoint subintervals of $0, \ldots, 2^n$. The vth subinterval covers $(C_v - 1)/2^n$ of the original time scale. Consequently $P_i\{A(n, N)\}$ is no more than the conditional P_i-probability that $\sum_{v=1}^{N-1} 2^{-n}(C_v - 1) \leq 1$, given there are N or more j-sequences. Given there are N or more j-sequences, (1.24) shows $C_1 - 1, \ldots, C_N - 1$ are conditionally P_i-independent and geometrically distributed, with common parameter

$$P(2^{-n}, j, j) = 1 - 2^{-n}q(j) + o(2^{-n}) \quad \text{as } n \to \infty.$$

By (7), the conditional P_i-distribution of $2^{-n}(C_v - 1)$ converges to the exponential distribution with parameter $q(j)$ as $n \to \infty$. As n increases to ∞, however, $A(n, N)$ increases to $A(N)$. Consequently, $A(N)$ is measurable, and $P_i\{A(N)\}$ is at most the probability that the sum of $N - 1$ independent exponential random variables with parameter $q(j)$ does not exceed 1. This is small for large N.   ★

For $\omega \in \Omega_v$ and nonnegative real t, there are only two possibilities as $r \in R$ decreases to t: either

(9) $$\omega(r) \to i \in I;$$

or

(10) $$\omega(r) \to \varphi.$$

If $t \in R$, only (9) can hold, with $i = \omega(t)$. For $\omega \in \Omega_v$ and positive real t, as $r \in R$ increases to t, the only two possibilities are still (9) and (10); if $t \in R$, only (9) can hold, with $i = \omega(t)$.

(11) Definition. *For $\omega \in \Omega_v$ and $t \geq 0$, let*

$$X(t, \omega) = \lim X(r, \omega) \quad \text{as } r \in R \text{ decreases to } t.$$

Plainly, $X(t)$ is a measurable function from Ω_v to $\bar{I}$, and for each $\omega \in \Omega_v$, the function $X(\cdot, \omega)$ is regular in the sense of (2).

(12) Theorem. *Define $X(t)$ by (11). For each $i \in I$, the process $\{X(t): 0 \leq t < \infty\}$ is a Markov chain on the probability triple (Ω_v, P_i), with stationary transitions P, starting from i, such that all sample functions are regular in the sense of (2).*

NOTE. The σ-field on Ω_v is the product σ-field on Ω relativized to Ω_v. And P_i is retracted to this σ-field.

PROOF. In (3), let $r_1, \ldots, r_n$ in R decrease to $t_1, \ldots, t_n$ respectively. By regularity,

$$\{X(r_m) = i_m \text{ for } m = 0, \ldots, n\} \to \{X(t_m) = i_m \text{ for } m = 0, \ldots, n\},$$

where $r_0 = t_0 = 0$. Use Fatou and (8). ★

(13) Lemma. *$t \to X(t)$ is continuous in P_i-probability.*

PROOF. Let $0 < s < t$. Then

$$P_i\{X(t) = X(s)\} = \Sigma_j P(s, i, j) P(t - s, j, j).$$

Let s, t tend to a common, finite limit. Use Fatou to check that the right side of the display tends to 1. ★

(14) Lemma. *Fix positive t. For P_i-almost all ω, the point t is interior to a j-interval for some j.*

PROOF. Suppose $P(t, i, j) > 0$. By (13), given that $X(t) = j$, for P_i-almost all ω, there is a sequence $r_n \in R$ increasing to t with $\omega(r_n) = j$. For $\omega \in \Omega_v$, each r_n is interior to a j-interval of ω, and there are only finitely many j-intervals of ω meeting $[0, t]$. Thus $X(s, \omega) = j$ for $t - \varepsilon \leq s \leq t$, where $\varepsilon = \varepsilon(\omega) > 0$. By (11), there is a sequence $r_n \in R$ decreasing to t, with $\omega(r_n) = j$. Only finitely many j-intervals of ω meet $[0, t + 1]$. Thus, $X(s, \omega) = j$ for $t \leq s \leq t + \varepsilon$, where $\varepsilon = \varepsilon(\omega) > 0$. ★

For (15) and (16), let $(\mathscr{X}, \mathscr{F}, \mathscr{P})$ be any probability triple, and $\{Y(t): 0 \leq t < \infty\}$ any Markov chain on $(\mathscr{X}, \mathscr{F}, \mathscr{P})$ with stationary transitions P.

(15) Corollary. *Let $\mathscr{X}_v$ be the set of $x \in \mathscr{X}$ such that $Y(\cdot, x)$ retracted to R is in Ω_v. Then $\mathscr{X}_v \in \mathscr{F}$ and $\mathscr{P}\{\mathscr{X}_v\} = 1$. For $x \in \mathscr{X}_v$, let $Y^*(t, x) = \lim Y(r, x)$ as $r \in R$ decreases to t. Then $Y^*(\cdot, x)$ is regular, and $\mathscr{P}\{Y(t) = Y^*(t)\} = 1$ for each t.*

PROOF. Without real loss, suppose $\mathscr{P}\{Y(0) = i\} = 1$. Let

$$\mathscr{X}_0 = \{x: Y(\cdot, x) \text{ is } I\text{-valued on } R\}.$$

Plainly, $\mathscr{X}_0 \in \mathscr{F}$ and $\mathscr{P}\{\mathscr{X}_0\} = 1$. Let M be the map which sends $x \in \mathscr{X}_0$

to the function $Y(\cdot, x)$ retracted to R. Then M is $\mathscr{F}$-measurable from $\mathscr{X}_0$ to Ω, the Borel set of functions from R to I. And $\mathscr{P}M^{-1} = P_i$. But $\mathscr{X}_v = M^{-1}\Omega_v$, so $\mathscr{X}_v \in \mathscr{F}$ and

$$\mathscr{P}\{\mathscr{X}_v\} = (\mathscr{P}M^{-1})\{\Omega_v\} = P_i\{\Omega_v\} = 1,$$

using (8).

I still have to argue that

$$\mathscr{P}\{Y(t) = Y^*(t)\} = 1.$$

The $\mathscr{P}$-distribution of $\{Y(s): s \in R \text{ or } s = t\}$ coincides with the P_i-distribution of $\{X(s): s \in R \text{ or } s = t\}$, by (12). And the set of functions ϕ from $R \cup \{t\}$ to $\bar{I}$ with

$$\phi(t) = \lim \phi(r) \quad \text{as } r \in R \text{ decreases to } t$$

is product measurable. Using (11),

$$
\begin{aligned}
1 &= P_i\{X(t) = \lim X(r) \quad \text{as } r \in R \text{ decreases to } t\} \\
&= \mathscr{P}\{Y(t) = \lim Y(r) \quad \text{as } r \in R \text{ decreases to } t\} \\
&= \mathscr{P}\{Y(t) = Y^*(t)\}.
\end{aligned}
$$

★

The Y^* process has an additional smoothness property: each sample function is I-valued and continuous at each $r \in R$. This property is an easy one to secure.

(16) Lemma. *Let Y be a Markov process on $(\mathscr{X}, \mathscr{F}, \mathscr{P})$ with stationary transitions P and regular sample functions. Fix a nonnegative real number t. Let $\mathscr{X}_t$ be the set of $x \in \mathscr{X}$ such that $Y(\cdot, x)$ is continuous and I-valued at t. Then $\mathscr{X}_t \in \mathscr{F}$ and $\mathscr{P}\{\mathscr{X}_t\} = 1$.*

PROOF.   As in (15), using (14).   ★

Return now to the X process of (11). Keep $\omega \in \Omega_v$. I say

(17a) $(t, \omega) \to X(t, \omega)$ is jointly measurable:

that is, with respect to the product of the Borel σ-field on $[0, \infty)$ and the relative σ-field on Ω_v. Indeed, $X(t, \omega) = j \in I$ iff for all n there is a binary rational r with

$$t < r < t + \frac{1}{n} \quad \text{and} \quad X(r, \omega) = j.$$

The sets of constancy

For $i \in \bar{I}$ and $\omega \in \Omega_v$, let

$$S_i(\omega) = \{t : 0 \leqq t < \infty \quad \text{and} \quad X(t, \omega) = i\}.$$

This is a *level set*, or a *set of constancy*.

For $i \in I$, the set $S_i(\omega)$ is either empty or a countable union of intervals $[a_n, b_n)$ with $a_1 < b_1 < a_2 < b_2 < \cdots$. If there are infinitely many such intervals, $a_n \to \infty$ as $n \to \infty$. In more picturesque language, X visits i for the nth time on $[a_n, b_n)$, with *holding time* $b_n - a_n$. Moreover, $S_\varphi(\omega)$ is a nowhere dense set closed from the right. In particular, $S_\varphi(\omega)$ is Borel. This follows more prosaically from Fubini, which also implies

(17b) $\{\omega : \text{Lebesgue } S_\varphi(\omega) = 0\}$ is measurable and has P_i-probability 1.

3. THE POST-EXIT PROCESS

The results of this section are essentially due to Lévy (1951). For another treatment, see Chung (1960, II.15). Here are some preliminaries (18–20). For now, drop the assumption (1) of stability. Let $\{Z(t) : 0 \leqq t < \infty\}$ be an $\bar{I}$-valued process on the probability triple $(\mathscr{X}, \mathscr{F}, \mathscr{P})$.

(18) Definition.   *Drop* (1). *Say Z is* Markov *on* $(0, \infty)$ *with stationary transitions P iff*:

$$\mathscr{P}\{Z(t_n) = i_n \text{ for } n = 0, \ldots, N\}$$
$$= \mathscr{P}\{Z(t_0) = i_0\} \, \Pi_{n=0}^{N-1} \, P(t_{n+1} - t_n, i_n, i_{n+1})$$

for all N and $i_n \in I$ and $0 \leqq t_0 < t_1 < \cdots < t_N$. If also

$$\mathscr{P}\{Z(t) \in I\} = 1 \quad \text{for all } t > 0,$$

then Z is finitary. *By convention, an empty product is* 1.

There is nothing here to prevent $\mathscr{P}\{Z(0) = \varphi\} > 0$, and this is the point of the generalization.

ILLUSTRATION.   Let X be the process of example (6.142). Let θ be the least t with $X(t) = \varphi$, so θ is finite and $X(\theta) = \varphi$ almost surely. Let $Z(s) = X(\theta + s)$. Then Z is finitary and Markov on $(0, \infty)$, with the same stationary transitions as X. But $Z(0) = \varphi$ almost surely.

(19) Criterion. Z is a finitary P-chain in $(0, \infty)$ iff $Z(\varepsilon + \cdot)$ is an ordinary P-chain on $[0, \infty)$ for a sequence of ε decreasing to 0. This holds without (1).

(20) Lemma.   *Let Z be Markov on $(0, \infty)$ with stationary transitions P. Suppose Z is jointly measurable and*

$$\text{Lebesgue } \{t : Z(t, \omega) = \varphi\} = 0$$

for almost all ω. Then Z is finitary, namely, $Z(t) \in I$ almost surely for all $t > 0$. This holds without (1).

PROOF. Let $f(t)$ be the probability that $Z(t) \in I$. I say f is nondecreasing on $[0, \infty)$, because P is stochastic:

$$f(t + s) \geq \Sigma_{i,j \in I} \mathscr{P}\{Z(t) = i \text{ and } Z(t + s) = j\}$$
$$= \Sigma_{i,j \in I} \mathscr{P}\{Z(t) = i\} \cdot P(s, i, j)$$
$$\geq \Sigma_{i \in I} \mathscr{P}\{Z(t) = i\}$$
$$= f(t).$$

By Fubini, $f(t) = 1$ except for a Lebesgue-null set of t. Now the monotonicity forces $f \equiv 1$.   ★

ASSUMPTION (1) IS BACK IN FORCE.

By convention, the inf of an empty set is ∞. Let

$$\tau(\omega) = \inf\{t : X(t, \omega) \neq X(0, \omega)\},$$

the *first holding time*. Check that τ is measurable, and positive everywhere. Let

$$Y(t, \omega) = X[\tau(\omega) + t, \omega] \quad \text{when } \tau(\omega) < \infty.$$

Call Y the *post-exit* process. Clearly, $Y(\cdot, \omega)$ is regular. Moreover, Y is jointly measurable: that is, with respect to the product of the Borel σ-field on $[0, \infty)$ and the relative σ-field on Ω_i. Indeed, Y is the composition of X with $(t, \omega) \to (\tau(\omega) + t, \omega)$. The last function is measurable, because its first component is the sum of two visibly measurable functions:

$$(t, \omega) \to t \quad \text{and} \quad (t, \omega) \to \tau(\omega)$$

And X is jointly measurable by (17a).

Fix $i \in I$ with $q(i) > 0$. Let Ω_i be the set of $\omega \in \Omega_v$ with $X(0, \omega) = i$ and $\tau(\omega) < \infty$. Then $P_i\{\Omega_i\} = 1$ by (4) and (8). Confine ω to Ω_i. The function $X(\tau)$, whose value at ω is $X[\tau(\omega), \omega]$, is measurable on Ω_i, and is either some element j of I other than i, or is φ. In the first case, say the process *jumps to j* on leaving i; in the second, say the process *jumps to φ* on leaving i.

To study this more closely, let

$$\Gamma(i, j) = Q(i, j)/q(i) \quad \text{for } i \neq j$$
$$= 0 \qquad\qquad \text{for } i = j.$$

Call Γ the *jump matrix* of Q.

(21) Theorem. *With respect to P_i, the post-exit process Y and the first holding time τ are independent; Y is Markov on $(0, \infty)$ with transitions P; and τ is exponential with parameter $q(i)$. Moreover,*

$$P_i\{Y(0) = j\} = \Gamma(i, j) \quad \text{for } j \in I.$$

Finally, Y is finitary;

$$P_i\{Y(t) \in I\} = 1 \quad \text{for all } t > 0.$$

PROOF. Keep ω in Ω_i. Let $0 \leq t < \infty$. Let $0 \leq s_0 < s_1 < \cdots < s_M$, and let $i_0, i_1, \ldots, i_M \in I$. The main thing to prove is

(22) $$P_i\{A\} = P_i\{\tau \geq t\} \cdot P_i\{Y(s_0) = i_0\} \cdot \pi,$$

where $A = \{\tau \geq t\} \cap B$ and

$$B = \{Y(s_m) = i_m \text{ for } m = 0, \ldots, M\}$$

and

$$\pi = \Pi_{m=0}^{M-1} P(s_{m+1} - s_m, i_m, i_{m+1}).$$

To begin with, suppose $t > 0$ and $t, s_0, \ldots, s_M \in R$. Let τ_n be the least $m/2^n$ with $X(m/2^n) \neq i$. So $\tau_n \geq 1/2^n$. Let

$$A_n = \{\tau_n \geq t \text{ and } X(\tau_n + s_m) = i_m \text{ for } m = 0, \ldots, M\}.$$

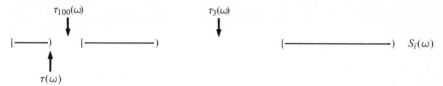

Figure 1.

Because the sample functions are regular, there is an interval to the right of τ free of $S_i = \{t:X(t) = i\}$, as in Figure 1. Check that τ_n is in this interval for large n, and $\tau_n \downarrow \tau$. So $\{\tau_n \geq t\} \downarrow \{\tau \geq t\}$. Using the regularity again,

$$\limsup A_n \subset A \subset \liminf A_n.$$

By Fatou,

(23) $$P_i\{A\} = \lim P_i\{A_n\},$$

and the problem is to compute $P_i\{A_n\}$.

Consider only n so large that $2^n t, 2^n s_0, \ldots, 2^n s_M$ are integers. Then

(24) $$A_n = \bigcup_N \{A_{n,N}: N \geq 2^n t\},$$

where $A_{n,N}$ is the event that $X(m/2^n) = i$ for $m = 0, \ldots, N-1$ and $X(N/2^n) \neq i$ and $X(N2^{-n} + s_m) = i_m$ for $m = 0, \ldots, M$. The problem is to compute $P_i\{A_{n,N}\}$. Let

(25a) $a(n) = P(2^{-n}, i, i) = 1 - q(i)2^{-n} + o(2^{-n})$
(25b) $b(n) = a(n)^{2^n t - 1} = P_i\{\tau_n \geq t\} \to P_i\{\tau \geq t\}$
(25c) $c(n) = [1 - a(n)]^{-1} P_i\{X(2^{-n}) \neq i \text{ and } X(2^{-n} + s_0) = i_0\}.$

Because $\{X(m/2^n): m = 0, 1, \ldots\}$ is Markov with transitions $P(2^{-n})$,

$$P_i\{A_{n,N}\} = a(n)^{N-1}[1 - a(n)]c(n)\pi.$$

Sum out $N = 2^n t, 2^n t + 1, \ldots$ and use (24):

(26) $$P_i\{A_n\} = b(n)c(n)\pi.$$

Put $t = 2^{-n}$ and $M = 0$ in (26):

(27) $$c(n) = P_i\{X(\tau_n + s_0) = i_0\} \to P_i\{Y(s_0) = i_0\}$$

by regularity. Let $n \to \infty$ in (26) and use (23, 25b, 27) to get (22) for $t > 0$ and $t, s_0, \ldots, s_M \in R$. Get the full (22) by a passage to the limit and regularity.

Put $t = 0$ in (22) to see that Y is Markov on $(0, \infty)$ with stationary transitions P. So you can rewrite (22):

(28) $$P_i\{C \text{ and } B\} = P_i\{C\} \cdot P_i\{B\}$$

for $C = \{\tau \geq t\}$. As M and $i_0, \ldots, i_M$ and $s_0, \ldots, s_M$ vary, the B's are closed under intersection, modulo the null set, and generate the same σ-field as Y. The sets C do similar things for τ. Both sides of (28) are countably additive in both arguments. So Y is independent of τ by (10.16). And τ is exponential with parameter $q(i)$ by (4).

From the definition, $P_i\{Y(0) = i\} = 0$. Let $j \neq i$. To compute $P_i\{Y(0) = j\}$, go back to (25–26). Put $s_0 = 0$ and $i_0 = j$. Then

$$\begin{aligned}
c(n) &= [1 - a(n)]^{-1} P_i\{X(2^{-n}) = j\} \\
&= [1 - a(n)]^{-1} P(2^{-n}, i, j) \\
&= [q(i)2^{-n} + o(2^{-n})]^{-1} [Q(i,j)2^{-n} + o(2^{-n})] \\
&= [q(i) + o(1)]^{-1} [Q(i,j) + o(1)] \\
&\to \Gamma(i,j).
\end{aligned}$$

Use (27).

I still have to prove $P_i\{Y(t) \in I\} = 1$ for all $t > 0$. But

$$\{t : Y(t, \omega) = \varphi\}$$

is the translate by $\tau(\omega)$ of a subset of

$$\{t : X(t, \omega) = \varphi\},$$

and has Lebesgue measure 0 for almost all ω by (17b). Use (20).   ★

(29) **Remark.** For each positive t, the function $Y(\cdot, \omega)$ is I-valued and continuous at t for P_i-almost all ω: use (21, 19, 16). So $Y(\cdot, \omega)$ is I-valued and continuous at all positive $r \in R$, for P_i-almost all ω.

(30) **Remark.** The post-exit process is studied again in Section 6, where it is shown that P' exists and is continuous, and

$$P_i\{Y(t) = j\} = P(t, i, j) + q(i)^{-1} P'(t, i, j).$$

This could also be deduced from the proof of (21), as follows. Abbreviate $t = s_0 \geqq 0$ and $j = i_0$ and $f(t) = P(t, i, j)$. Let $\varepsilon_n = 2^{-n}$. Recall that $a(n) = P(\varepsilon_n, i, i)$ from (25). Now

$$P_i\{X(\varepsilon_n) \neq i \text{ and } X(\varepsilon_n + t) = j\} = f(\varepsilon_n + t) - P_i\{X(\varepsilon_n) = i \text{ and } X(\varepsilon_n + t) = j\}$$
$$= [1 - a(n)]f(t) + f(t + \varepsilon_n) - f(t).$$

Recall $c(n)$ from (25). Check

$$c(n) = f(t) + \frac{\varepsilon_n}{1 - a(n)} \cdot \frac{f(t + \varepsilon_n) - f(t)}{\varepsilon_n}.$$

But $c(n) \to P_i\{Y(t) = j\}$ by (27). Consequently,

$$f^*(t) = \lim_{n \to \infty} \frac{f(t + \varepsilon_n) - f(t)}{\varepsilon_n}$$

exists, and

(31) $$P_i\{Y(t) = j\} = f(t) + q(i)^{-1}f^*(t).$$

By regularity, Y is continuous at 0. And Y is continuous with P_i-probability 1 at each $t > 0$ by (29). Consequently, $t \to P_i\{Y(t) = j\}$ is continuous, and therefore f^* is continuous. But ε_n can be replaced by any sequence tending to 0, without affecting the argument much: τ_n is the least $m\varepsilon_n$ with $X(m\varepsilon_n) \neq i$, and $\tau_n \to \tau$ from the right but not monotonically. The limit of the difference quotient does not depend on the sequence, in view of (31). Thus, the right derivative of f exists, and is continuous, being f^*. Use (10.67) to see that f^* is the calculus derivative of f.   ★

The global structure of $\{X(t) : 0 \leqq t < \infty\}$ is not well understood. But the local behavior is no harder than in the uniform case. To explain it, introduce the following notation. At time 0, the process is in some state ξ_0; it remains there for some time τ_0, then jumps to φ or to a new state ξ_1. If the latter, it remains in ξ_1 for some time τ_1, then jumps, and so on. See Figure 2.

More formally, let $\xi_0 = X(0)$ and let

$$\tau_0 = \inf\{t : X(t) \neq \xi_0\}.$$

The inf of an empty set is ∞. Suppose $\xi_0, \ldots, \xi_n$ and $\tau_0, \ldots, \tau_n$ are defined. If $\tau_n = \infty$ or $\tau_n < \infty$ but $X(\tau_0 + \cdots + \tau_n) = \varphi$, then $\xi_{n+1}, \xi_{n+2}, \ldots$ as well as $\tau_{n+1}, \tau_{n+2}, \ldots$ are undefined. If $\tau_n < \infty$ and $X(\tau_0 + \cdots + \tau_n) \in I$, then $\xi_{n+1} = X(\tau_0 + \cdots + \tau_n)$ and

$$\tau_1 + \cdots + \tau_{n+1} = \inf\{t : \tau_1 + \cdots + \tau_n \leqq t \text{ and } X(t) \neq \xi_{n+1}\}.$$

Define the substochastic jump matrix Γ on I by

(32) $\Gamma(i, j) = Q(i, j)/q(i)$ for $q(i) > 0$ and $i \neq j$
                              $= 0$ elsewhere.

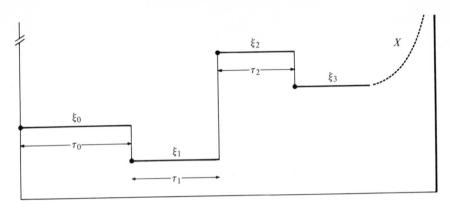

Figure 2.

(33) Theorem.   *With respect to P_i; the process $\xi_0, \ldots$ is a partially defined discrete time Markov chain, with stationary transitions Γ, and starting state i. Given $\xi_0, \ldots$, the holding times $\tau_0, \ldots$ are conditionally P_i-independent and exponential with parameters $q(\xi_0), \ldots$. In other terms, let $i_0 = i, \ldots, i_n \in I$. Given that $\xi_0, \ldots, \xi_n$ are defined, and $\xi_0 = i_0, \ldots, \xi_n = i_n$, the random variables $\tau_0, \ldots, \tau_n$ are conditionally P_i-independent and exponentially distributed, with parameters $q(i_0), \ldots, q(i_n)$.*

PROOF.   Let $t_0, \ldots, t_n$ be positive real numbers. The thing to prove is

$$P_i\{A\} = e^{-s}\,\pi,$$

where

$$A = \{\xi_0 = i_0, \xi_1 = i_1, \ldots, \xi_n = i_n \text{ and } \tau_0 \geqq t_0, \tau_1 \geqq t_1, \ldots, \tau_n \geqq t_n\}$$

and

$$s = q(i_0)t_0 + q(i_1)t_1 + \cdots + q(i_n)t_n$$

and

$$\pi = \Pi_{m=0}^{n-1}\,\Gamma(i_m, i_{m+1}).$$

Let

$$B = \{\xi_0 = i_1, \ldots, \xi_{n-1} = i_n \text{ and } \tau_0 \geqq t_1, \ldots, \tau_{n-1} \geqq t_n\}.$$

Put $i = i_0$ and $j = i_1$ and $t = t_0$. Let Y be the post-exit process. Let $Y_R(\cdot, \omega)$ be $Y(\cdot, \omega)$ retracted to R. I claim

$$P_i\{Y(0) = j \text{ but } Y_R \notin \Omega_v\} = 0.$$

Indeed, $Y(\cdot, \omega)$ is the translate of part of $X(\cdot, \omega)$, and has only finitely many k-intervals in any finite time interval, and is continuous at 0. If $Y(\cdot, \omega)$ is continuous and I-valued at positive $r \in R$, and $Y(\cdot, \omega)$ is I-valued at 0, then $Y_R(\cdot, \omega) \in \Omega_v$. So (29) gets the claim. Exclude this null set. Then

$$A = \{\tau_0 \geqq t \text{ and } X(0) = i \text{ and } Y(0) = j\} \cap \{Y_R \in B\}.$$

Use (21) to see

$$P_i\{A\} = e^{-q(i)t}\,\Gamma(i,j)\,P_j\{B\}.$$

Induct.   ★

NOTE. This result and (5.38) prove (5.45). For $\sup_i q(i) < \infty$ implies $\Sigma\, 1/q(\xi_n) = \infty$.

There is a useful way to restate (33). Define

$$d = \inf\,\{n: \xi_n \notin I\} \quad \text{on } \Omega_v.$$

As in Section 5.6, let $\partial \notin I$. Let $\mathscr{X}$ be the set of pairs (ω, w), where ω is a sequence of elements of $I \cup \{\partial\}$, and w is a sequence of elements of $(0, \infty]$. Let

$$\xi_n(\omega, w) = \omega(n) \quad \text{and} \quad \tau_n(\omega, w) = w(n).$$

Let

$$d = \inf\,\{n: \xi_n = \partial\} \quad \text{on } \mathscr{X}.$$

Define the probability π_i in $\mathscr{X}$ by (5.35). Relative to π_i,

> ξ_n is a discrete time Markov chain with stationary transitions Γ, extended to be stochastic on $I \cup \{\partial\}$, and starting state i; given $\{\xi_n\}$, the holding times τ_n are conditionally independent and exponential, the parameter for τ_n being $q(\xi_n)$.

Then

> the P_i-distribution of $\{(\xi_n, \tau_n): 0 \leqq n < d\}$ coincides with the π_i-distribution of $\{(\xi_n, \tau_n): 0 \leqq n < d\}$.

Suppose $q(i) > 0$ and $\Sigma_j\, Q(i, j) = 0$ for all i. Then $P_i\{d = \infty\} = 1$, so

(33*)   the P_i-distribution of $\{(\xi_n, \tau_n): 0 \leqq n < \infty\}$ is π_i.

4. THE STRONG MARKOV PROPERTY

For a moment, let τ be the first holding time in X. According to (21), the post-τ process Y is a finitary P-chain on $(0, \infty)$, and is independent of τ. The strong Markov property (41) makes a similar but weaker assertion for a much more general class of times τ. In suggestive but misleading language: the post-τ process is conditionally a finitary P-chain on $(0, \infty)$, given the process to time τ. By example (6.182), the post-τ process need not be independent of the process to time τ. Here is the program for proving strong Markov. First, I will prove a weak but computational form of the assertion for constant τ, in (34). Using (34), I will get this computational assertion for random τ in (35). I can then get strong Markov (38) on the set $\{X(\tau) \in I\}$. General strong Markov finally appears in (41).

You may wish to look at Sections 6.6 and 6.7 when you think about the present material. David Gilat forced me to rewrite this section one more time than I had meant to.

Let $\mathscr{F}(t)$ be the σ-field generated by $X(s)$ for $0 \leqq s \leqq t$.

(34) Lemma   *Let* $0 \leqq s_0 < s_1 < \cdots < s_M$ *and let* $i_0, i_1, \ldots, i_M \in I$. *Let* $0 \leqq t < \infty$ *and let* $D \in \mathscr{F}(t)$. *Then*

$$P_i\{D \text{ and } X(t + s_m) = i_m \text{ for } m = 0, \ldots, M\}$$
$$= P_i\{D \text{ and } X(t + s_0) = i_0\} \cdot \pi,$$

where

$$\pi = \Pi_{m=0}^{M-1} P(s_{m+1} - s_m, i_m, i_{m+1}).$$

PROOF.   This is clear from (12) for special D, of the form

$$\{X(t_n) = j_n \text{ for } n = 0, \ldots, N\},$$

where $0 \leqq t_0 < t_1 < \cdots < t_N \leqq t$ and $j_0, j_1, \ldots, j_N \in I$. Now use (10.16). ★

Call a nonnegative random variable τ on Ω_v a *Markov time*, or *Markov*, iff $\{\tau < t\}$ is in $\mathscr{F}(t)$ for all t. Let $\mathscr{F}(\tau+)$ be the σ-field of all measurable sets A such that $A \cap \{\tau < t\}$ is in $\mathscr{F}(t)$ for all t. Call this the *pre-τ sigma field*. Let $\Delta = \{\tau < \infty\}$. Now $\Delta \subset \Omega_v$, for ω is confined to Ω_v. Let $Y(t) = X(\tau + t)$ on Δ. More explicitly,

$$Y(t, \omega) = X[\tau(\omega) + t, \omega] \quad \text{for } \tau(\omega) < \infty.$$

Call Y the *post-τ process*. Clearly, Y is a jointly measurable process with regular sample functions.

WARNING. τ and $\mathscr{F}(\tau+)$ both peek at the future; this point is discussed in Sections 6.6 and 6.7.

Here is a preliminary version of the strong Markov property.

(35) Proposition.   *Suppose* τ *is a Markov time. Let* Y *be the post-τ process. Let*

$$0 \leqq s_0 < s_1 < \cdots < s_M$$

and let $i_0, i_1, \ldots, i_M \in I$. *Let*

$$B = \{Y(s_m) = i_m \text{ for } m = 0, \ldots, M\}$$

and

$$\pi = \Pi_{m=0}^{M-1} P(s_{m+1} - s_m, i_m, i_{m+1}).$$

Let $A \in \mathscr{F}(\tau+)$ *with* $A \subset \Delta$. *Then*

$$P_i\{A \cap B\} = P_i\{A \text{ and } Y(s_0) = i_0\} \cdot \pi.$$

PROOF. Let τ_n be the least $m/2^n$ greater than τ. This approximation differs from the one in (21). Let B_n be the event that $X(\tau_n + s_m) = i_m$ for $m = 0, \ldots, M$. By regularity,

$$\limsup (A \cap B_n) \subset A \cap B \subset \liminf (A \cap B_n),$$

so $P_i\{A \cap B_n\} \to P_i\{A \cap B\}$ by Fatou. Let

$$C_{n,m} = \{(m - 1)/2^n \leq \tau < m/2^n\}.$$

Then

$$P_i\{A \cap B_n\} = \Sigma_{m=1}^{\infty} P_i\{A \cap B_n \cap C_{n,m}\}.$$

But by (34),

$$P_i\{A \cap B_n \cap C_{n,m}\} = P_i\{A \cap C_{n,m} \cap [X(2^{-n}m + s_0) = i_0]\} \cdot \pi,$$

because

$$A \cap C_{n,m} \in \mathcal{F}(m/2^n).$$

Consequently,

$$P_i\{A \cap B_n\} = P_i\{A \text{ and } X(\tau_n + s_0) = i_0\} \cdot \pi.$$

Let $n \to \infty$ and use regularity again. ★

Let Δ_v be the set of $\omega \in \Delta$ such that $Y(\cdot, \omega)$ is continuous when retracted to R, and I-valued when retracted to positive R.

(36) Corollary. (a) *Given Δ and $X(\tau) = j \in I$, the σ-field $\mathcal{F}(\tau+)$ and the post-τ process Y are conditionally P_i-independent, Y being Markov with stationary transitions P and starting state j.*

(b) *With respect to $P_i\{\cdot \mid \Delta\}$, the post-τ process Y is a Markov chain on $(0, \infty)$, with stationary transitions P.*

(c) $P_i\{Y(t) \in I \mid \Delta\} = 1$ *for $t > 0$, so Y is finitary.*
(d) $P_i\{\Delta_v \mid \Delta\} = 1.$

PROOF. Use (35) to get (a). Put $A = \Delta$ in (35) to get (b). Then use (20) to get (c): the set $\{t: Y(t, \omega) = \varphi\}$ is the translate by $\tau(\omega)$ of a subset of $\{t: X(t, \omega) = \varphi\}$, and the latter set is typically Lebesgue null by (17b). Let $r \in R$ be positive. Then $Y(r + \cdot)$ is an ordinary Markov chain on $[0, \infty)$ relative to $P_i\{\cdot \mid \Delta\}$ by (b–c). So $Y(r + \cdot)$ is almost surely continuous and I-valued on R, by (16). Intersect on r to get (d): the continuity at 0 is forced by regularity. ★

I would now like to explain why (35–36) are inadequate. Fix $i \neq j$. Let θ be the length of the first j-interval in X. According to (46) below, the P_i-distribution of θ is exponential with parameter $q(j)$.

PSEUDO PROOF.   Let τ be the least t if any with $X(t) = j$, and $\tau = \infty$ if none. Then τ is a Markov time. Suppose $P_i\{\tau < \infty\} = 1$ for a moment. Let Y be the post-τ process. Let σ be the first holding time in X.

(a) $\theta = \sigma \circ Y$.

(b) The P_i-distribution of Y is P_j.

(c) The P_i-distribution of θ coincides with the P_j-distribution of σ.

(d) The last distribution is exponential with parameter $q(j)$, by (21).   ★

This proof is perfectly sound in principle, but it breaks down in detail. The time domain of a Y sample function is $[0, \infty)$. But σ is defined only on a space of functions with time domain R. And P_j also acts only in this space. So (a) and (b) stand discredited, for the most sophistical of reasons. For a quick rescue, let S be the retraction of Y to time domain R.

REAL PROOF.   (a) $\theta = \sigma \circ S$ on $\{S \in \Omega_v\}$, and $P_i\{S \in \Omega_v\} = 1$.

(b) The P_i-distribution of S is P_j.

(c, d) stay the same.   ★

I would now like to set this argument up in a fair degree of generality. Let $\bar{\Omega}$ be the set of all functions ω from R to $\bar{I}$. Let $\bar{X}(r, \omega) = \omega(r)$ for $\omega \in \bar{\Omega}$ and $r \in R$. Give $\bar{\Omega}$ the product σ-field, namely the smallest σ-field over which all $\bar{X}(r)$ are measurable. Here is the *shift* mapping S from Δ to $\bar{\Omega}$:

$$(37) \qquad \bar{X}(r, S\omega) = Y(r, \omega) = X[\tau(\omega) + r, \omega].$$

You should check that S is measurable. Here is the *strong Markov property* on $\{X(\tau) \in I\}$; the set $\{X(\tau) = \varphi\}$ is tougher, and I postpone dealing with it until (41).

(38) Theorem. *Suppose τ is Markov. Let $\Delta = \{\tau < \infty\}$, and let Y be the post-τ process. Define the shift S by (37).*

(a) $P_i\{\Delta$ *and* $X(\tau) \in I$ *but* $S \notin \Omega_v\} = 0$.

(b) *If* $\omega \in \{\Delta$ *and* $X(\tau) \in I$ *and* $S \in \Omega_v\}$, *then* $Y = X \circ S$; *that is*, $Y(t, \omega) = X(t, S\omega)$ *for all* $t \geq 0$.

(c) *Suppose* $A \in \mathscr{F}(\tau+)$ *and* $A \subset \{\Delta$ *and* $X(\tau) = j \in I\}$. *Suppose B is a measurable subset of* Ω. *Then*

$$P_i\{A \text{ and } S \in B\} = P_i(A) \cdot P_j(B).$$

(d) *Given* Δ *and* $X(\tau) = j \in I$, *the pre-τ sigma field* $\mathscr{F}(\tau+)$ *is conditionally P_i-independent of the shift S, and the conditional P_i-distribution of S is P_j.*

NOTE.   Claims (b–d) make sense, if you visualize Ω as this subset of $\bar{\Omega}$:

$$\{\omega : \omega \in \bar{\Omega} \text{ and } \omega(r) \in I \text{ for all } r \in R\}.$$

PROOF.   *Claim (a).* Suppose $\omega \in \Omega_v$. Then $Y(\cdot, \omega)$ is the translate of part of $X(\cdot, \omega)$, is continuous at 0, and has only finitely many k-intervals on finite intervals. Now use (36d).

Claim (b).   Use the definitions.

Claim (c).   Use (35) to handle the special B, of the form

$$\{X(s_m) = i_m \text{ for } m = 0, \ldots, M\},$$

with $0 \leq s_0 < \cdots < s_M$ and $i_0, \ldots, i_M$ in I. Then use (10.16).

Claim (d).   Use claim (c).   ★

The general statement (41) of strong Markov is quite heavy. Here is some explanation of why a light statement doesn't work. Suppose for a bit that

$$P_i\{\tau < \infty \text{ and } X(\tau) = \varphi\} = 1.$$

Then $\mathcal{F}(\tau+)$ and Y are usually dependent: see (6.182) for an example. At the beginning, I said that Y is conditionally a finitary P-chain on $(0, \infty)$, given $\mathcal{F}(\tau+)$. This is much less crisp than it sounds. To formalize it, I would have to introduce the conditional distribution of Y given $\mathcal{F}(\tau+)$. Unless one takes precautions, this distribution would act only on the meager collection of product measurable subsets in $I^{[0,\infty)}$; it loses the fine structure of the sample functions. Also, to check that a probability on $I^{[0,\infty)}$ is Markov, one would have to look at an uncountable number of conditions; it's hard to organize the work in a measurable way. So, the charming informal statement leads into a morass. My way around is dividing the problem in two. First, retract the time domain of Y to R; call this retract S, for shift. Now S takes its values in the Borel space $\bar{\Omega} = I^R$; we can discuss its conditional distribution with a clear conscience, and in a countable number of moves force these distributions to be finitary P-chains on positive R. Second, we can relate Y to S: except for a negligible set of ω, the sample function $Y(\cdot, \omega)$ is obtained by filling in $S(\omega)(\cdot)$.

The detail work begins here. Remember that $\bar{\Omega}$ is the set of all functions ω from R to $\bar{I}$; and $\bar{X}(r, \omega) = \omega(r)$ for $\omega \in \bar{\Omega}$ and $r \in R$; and $\bar{\Omega}$ is endowed with the product σ-field, namely the smallest σ-field over which all $\bar{X}(r)$ are measurable. Let $\bar{\Omega}_v$ be the set of $\omega \in \bar{\Omega}$ such that:

$\omega(\cdot)$ is continuous at all $r \in R$;

$\omega(\cdot)$ is I-valued at all positive $r \in R$;

$\omega(\cdot)$ has only finitely many j-intervals on finite time intervals, for all $j \in I$.

CRITERION.   Let $\omega \in \bar{\Omega}$. Then $\omega \in \bar{\Omega}_v$ iff $\omega(r + \cdot) \in \bar{\Omega}_v$ for all positive $r \in R$, and $\omega(0) = \lim \omega(r)$ as $r \in R$ decreases to 0.

As (8) implies, $\bar{\Omega}_v$ is measurable. For $\omega \in \bar{\Omega}_v$, let

(39) $\bar{X}(t, \omega) = \lim \bar{X}(r, \omega)$ as $r \in R$ decreases to t.

Remember the mapping S from (37). What does it mean to say that the conditional distribution of S is a finitary P-chain on positive R? To answer this question, introduce the class $\mathbf{P}$ of probabilities μ on $\bar{\Omega}$, having the properties:

(40a) $\mu\{\bar{X}(r) \in I\} = 1$ for all positive $r \in R$;

(40b) $\mu\{\bar{X}(r_n) = i_n \text{ for } n = 0, \ldots, N\}$

$$= \mu\{\bar{X}(r_0) = i_0\} \, \Pi_{n=0}^{N-1} \, P(r_{n+1} - r_n, i_n, i_{n+1})$$

for nonnegative integers N, and $0 \leq r_0 < r_1 < \cdots < r_N$ in R and $i_0, \ldots, i_N$ in I. By convention, an empty product is 1.

CRITERION. $\mu \in \mathbf{P}$ iff for all positive $r \in R$,

$$\mu\{\bar{X}(r) \in I\} = 1,$$

and the μ-distribution of $\{\bar{X}(r + s) : s \in R\}$ is

$$\Sigma_{j \in I} \, \mu\{\bar{X}(r) = j\} \cdot P_j.$$

This makes sense, because $\{s \to \bar{X}(r + s, \omega) : s \in R\}$ is an element of Ω for μ-almost all $\omega \in \bar{\Omega}$. And P_j acts on Ω.

The results of this chapter extend to all $\mu \in \mathbf{P}$ in a fairly obvious way, with $\bar{X}$ replacing X. In particular, $\mu \in \mathbf{P}$ concentrates on $\bar{\Omega}_v$ by (8). And $\mu \in \mathbf{P}$ iff relative to μ, the process $\{\bar{X}(t) : 0 \leq t < \infty\}$ is finitary and Markov on $(0, \infty)$ with stationary transitions P, by (12): the definition is (18).

Here is the *strong Markov property*. Regular conditional distributions are discussed in Section 10.10.

(41) Theorem.   *Suppose τ is Markov. Let $\Delta = \{\tau < \infty\}$, and let Y be the post-τ process. Define S by (37).*

   (a) $P_i\{S \in \bar{\Omega}_v \mid \Delta\} = 1$.

Define $\bar{X}$ by (39).

   (b) *If $\omega \in \{\Delta \text{ and } S \in \bar{\Omega}_v\}$, then $Y = \bar{X} \circ S$; that is,*

$$Y(t, \omega) = \bar{X}(t, S\omega) \quad \text{for all } t \geq 0.$$

On Δ, let $Q(\cdot, \cdot)$ be a regular conditional P_i-distribution for S given $\mathscr{F}(\tau+)$. Remember that $\mathbf{P}$ is the set of probabilities μ on $\bar{\Omega}$ which satisfy (40). Let Δ_P be the set of $\omega \in \Delta$ such that $Q(\omega, \cdot) \in \mathbf{P}$.

   (c) $\Delta_P \in \mathscr{F}(\tau+)$, and $P_i\{\Delta_P \mid \Delta\} = 1$.

PROOF.   *Claim (a).*   As in (38a).
Claim (b).   Use the definitions.

Claim (c). Let $r \in R$ be positive. Let $G(r)$ be the set of $\omega \in \Delta$ satisfying

(42) $$Q(\omega, [\bar{X}(r) \in I]) = 1.$$

I say

(43) $$G(r) \in \mathscr{F}(\tau+) \quad \text{and} \quad P_i\{\Delta \backslash G(r)\} = 0.$$

The measurability is clear. Furthermore,

$$Q(\omega, [\bar{X}(r) \in I]) \leq 1$$

for all $\omega \in \Delta$. Integrate both sides of this inequality over Δ. On the right, you get $P_i\{\Delta\}$. On the left, you get

$$P_i\{\Delta \text{ and } S \in [\bar{X}(r) \in I]\} = P_i\{\Delta \text{ and } Y(r) \in I\} = P_i\{\Delta\}$$

by (36c). So, strict inequality holds almost nowhere, proving (43).

Let $s = (s_0, \ldots, s_M)$ be an $(M+1)$-tuple of elements of R, with $0 \leq s_0 < \cdots < s_M$. Let $i = (i_0, \ldots, i_M)$ be an $(M+1)$-tuple of elements of I. Let

$$B = B(s, i) = \{\bar{X}(s_m) = i_m \text{ for } m = 0, \ldots, M\}$$
$$C = C(s, i) = \{\bar{X}(s_0) = i_0\}$$
$$\pi = \pi(s, i) = \Pi_{m=0}^{M-1} P(s_{m+1} - s_m, i_m, i_{m+1}).$$

Let $G(s, i)$ be the set of $\omega \in \Delta$ satisfying

(44) $$Q(\omega, B) = Q(\omega, C) \cdot \pi.$$

I say

(45) $$G(s, i) \in \mathscr{F}(\tau+) \quad \text{and} \quad P_i\{\Delta \backslash G(s, i)\} = 0.$$

The measurability is clear. To proceed, integrate both sides of (44) over an arbitrary $A \in \mathscr{F}(\tau+)$ with $A \subset \Delta$. On the left, you get

$$P_i\{A \text{ and } Y(s_m) = i_m \text{ for } m = 0, \ldots, M\}.$$

On the right you get

$$P_i\{A \text{ and } Y(s_0) = i_0\} \cdot \pi.$$

These two expressions are equal by (35). Now (10.10a) settles (45). But

$$\Delta_P = [\cap_r G(r)] \cap [\cap_{s,i} G(s, i)]. \qquad \bigstar$$

NOTE. Strong Markov (38 and 41) holds for any $\mu \in \mathbf{P}$ in place of P_i; review the proofs.

Given the ordering of the states, the holding time on each visit to state i is exponential with parameter $q(i)$, independent of other holding times. This can be made precise in various ways. For example, let D be a finite set of states. Let $i_1, \ldots, i_n \in D$, not necessarily distinct. Suppose $q(i_1), \ldots, q(i_n)$ positive. You may wish to review (1.24) before tackling the next theorem.

(46) Theorem. *Let $\mu \in \mathbf{P}$. Given that $\{\bar{X}(t) : 0 \leq t < \infty\}$ pays at least n visits to D, the 1st being to $i_1, \ldots$, the nth being to i_n, the holding times on these visits are conditionally μ-independent and exponential with parameters $q(i_1), \ldots, q(i_n)$.*

PROOF. Let A be the event $\bar{X}$ visits D at least once, the 1st visit being to i_1. Let σ_1 be the holding time on this visit. Let B be the event $\bar{X}$ visits D at least $n + 1$ times, the 2nd visit being to $i_2, \ldots$, the $n + 1$st to i_{n+1}. On B, let $\sigma_2, \ldots, \sigma_{n+1}$ be the holding times on visits 2 through $n + 1$. Let C be the event $\bar{X}$ visits D at least n times, the first visit being to $i_2, \ldots$, the nth to i_{n+1}. On C, let $\tau_1, \ldots, \tau_n$ be the holding times on the n visits. Let $t_1, \ldots, t_{n+1}$ be nonnegative numbers. Let

$$G = \{A \text{ and } B \text{ and } \sigma_m > t_m \text{ for } m = 1, \ldots, n + 1\}$$

$$H = \{C \text{ and } \tau_m > t_{m+1} \text{ for } m = 1, \ldots, n\}.$$

If $\mu \in \mathbf{P}$, I claim

(a) $$\mu\{G \mid A \cap B\} = P_{i_1}\{G \mid B\}.$$

Argument for (a). Let ϕ be the time of first visiting D. Then ϕ is Markov. Let S_ϕ be $\bar{X}(\phi + \cdot)$ retracted to R. Confine ω to $\bar{\Omega}_v \cap S_\phi^{-1}\Omega_v$, a set of μ-probability 1 by (38 on μ). Then

$$A \cap B = A \cap \{\bar{X}(\phi) = i_1 \text{ and } S_\phi \in B\}$$

$$G = A \cap \{\bar{X}(\phi) = i_1 \text{ and } S_\phi \in G\}.$$

So (38 on μ) implies

$$\mu\{A \cap B\} = \mu\{A\} \cdot P_{i_1}\{B\}$$

$$\mu\{G\} = \mu\{A\} \cdot P_{i_1}\{G\}.$$

Divide to get (a).

Confine ω to $\{\Omega_v \text{ and } X(0) = i_1\}$. There, σ_1 coincides with the first holding time in X, which is Markov. Let S_1 be $X(\sigma_1 + \cdot)$ retracted to R. Let ν be the P_{i_1}-distribution of S_1; so ν depends on i_1. I claim

(b) $$P_{i_1}\{G \mid B\} = \nu\{H \mid C\} \cdot e^{-q(i_1)t_1}.$$

Argument for (b). Confine ω to $\{\Omega_v \text{ and } X(0) = i_1 \text{ and } S_1 \in \bar{\Omega}_v\}$, which has P_{i_1}-probability 1 by (41a). There,

$$B = \{S_1 \in C\}$$

$$G = \{S_1 \in H\} \cap \{\sigma_1 > t_1\}.$$

Using (21),

$$P_{i_1}\{B\} = \nu\{C\}$$
$$P_{i_1}\{G\} = \nu\{H\} \cdot e^{-q(i_1)t_1}.$$

Divide to clinch (b).

Combine (a) and (b):

$$\mu\{G \mid A \cap B\} = \nu\{H \mid C\} \cdot e^{-q(i_1)t_1}.$$

But $\nu \in \mathbf{P}$ by (21), so induction wins again. ★

NOTE. Specialize $D = \{j\}$. Then (46) asserts: given X visits j at least n times, the first n holding times in j are conditionally independent and exponential, with common parameter $q(j)$. This is the secret explanation for the proof of (8).

REFERENCES. Strong Markov was first treated formally by Blumenthal (1957) and Hunt (1956), but was used implicitly by Lévy (1951) to prove (46). For another discussion, see (Chung, 1960, II.9).

5. THE MINIMAL SOLUTION

The results of this section, which can be skipped on a first reading of the book, are due to Feller (1945, 1957). For another treatment, see Chung (1960, II.18). Let Q be a real-valued matrix on the countable set I, with

(47a) $$q(i) = -Q(i, i) \geqq 0;$$

(47b) $$Q(i, j) \geqq 0 \quad \text{for } i \neq j;$$

(47c) $$\Sigma_j \, Q(i, j) \leqq 0.$$

NOTE. Any generator Q with all its entries finite satisfies (47), by (5.10, 14). If I is finite, or even if $\sup_i q(i) < \infty$, then (47) makes Q a generator by (5.29).

When is there a standard stochastic or substochastic semigroup P with $P'(0) = Q$? When is P unique? To answer these questions, at least in part, define Γ by

$$\Gamma(i, j) = Q(i, j)/q(i) \quad \text{for } i \neq j \quad \text{and} \quad q(i) > 0$$
$$= 0 \qquad \qquad \text{elsewhere.}$$

Define a *minimal Q-process starting from i* as follows. The order of states is a partially defined discrete time Markov chain with stationary substochastic transitions Γ; the holding time on each visit to j is exponential with parameter $q(j)$, independent of other holding times and the order of states; the sample function is continuous from the right where defined. In general, there is positive probability that a sample function will be defined so far only on a finite interval. The most satisfying thing to do is simply to leave the process partially defined. To avoid new definitions, however, introduce an absorbing state $\partial \notin I$. When the sample function is still undefined, set it equal to ∂. The minimal Q-process then has state space $I \cup \{\partial\}$; starting from ∂, it remains in ∂ forever. The minimal Q-process is Markov, with stationary standard transitions, say $\bar{P}_\partial$. And $\bar{P}_\partial$ is a stochastic semigroup on $I \cup \{\partial\}$, with ∂ absorbing.

NOTE. All minimal Q-processes starting from i have the same distribution.

One rigorous minimal Q-process can be constructed by (5.39): just plug in the present matrix Q for the (5.39) matrix Q; more exactly, plug in the present Γ for the (5.39) matrix Γ, and the present q for the (5.39) function q. Check that the (5.39) matrix Q coincides with the present Q. The construction embodied in (5.39) produces a process, which I just described informally; and (5.39) asserts that the formal process is Markov with stationary transitions $\bar{P}_\partial$, which are standard and stochastic on $I \cup \{\partial\}$. Let $\bar{P}$ be the retraction of $\bar{P}_\partial$ to I:

$$\bar{P}(t, i, j) = \bar{P}_\partial(t, i, j) \quad \text{for } t \geq 0 \quad \text{and} \quad i, j \text{ in } I.$$

Now $\bar{P}$ is a substochastic semigroup on I, because ∂ is absorbing. And P is standard because $\bar{P}_\partial$ is. And (5.39) shows $\bar{P}'(0) = Q$.

(48) Lemma. $\bar{P}$ *is stochastic iff the minimal Q-process starting from any $i \in I$ has almost all its sample functions I-valued on $[0, \infty)$. Indeed, $\Sigma_{j\in I} \bar{P}(t, i, j)$ is just the probability that a minimal Q-process starting from i is I-valued on $[0, t]$.*

PROOF. If the process hits ∂ at all on $[0, t]$, it does so at a rational time, and is then stuck in ∂ at time t.   ★

It is convenient to generalize the notion of minimal Q-process slightly. Let the stochastic process $\{Z(t) : 0 \leq t < \infty\}$ on a triple $(\mathscr{X}, \mathscr{F}, \mathscr{P})$ be a *regular Q-process* starting from $i \in I$, namely: Z is a Markov process starting from i, with state space $I \cup \{\partial\}$, and stationary standard stochastic transitions P on $I \cup \{\partial\}$; moreover, ∂ is absorbing for P, and the retraction of P to I has generator Q; finally, the sample functions of Z are $\bar{I} \cup \{\partial\}$-valued and regular: ∂ is an isolated point of $\bar{I} \cup \{\partial\}$. Let τ be the least t if any with

$$Z(t) \notin I \quad \text{or} \quad \lim_{s \uparrow t} Z(s) \notin I;$$

if none, $\tau = \infty$. Let $Z^*(t) = Z(t)$ if $t < \tau$ and $Z^*(t) = \partial$ if $t \geq \tau$.

(49) Lemma. Z^* *is a minimal Q-process starting from i.*

PROOF. Use (33).   ★

NOTE. Different regular Q-processes can have different distributions. As (49) shows, however, the distribution of Z^* is determined by Q for any regular Q-process Z.

These considerations answer the existence question for substochastic P.

(50) Theorem. *The standard substochastic semigroup $\bar{P}$ on I satisfies $\bar{P}'(0) = Q$. If a standard substochastic semigroup P on I satisfies $P'(0) = Q$, then $P \geq \bar{P}$.*

PROOF. Only the second claim has to be argued. Define the standard stochastic semigroup P_∂ on $I \cup \{\partial\}$ by:

$$P_\partial(t, i, j) = P(t, i, j) \qquad \text{for } i, j \in I;$$
$$P_\partial(t, \partial, i) = 0 \qquad \text{for } i \in I;$$
$$P_\partial(t, i, \partial) = 1 - \Sigma_{j \in I} P(t, i, j) \quad \text{for } i \in I;$$
$$P_\partial(t, \partial, \partial) = 1.$$

Now construct a Markov chain $\{Z(t) : 0 \leq t < \infty\}$ on some triple $(\mathcal{X}, \mathcal{F}, \mathcal{P})$, with state space $I \cup \{\partial\}$, regular $\bar{I} \cup \{\partial\}$-valued sample functions, stationary transitions P_∂, starting from $i \in I$. In particular, Z is a regular Q-process. Use the notation and result of (49). For i and j in I,

$$\begin{aligned}
P(t, i, j) &= P_\partial(t, i, j) \\
&= \mathcal{P}\{Z(t) = j\} \\
&\geq \mathcal{P}\{Z(t) = j \text{ and } \tau > t\} \\
&= \mathcal{P}\{Z^*(t) = j\} \\
&= \bar{P}(t, i, j).
\end{aligned} \qquad \bigstar$$

Suppose $\Sigma_j Q(i, j) = 0$ for all i. Then (51) below answers the uniqueness question. But I believe the most interesting development in (51) to be this: if $\bar{P}$ is not stochastic, then there is a continuum of stochastic P with $P'(0) = Q$.

(51) Theorem. *Suppose $\Sigma_j Q(i, j) = 0$ for all i. Then conditions (a)–(g) are all equivalent.*

(a) *$\bar{P}$ is stochastic.*

(b) *Any regular Q-process has for almost all its sample functions I-valued step functions on $[0, \infty)$.*

(c) *Any discrete time Markov chain $\xi_0, \ldots$ with stationary transitions Γ satisfies $\Sigma_n 1/q(\xi_n) = \infty$ almost surely.*

(d) *There is at most one standard substochastic semigroup P with $P'(0) = Q$.*

(e) *There is at most one standard stochastic semigroup P with $P'(0) = Q$.*

Claim A. *If $\bar{P}$ is not stochastic, there are a continuum of different standard stochastic semigroups P on I with $P'(0) = Q$.*

To state (f) and (g), let z be a function from I to $[0, 1]$. Then Qz is this function on I:

$$(Qz)(i) = \Sigma_j \, Q(i, j) z(j).$$

For $\lambda > 0$, let (S_λ) be this system:

(S_λ) *z is a function from I to $[0, 1]$ and $Qz = \lambda z$.*

(f) *For all $\lambda > 0$, the system (S_λ) has only the trivial solution $z \equiv 0$.*

(g) *For one $\lambda > 0$, the system (S_λ) has only the trivial solution $z \equiv 0$.*

Let θ be the first time if any at which the minimal Q-process starting from i hits ∂, and $\theta = \infty$ if none. Let $z(\lambda, i)$ be the expectation of $e^{-\lambda \theta}$ for $\lambda > 0$. Then

Claim B. *$z(\lambda, \cdot)$ is the maximal solution of (S_λ) for each $\lambda > 0$.*

I learned about (S_λ) from Harry Reuter.

PROOF.   (a) iff (b).   Use (48) and (49).
(a) iff (c).   Use (5.38).
(a) implies (d).   Suppose (a), and suppose P is a standard substochastic semigroup with $P'(0) = Q$. By (50).

$$P(t, i, j) \geqq \bar{P}(t, i, j).$$

Sum out j to see that equality holds; namely, $P = \bar{P}$.
(d) implies (e).   Logic.
(e) implies (a).   Follows from claim A.
Claim A.   The construction in (6.132) had a parameter p, which ranged over the probabilities on I. The construction produced a triple $(\mathscr{X}_0, \pi_i^p)$ and a process X, which was Markov with stationary transitions P^p and starting state i. The transitions formed a standard stochastic semigroup on I, with generator Q, by (6.37). The sample functions of X were I-valued and regular, by (6.136). So X is a regular Q-process. Remember that θ in (6.132) was the least time to the left of which there were infinitely many jumps.
Suppose

$$\Sigma_j \, \bar{P}(t, i, j) < 1.$$

By (48–49),

$$\pi_i^p\{\theta \leqq t\} = 1 - \Sigma_j \, \bar{P}(t, i, j) > 0.$$

By (6.135),

$$\pi_i^p\{\xi(1, 0) = j \mid \theta < \infty\} = p(j).$$

By (6.140), this conditional probability can be computed from the π_i^p-distribution of $\{X(r): \text{rational } r\}$. This distribution can be computed from i and P^p. So P^p determines p, and there are continuum many distinct P^p.

Underground explanation. The minimal Q-process starting from j cannot reach ∂ by jumping into it, but only by visiting an infinite sequence of states in a finite amount of time. This protects me from increasing $Q(j, k)$ when I replace ∂ by k.

(*f*) *implies* (*g*).   Logic.
(*a*) *implies* (*f*) *and* (*g*) *implies* (*a*).   Use claim B.
Claim B.   Let X be the minimal Q-process constructed in (5.39), with visits ξ_n, and holding times τ_n. The probability controlling X is π_i; write E_i for π_i-expectation. Relative to π_i, the visits $\{\xi_n\}$ are a $\hat{\Gamma}$-chain starting from i, where $\hat{\Gamma}$ is Γ extended to be stochastic on $I \cup \{\partial\}$ and absorbing at ∂; given $\{\xi_n\}$, the holding times $\{\tau_n\}$ are conditionally independent and exponentially distributed, with parameters $\{q(\xi_n)\}$, where $q(\partial) = 0$. By deleting a set which is null for all π_i, suppose that for all n,

$$\tau_n = \infty \quad \text{iff} \quad q(\xi_n) = 0 \quad \text{iff} \quad \xi_{n+1} = \partial.$$

Let d be the least n if any with $\xi_n = \partial$, and $d = \infty$ if none. Let

$$\theta = \Sigma \{\tau_n : 0 \leqq n < d\} = \Sigma_n \tau_n,$$

because $d < \infty$ makes $\tau_{d-1} = \infty$, so $\theta = \infty$ either way. Then θ is the least t if any with $X(t) = \partial$, and $\theta = \infty$ if none.

Fix $\lambda > 0$. Abbreviate $\exp x = e^x$. Let

$$z(\lambda, i) = E_i\{\exp(-\lambda\theta)\};$$
$$z_0(\lambda, i) = 1;$$
$$z_n(\lambda, i) = E_i\{\exp[-\lambda(\tau_0 + \cdots + \tau_{n-1})]\} \quad \text{for } n \geqq 1.$$

I claim

$(S_{\lambda,n})$ $[q(i) + \lambda]z_{n+1}(\lambda, i) = \Sigma_{j \neq i} Q(i, j)z_n(\lambda, j).$

Indeed, with respect to π_i, the time τ_0 is independent of $(\xi_1, \tau_1, \ldots, \tau_n)$ and is exponential with parameter $q(i)$, by (5.37). So (5.31) makes

$$z_1(\lambda, i) = E_i\{\exp(-\lambda\tau_0)\} = \frac{q(i)}{q(i) + \lambda}.$$

This is $(S_{\lambda,0})$. For $n \geqq 1$, the argument shows

$$z_{n+1}(\lambda, i) = E_i\{\exp(-\lambda\tau_0) \cdot \exp[-\lambda(\tau_1 + \cdots + \tau_n)]\}$$

$$= \frac{q(i)}{q(i) + \lambda} \cdot E_i\{\exp[-\lambda(\tau_1 + \cdots + \tau_n)]\}.$$

Let ϕ be a nonnegative, measurable function of n real variables. Then

$$E_i\{\phi(\tau_1, \ldots, \tau_n)\} = \Sigma_{j \neq i} \Gamma(i,j) E_j\{\phi(\tau_0, \ldots, \tau_{n-1})\}:$$

to check this, split the left side over the sets

$$\{\xi_1 = j, \xi_2 = j_2, \ldots, \xi_n = j_n\};$$

split the jth term on the right over

$$\{\xi_1 = j_2, \ldots, \xi_{n-1} = j_n\};$$

and use (5.37). Consequently,

$$z_{n+1}(\lambda, i) = \frac{q(i)}{q(i) + \lambda} \cdot \Sigma_{j \neq i} \Gamma(i,j) z_n(\lambda, j).$$

This settles $(S_{\lambda,n})$ for $n \geq 1$.

Check $z_n(\lambda, i) \downarrow z(\lambda, i)$. Let $n \to \infty$ in $(S_{\lambda,n})$:

$$[q(i) + \lambda]z(\lambda, i) = \Sigma_{j \neq i} Q(i,j)z(\lambda, j).$$

This shows $z(\lambda, \cdot)$ to be a solution of (S_λ). Why is $z(\lambda, \cdot)$ maximal? Let z be a competitive solution. Then

$$z(j) \leq 1 = z_0(\lambda, j).$$

So

$$\begin{aligned}[q(i) + \lambda]z(i) = \Sigma_{j \neq i} Q(i,j)z(j) &\quad \text{by } (S_\lambda) \\ \leq \Sigma_{j \neq i} Q(i,j)z_0(\lambda, j) \\ = [q(i) + \lambda]z_1(\lambda, i) &\quad \text{by } (S_{\lambda,0}).\end{aligned}$$

That is,

$$z(i) \leq z_1(\lambda, i) \qquad \text{for all } i.$$

Persevering,

$$z(i) \leq z_n(\lambda, i) \quad \text{for all } n \text{ and } i.$$

Let $n \to \infty$ and remember $z_n(\lambda, i) \downarrow z(\lambda, i)$ to get

$$z(i) \leq z(\lambda, i) \qquad \text{for all } i,$$

as advertised. ★

NOTE. Suppose $\Sigma_j Q(i,j) = 0$, and suppose the minimal solution $\bar{P}$ is really substochastic. If P is a standard substochastic semigroup with $P \geq \bar{P}$ coordinatewise, this forces $P'(0) = Q$, as suggested by Volker Strassen. Indeed, $P'(0) \geq Q$ coordinatewise by calculus, and the row sums of $P'(0)$ are nonpositive by (5.14).

NOTE. If $\Sigma_j Q(i,j) < 0$, then $\bar{P}$ cannot be stochastic. Indeed, $q(i) > 0$ and $\Gamma(i, \cdot)$ has mass strictly less than 1. So the minimal Q-process starting from i can reach ∂ on first leaving i. However, $\bar{P}$ may be the only standard substochastic semigroup with generator Q, as is the case when $\sup_i q(i) < \infty$. It is not known when Q generates a standard stochastic semigroup.

6. THE BACKWARD AND FORWARD EQUATIONS

The main results of this section, which can be skipped on a first reading of the book, are due to Doob (1945). For another treatment, see Chung (1960, II.17). Let P be a standard stochastic semigroup on the countable set I; the finite I case has already been dealt with, in (5.29). Let $P'(0) = Q$; let $q(i) = -Q(i, i)$, and suppose $q(i)$ finite for all i. The problem is to decide when the following two equations hold:

(Backward equation) $P'(t) = QP(t)$

(Forward equation) $P'(t) = P(t)Q.$

It is clear from Fatou and the existence of P', to be demonstrated below, that both relations hold with = replaced by $\geqq$. Let X be a Markov process on the triple (Ω_v, P_j), with stationary transitions P, starting state j, and regular sample functions which are finite and continuous at the binary rationals; for a discussion, see (12).

The backward equation

Recall the definition and properties of the post-exit process Y from Section 3. Let $b(s, i, j)$ be $q(i)$ times the P_i-probability that the post-exit process starts in φ, and is in j at time s:

$$b(s, i, j) = q(i)P_i\{Y(0) = \varphi \text{ and } Y(s) = j\}.$$

By (57) below, $b(\cdot, i, j)$ is continuous on $[0, \infty)$; here is a more interesting argument. Suppose $s_n \to s$. Using (29),

$$\{Y(0) = \varphi \text{ and } Y(s_n) = j\} \to \{Y(0) = \varphi \text{ and } Y(s) = j\}, \quad a.c.$$

Fatou implies $b(s_n, i, j) \to b(s, i, j)$.
    Let

$$\sigma(s, i, j) = \Sigma_{k \neq i} \, Q(i, k)P(s, k, j).$$

As (5.14) implies, $\Sigma_{k \neq i} Q(i, k) < \infty$. As (5.9) implies, $P(\cdot, k, j)$ is continuous. By dominated convergence, $\sigma(\cdot, i, j)$ is finite and continuous on $[0, \infty)$. Let $\Delta(i, j)$ be 1 or 0, according as $i = j$ or $i \neq j$.

(52) Theorem.

(a) $P(t, i, j) = \Delta(i, j)e^{-q(i)t} + \int_0^t e^{-q(i)u} \left[b(t - u, i, j) + \sigma(t - u, i, j) \right] du.$

(b) $P'(t, i, j)$ *exists and is continuous on* $[0, \infty)$; *namely*,

$$P'(t, i, j) = b(t, i, j) - q(i)P(t, i, j) + \sigma(t, i, j).$$

(c)   *In particular,*

(53) $P'(t, i, j) = \Sigma_k \, Q(i, k)P(t, k, j)$

is equivalent to

(54) $b(t, i, j) = 0.$

PROOF.   *Claim* (a).   Condition on the first holding time and use (21). More prosaically, suppose $q(i) > 0$; the opposite case is trivial. And suppose $i \neq j$; the case $i = j$ is very similar. Confine ω to the set $\{X(0) = i\}$. Let τ be the first holding time. Confine ω further to the set $\{X(0) = i$ and $\tau < \infty\}$, which has P_i-probability 1. Clearly,

$$\{X(t) = j\} = \{\tau \leqq t \text{ and } Y(t - \tau) = j\}.$$

But (21) makes τ and Y independent, τ being exponential with parameter $q(i)$. Fubini this:

$$P_i\{X(t) = j\} = \int_0^t e^{-q(i)u} \, q(i)P_i\{Y(t - u) = j\} \, du.$$

Clearly,

$$\{Y(s) = j\} = \mathsf{U}_{k \in \bar{I}} \, \{Y(0) = k \text{ and } Y(s) = j\}.$$

So

$$P_i\{Y(s) = j\} = \Sigma_{k \in \bar{I}} \, P_i\{Y(0) = k \text{ and } Y(s) = j\}.$$

Use (21) again for $k \in I$:

$$q(i)P_i\{Y(0) = k \text{ and } Y(s) = j\} = q(i)\Gamma(i, k)P(s, k, j)$$
$$= Q(i, k)P(s, k, j).$$

By definition,

$$q(i)P_i\{Y(0) = \varphi \text{ and } Y(s) = j\} = b(s, i, j).$$

Claim (b).   Put $s = t - u$ in (a) and differentiate with respect to t, using the continuity of b and σ. Then use (a) again.

Claim (c).   Use (b) and the definition of σ.   ★

(55) Theorem.   **(a)** *If* (53) *holds for any* $t > 0$, *it holds for all* t; *and then*

(56) $P(t, i, j) = \Delta(i, j)e^{-q(i)t} + \int_0^t e^{-q(i)u} \, \sigma(t - u, i, j) \, du.$

(b) *If* (56) *holds for any* $t > 0$, *then* (53) *holds.*

PROOF. *Claim (a).* To begin with, I say $t > 0$ and $s \geqq 0$ imply

(57) $b(t + s, i, j) = \Sigma_k\, b(t, i, k)P(s, k, j).$

Indeed, (21) makes $P_i\{Y(t) \in I\} = 1$ and

$$P_i\{Y(0) = \varphi \text{ and } Y(t) = k \text{ and } Y(t + s) = j\}$$
$$= P_i\{Y(0) = \varphi \text{ and } Y(t) = k\} \cdot P(s, k, j).$$

So,

$$
\begin{aligned}
b(t + s, i, j) &= q(i)P_i\{Y(0) = \varphi \text{ and } Y(t + s) = j\} \\
&= q(i)P_i\{Y(0) = \varphi \text{ and } Y(t) \in I \text{ and } Y(t + s) = j\} \\
&= \Sigma_k\, q(i)P_i\{Y(0) = \varphi \text{ and } Y(t) = k \text{ and } Y(t + s) = j\} \\
&= \Sigma_k\, q(i)P_i\{Y(0) = \varphi \text{ and } Y(t) = k\} \cdot P(s, k, j) \\
&= \Sigma_k\, b(t, i, k)P(s, k, j).
\end{aligned}
$$

This proves (57).

By a theorem of Lévy, to be proved in Section 2.3 of *ACM.*, the function $s \to P(s, k, j)$ is identically 0 or strictly positive on $(0, \infty)$; there is a really clever analytic proof to this theorem, due to Don Ornstein, in Chung (1960, pp. 121–122). By (57), the same dichotomy applies to $b(t, i, j)$. Indeed, suppose $b(t, i, j) > 0$ for some $t > 0$. Use (5.7):

$$b(t + s, i, j) \geqq b(t, i, j)P(s, j, j) > 0,$$

so $b(u, i, j) > 0$ for $u \geqq t$. Next, let $0 < s < t$. Then

$$b(t, i, j) = \Sigma_k\, b(s, i, k)P(t - s, k, j),$$

so $b(s, i, k)P(t - s, k, j) > 0$ for some k. Then $P(u, k, j) > 0$ for all $u > 0$ and $b(s + u, i, j) \geqq b(s, i, k)P(u, k, j) > 0$. This proves the dichotomy. The dichotomy and (52c) prove the first part of (a). Then (52a) gets the rest of (a).

Claim (b). If (56) holds, then (52a) implies

$$\int_0^t e^{-q(i)u}\, b(t - u, i, j)\, du = 0.$$

This forces b to vanish somewhere, so everywhere. Use (52c).   ★

(58) Theorem. *The following three relations are all equivalent:*

    (a) *Relation (53) holds for all j.*

    (b) *The X sample function jumps from i to φ with P_i-probability 0.*

    (c) $\Sigma_j\, Q(i, j) = 0.$

PROOF. *(a) iff (b).* As usual, suppose $q(i) > 0$. Let τ be the first holding time, and confine ω to

$$\{X(0) = i \text{ and } \tau < \infty \text{ and } X(\tau + 1) \in I\},$$

which has P_i-probability 1 by (29). But

$$\{X(0) = i \text{ and } X(\tau) = \varphi\} = \bigcup_j \{X(0) = i \text{ and } X(\tau) = \varphi \text{ and } X(\tau + 1) = j\};$$

so

$$q(i)P_i\{X(\tau) = \varphi\} = \Sigma_j b(1, i, j).$$

Use (52c).

(b) *iff* (c) is clear from (21).   ★

The forward equation

The next difficulty is proving that $\rho(t, i, j)$ is finite and continuous on $[0, \infty)$, where

$$\rho(t, i, j) = \Sigma_{k \neq j} P(t, i, k)Q(k, j).$$

Suppose without real loss of generality that $P(t, i, k) > 0$ for all $t > 0$ and $k \in I$: you can reduce I to this set of k, and then retract P to the smaller I. Let $\mu(k) = \int_0^\infty P(t, i, k)\, dt$, and suppose first

(59) $\mu(k) < \infty$ for all k.

Check that

$$\Sigma_k \mu(k)P(t, k, j) \leqq \mu(j).$$

Let

$$\hat{P}(t, j, k) = \mu(k)P(t, k, j)/\mu(j).$$

Then $\hat{P}$ is a standard substochastic semigroup with generator $\hat{Q}$, where

$$\hat{Q}(j, k) = \mu(k)Q(k, j)/\mu(j).$$

But

$$\rho(t, i, j) = \mu(j) \cdot \Sigma_{k \neq j} \hat{Q}(j, k)\hat{P}(t, k, i)/\mu(i),$$

and is now finite and continuous, by the argument for σ. To remove condition (59), use the argument on the standard substochastic semigroup $t \rightarrow e^{-t} P(t)$.

Let

$$f(t, i, j) = P'(t, i, j) + q(j)P(t, i, j) - \rho(t, i, j),$$

a continuous function on $[0, \infty]$. By Fatou, $f \geqq 0$.

(60) Lemma.   *For $t \geqq 0$ and $s > 0$,*

$$f(t + s, i, j) = \Sigma_k P(t, i, k)f(s, k, j).$$

PROOF.   First, suppose (59). As (52b) implies, $\hat{P}$ is continuously differentiable. Let

$$\hat{b}(t, i, j) = \hat{P}'(t, i, j) - \Sigma_k \hat{Q}(i, k)\hat{P}(s, k, j).$$

Informally, (52b) reveals $\hat{b}$ as the b of $\hat{P}$. That is, $\hat{b}(t, i, j)$ is $\hat{q}(i)$ times the probability that a $\hat{P}$-chain with regular sample functions starting from i jumps to φ on leaving i, and is in j at time t after the jump. So, (57) holds with hats on. By algebra,

$$\mu(i)f(t, i, j)/\mu(j) = \hat{b}(t, j, i);$$

By more algebra, hatted (57) is (60). To remove condition (59), apply the argument to the standard substochastic semigroup $t \to e^{-t} P(t)$. The fudge factors cancel.   ★

Of course, (52) and (57) work for substochastic P, by the usual maneuver of adding ∂. The argument in (55) shows

(61) Corollary. $f(t, i, j)$ *is identically* 0 *or strictly positive on* $(0, \infty)$.

If p is continuously differentiable on $[0, \infty)$, and $0 \leq q < \infty$ is a real number, integration by parts shows

(62) $$\int_0^t [p'(s) + qp(s)]e^{-q(t-s)} ds = p(t) - e^{-qt} p(0).$$

Recall that

$$f(t, i, j) = P'(t, i, j) + q(j)P(t, i, j) - \rho(t, i, j),$$

where

$$\rho(t, i, j) = \Sigma_{k \neq j} P(t, i, k)Q(k, j)$$

is finite and continuous.

(63) Theorem.

 (a) $P'(t, i, j) = f(t, i, j) - q(j)P(t, i, j) + \rho(t, i, j)$.

 (b) $P(t, i, j) = \Delta(i, j)e^{-q(j)t} + \int_0^t [f(s, i, j) + \rho(s, i, j)] e^{-q(j)(t-s)} ds$.

 (c) *In particular,*

(64) $$P'(t, i, j) = \Sigma_k P(t, i, k)Q(k, j)$$

is equivalent to

(65) $$f(t, i, j) = 0.$$

 (d) *If* (64) *holds for any* $t > 0$, *it holds for all* t; *and this is equivalent to*

$$P(t, i, j) = \Delta(i, j)e^{-q(j)t} + \int_0^t \rho(s, i, j)e^{-q(j)(t-s)} ds.$$

PROOF.   *Claim* (a) rearranges the definition of f.
Claim (b).   Use (62), with $P(\cdot, i, j)$ for p and $q(j)$ for q:

$$P(t, i, j) = \Delta(i, j)e^{-q(j)t} + \int_0^t [P'(s, i, j) + q(j)P(s, i, j)] e^{-q(j)(t-s)} ds.$$

Substitute claim (a) into this formula.

Claim (c) is immediate from (a).

Claim (d). Use (b) and (c) and (61). ★

I will now obtain an analog of (58a ↔ 58b). It is even possible to throw in (58c), by working with $\hat{P}$ and $\hat{Q}$; but I will not do this. Informally, for $k \in I$ and $k \neq j$,

$$P(s, i, k)Q(k, j)\,ds$$

is the P_i-probability that a jump from k to j occurs in $(s, s + ds)$. Thus, $\rho(s, i, j)\,ds$ is the P_i-probability that X jumps from some state to j in $(s, s + ds)$, and

$$\int_0^t \rho(s, i, j)e^{-q(j)(t-s)}\,ds$$

is the P_i-probability that the sample function experiences at least one discontinuity on $[0, t]$, and the last discontinuity is a jump from some real state to j. Now (63b) reveals $f(s, i, j)\,ds$ as the P_i-probability of a jump from φ to j in $(s, s + ds)$. All these statements are rigorous in their way. To begin checking this out, let γ be the time of the last discontinuity of X on or before time t, on the set D where X has at least one such discontinuity. That is, D is the complement of $\{X(s) = X(t) \text{ for } 0 \leq s \leq t\}$. On D, the random variable γ is the sup of $s < t$ with $X(s) \neq X(t)$. By regularity,

$$X(\gamma) = \lim_{s \downarrow \gamma} X(s) = X(t);$$
$$X(\gamma-) = \lim_{s \uparrow \gamma} X(s) \neq X(t)$$

is a random element of $\bar{I}$.

NOTE. D and γ depend on t.

(66) Proposition. *Let* $j, k \in I$ *and* $k \neq j$.

(a) $P_i\{D \text{ and } X(\gamma-) = k \text{ and } X(t) = j\} = \int_0^t P(s, i, k)Q(k, j)e^{-q(j)(t-s)}\,ds$.

(b) $P_i\{D \text{ and } X(\gamma-) \in I \text{ and } X(t) = j\} = \int_0^t \rho(s, i, j)e^{-q(j)(t-s)}\,ds$.

(c) $P_i\{D \text{ and } X(\gamma-) = \varphi \text{ and } X(t) = j\} = \int_0^t f(s, i, j)e^{-q(j)(t-s)}\,ds$.

PROOF. *Claim* (a). Without real loss, put $t = 1$. Let D_n be the event $X(m/2^n) \neq X(1)$ for some $m = 0, \ldots, 2^n - 1$. On D_n, let γ_n be the greatest $m/2^n < 1$ with $X(m/2^n) \neq X(1)$. Using the regularity, check $D_n \uparrow D$ and $\gamma_n \uparrow \gamma$ and

$$\{D_n \text{ and } X(\gamma_n) = k \text{ and } X(1) = j\} \rightarrow \{D \text{ and } X(\gamma-) = k \text{ and } X(1) = j\}.$$

By Fatou, the P_i-probability of the left side converges to the P_i-probability of the right side. But the P_i-probability of the left side is

(67) $\sum_{m=0}^{2^n-1} P\left(1 - \frac{m+1}{2^n}, i, k\right) \cdot P\left(\frac{1}{2^n}, k, j\right) \cdot P\left(\frac{1}{2^n}, j, j\right)^m.$

As $n \to \infty$,

$$P\left(\frac{1}{2^n}, k, j\right) = \frac{1}{2^n} Q(k, j) + o\left(\frac{1}{2^n}\right);$$

and

$$P\left(\frac{1}{2^n}, j, j\right) = 1 - \frac{1}{2^n} q(j) + o\left(\frac{1}{2^n}\right),$$

so

$$P\left(\frac{1}{2^n}, j, j\right)^m \to e^{-q(j)u} \quad \text{as} \quad \frac{m}{2^n} \to u, \quad \text{uniformly in } 0 \leqq u \leqq 1.$$

Consequently, (67) converges to the right side of claim (a).

  Claim (b).   Sum claim (a) over $k \in I \backslash \{j\}$.

  Claim (c).   Subtract claim (b) from (63b).   ★

If $X(\gamma-) = k$ and $X(t) = j$, call the last discontinuity of X on $[0, t]$ a *jump from k to j*; even if $k = \varphi$. Let $\mu_{ij}(t)$ be the P_i-mean number of jumps from φ to j in $[0, t]$.

(68) Proposition. $\mu_{ij}(t) = \int_0^t f(s, i, j) \, ds.$

  PROOF.   This is a rehash of (66c). Jumps from φ to j occur at the beginning of j-intervals; so there are only finitely many on finite time intervals:

$$\omega \in \Omega_v.$$

Let $0 < \gamma_1 < \gamma_2 < \cdots$ be the times of the first, second, ... jumps from φ to j. If there are fewer than n jumps, put $\gamma_n = \infty$. Thus $\gamma_n \to \infty$ as $n \to \infty$. If $\gamma_n < \infty$, let τ_n be the length of the j-interval whose left endpoint is γ_n. Now $\gamma_n + \tau_n \leqq \gamma_{n+1}$; while

$$\{D \text{ and } X(\gamma-) = \varphi \text{ and } X(t) = j\} = \bigcup_{n=1}^\infty \{\gamma_n \leqq t < \gamma_n + \tau_n\}.$$

So,

$$P_i\{D \text{ and } X(\gamma-) = \varphi \text{ and } X(t) = j\} = \sum_{n=1}^\infty P_i\{\gamma_n \leqq t < \gamma_n + \tau_n\}.$$

Fix a positive real number t. I say that $\{\gamma_n \leqq t\} \in \mathscr{F}(t)$. Indeed, let D be a countable dense subset of $[0, t]$, with $t \in D$. For $a < b$ in D, let $E(a, b)$ be the event that $X(b) = j$, and for all finite subsets J of I there are binary rational $r \in (a, b)$ with $X(r) \notin J$. Let $F(s)$ be the event that for all positive integer m, there are a and b in D with

$$s < a < b \leqq t \quad \text{and} \quad b - a < 1/m \quad \text{and} \quad E(a, b).$$

Then
$$\{\gamma_n \leqq t\} = \bigcup_s \{\gamma_{n-1} \leqq s \text{ and } F(s):s \in D\},$$
proving that γ_n is a Markov time. Clearly $X(\gamma_n) = j$ on $\{\gamma_n < \infty\}$. By (21) and strong Markov (38), given $\{\gamma_n < \infty\}$, the variable τ_n is conditionally P_i-exponential with parameter $q(j)$, independent of γ_n. Let $\nu_n(t) = P_i\{\gamma_n \leqq t\}$. By Fubini,
$$P_i\{\gamma_n \leqq t < \gamma_n + \tau_n\} = \int_0^t e^{-q(j)(t-s)}\nu_n(ds).$$
But $\mu_{ij} = \Sigma_{n=1}^\infty \nu_n$, so
$$P_i\{D \text{ and } X(\gamma-) = \varphi \text{ and } X(t) = j\} = \int_0^t e^{-q(j)(t-s)}\mu_{ij}(ds).$$
Compare this with (66c) to get
$$\mu_{ij}(ds) = f(s, i, j)\, ds. \qquad\bigstar$$

Aside. With respect to P_i, given $\{\gamma_n < \infty\}$, the variable $\gamma_{n+1} - \gamma_n$ is independent of $\gamma_1, \ldots, \gamma_n$ and its distribution coincides with the P_j-distribution of γ_1.

(69) Theorem. *The following two relations are equivalent:*

(a) *Relation (64) holds for some $t > 0$.*

(b) *The X sample function jumps from φ to j with P_i-probability 0.*

PROOF. Suppose (a). By (63c, d), relation (65) holds for all t. Then (68) shows $\mu_{ij} = 0$. So (b) holds. Conversely, suppose (b). Remember $f \geqq 0$. Then $f(t, i, j) = 0$ for some $t > 0$ by (68), so for all t by (61). Then (a) holds by (63c). $\qquad\bigstar$

(70) Theorem. *For positive t and s,*
$$\Sigma_j |P'(t, i, j)| < \infty \quad \text{and} \quad \Sigma_j P'(t, i, j) = 0;$$
moreover,
$$P'(t + s) = P'(t)P(s) = P(t)P'(s).$$
PROOF. Use (21) and (52b) for the first display. Use (57) and (60) for the second, with b and f expressed in terms of P using (52b) and (63b). $\qquad\bigstar$

Reversing the sample function

The semigroup $\hat{P}$, introduced after (59), has a sample function interpretation. Temporarily, renormalize X to be continuous from the left. Fix P_i and compute relative to it. Suppose (59). For finite subsets J of I, let τ_J be the time of the last visit to J. The process
$$\{X(\tau_J - t):0 \leqq t < \tau_J\}$$

is Markov with stationary transitions $\hat{P}$, and regular (partially defined) sample functions.

What happens without (59)? Keep looking at P_i. If τ is independent of X and exponentially distributed with parameter 1, then

$$\{X(t): 0 \leqq t < \tau\}$$

is Markov with stationary transitions $t \to e^{-t}P(t)$; and

$$\{X(\tau - t): 0 \leqq t < \tau\}$$

is Markov with stationary transitions $\hat{P}$, where

$$\hat{P}(t, j, k) = \mu(k)e^{-t} P(t, k, j)/\mu(j)$$

$$\mu(j) = \int_0^\infty e^{-t} P(t, i, j) \, dt.$$

8

MORE EXAMPLES FOR THE STABLE CASE

1. AN OSCILLATING SEMIGROUP

Let P be a standard stochastic semigroup on the countably infinite set I. Suppose a, b, c are distinct elements of I. As usual, $Q = P'(0)$ exists by (5.21). If $Q(a, b) > 0$ or if $Q(a, c) > 0$, then $P(t, a, b)/P(t, a, c)$ converges as $t \to 0$, namely to $Q(a, b)/Q(a, c)$. If P is uniform, then P is analytic by (5.29), so the convergence holds by l'Hôpital (10.78 and 80). Lester Dubins asked me whether the convergence held in general. The object in this section is to provide a counterexample:

(1) Theorem. *There is a countable set I, containing a, b, c, and a standard stochastic semigroup P on I, such that;*

    (a) *all elements of $Q = P'(0)$ are finite;*

    (b) *there is a Markov chain with stationary transitions P, all of whose sample functions are step functions;*

    (c) $\qquad \limsup_{t \to 0} P(t, a, b)/P(t, a, c) = \infty, \quad and$

                   $\liminf_{t \to 0} P(t, a, b)/P(t, a, c) = 0.$

The construction

OUTLINE. The state space I consists of a, b, c and (d, n, m) for $d = b$ or c and $n = 1, 2, \ldots$ and $m = 1, \ldots, f(n)$. Here $f(n)$ is a positive

---

I want to thank Isaac Meilijson for checking the final draft of this chapter, which can be skipped on a first reading of the book.

integer to be chosen later. Think of it as large. Let b_n and c_n be positive, with $\sum_{n=1}^{\infty} (b_n + c_n) = 1$. Suppose:

(2) $d_{n+1} + d_{n+2} + \cdots = o(d_n)$ for $d = b$ or c;

(3) $\lim \sup_{n \to \infty} b_n/c_n = \infty$ and $\lim \inf_{n \to \infty} b_n/c_n = 0$.

For instance. If n is even, let

$$b_n = \beta\, n^{-1}\, 2^{-n} \quad \text{and} \quad c_n = \gamma\, n^{-2}\, 2^{-n}.$$

If n is odd, let

$$b_n = \beta\, n^{-2}\, 2^{-n} \quad \text{and} \quad c_n = \gamma\, n^{-1}\, 2^{-n}.$$

Choose the positive constants β and γ so $\sum_{n=1}^{\infty} (b_n + c_n) = 1$.

Let $0 < q_{n,m} < \infty$. These numbers will be chosen later; think of them as large. On a convenient probability triple $(\Omega, \mathscr{F}, \mathscr{P})$, let T_0 be exponential with parameter 1; and let $T_{n,m}$ be exponential with parameter $q_{n,m}$ for $n = 1, 2, \ldots$ and $m = 1, \ldots f(n)$. Suppose the $f(n) + 1$ variables

$$T_0, T_{n,1}, \ldots, T_{n,f(n)}$$

are mutually independent, for each n.

Construct an informal stochastic process as in Figure 1, with $d = b$ or c. The process starts in a, stays there time T_0, then jumps to $(d, n, 1)$ with probability d_n. Having reached (d, n, m), the process stays there time $T_{n,m}$, and then jumps to $(d, n, m + 1)$ unless $m = f(n)$, in which case the process jumps to d. Having reached d, the process stays there. By (5.39), this process is a Markov chain with stationary transitions P, which are standard and stochastic on I; and (a) holds. Later, I will choose $q_{n,m}$ and $f(n)$, and argue (c).

Formal use of (5.39)

The elements of (5.39) are I, Γ, q. The state space I has already been defined. Define the substochastic matrix Γ on I as follows, with $d = b$ or c and with $n = 1, 2, \ldots$:

$$\Gamma[a, (d, n, 1)] = d_n;$$

$$\Gamma[(d, n, m), (d, n, m + 1)] = 1 \quad \text{for } m = 1, \ldots, f(n) - 1;$$

$$\Gamma[(n, f(n)), d] = 1;$$

all other entries in Γ vanish. Define the function q on I as follows, with $d = b$ or c and $n = 1, 2, \ldots$:

$$q(a) = 1;$$

$$q(d, n, m) = q_{n,m};$$

$$q(d) = 0.$$

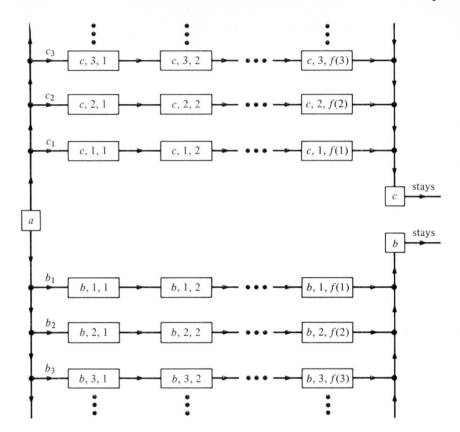

Figure 1.

Now (5.39) yields a process X and a probability π_i which makes X Markov with stationary transitions P and starting state i. The semigroup P is standard and stochastic on $I \cup \{\partial\}$, where ∂ is absorbing. As you will agree in a minute, X cannot really reach ∂ starting from $i \in I$; so P is standard and stochastic when retracted to I. Use (5.39) to check (1a-b).

The visiting process in (5.39) was called $\{\xi_n\}$, and the holding time process, $\{\tau_n\}$. Let $\mathcal{X}_0$ be the set where $\xi_0 = a$, and $\xi_1 = (d, n, 1)$ for some d and n, while $\xi_1 = (d, n, 1)$ implies:

$$\tau_m < \infty \quad \text{and} \quad \xi_m = (d, n, m) \quad \text{for } m = 1, \ldots, f(n);$$

$$\tau_m = \infty \quad \text{and} \quad \xi_m = d \qquad\qquad \text{for } m = f(n) + 1;$$

$$\tau_m = \infty \quad \text{and} \quad \xi_m = \partial \qquad\qquad \text{for } m > f(n) + 1.$$

Use (5.37) to check (4–6):

(4) $\pi_a\{\mathscr{X}_0\} = 1;$

(5) $\pi_a\{\xi_1 = (d, n, 1)\} = d_n;$

(6)   given $\{\xi_1 = (d, n, 1)\}$, the conditional π_a-distribution of

$$\tau_0, \tau_1, \ldots, \tau_{f(n)}$$

coincides with the $\mathscr{P}$-distribution of

$$T_0, T_{n,1}, \ldots, T_{n,f(n)}. \qquad \qquad \bigstar$$

Aside.   It is easy to make the chain more attractive. Let it stay in b or c for an independent, exponential time, and then return to a. This complicates the argument, but not by much. Some of the details for this modification are presented later in the section.

Lemmas

Here are some preliminaries to choosing $f(n)$ and $q_{n,m}$. For (7), let U_n and V_n be random variables on a probability triple $(\Omega_n, \mathscr{F}_n, \mathscr{P}_n)$. Suppose U_n has a continuous $\mathscr{P}_n$-distribution function F, which does not depend on n. Suppose V_n converges in $\mathscr{P}_n$-probability to the constant v, as $n \to \infty$. Let G be the $\mathscr{P}_n$-distribution function of $U_n + v$, so

$$G(t) = F(t - v).$$

(7) Lemma. $\mathscr{P}_n\{U_n + V_n \leq t\} \to G(t)$ *uniformly in* t, *as* $n \to \infty$.

PROOF.   Let $\varepsilon > 0$. Check

$$\{U_n + V_n \leq t\} \subset \{U_n + v \leq t + \varepsilon\} \cup \{V_n < v - \varepsilon\}:$$

because $U_n(\omega) + V_n(\omega) \leq t$ and $V_n(\omega) \geq v - \varepsilon$ imply

$$U_n(\omega) + v \leq U_n(\omega) + V_n(\omega) + \varepsilon \leq t + \varepsilon.$$

Similarly,

$$\{U_n + v \leq t - \varepsilon\} \subset \{U_n + V_n \leq t\} \cup \{V_n > v + \varepsilon\}.$$

So,

$$\mathscr{P}_n\{U_n + V_n \leq t\} \leq \mathscr{P}_n\{U_n + v \leq t + \varepsilon\} + \mathscr{P}_n\{V_n < v - \varepsilon\}$$

and

$$\mathscr{P}_n\{U_n + v \leq t - \varepsilon\} \leq \mathscr{P}_n\{U_n + V_n \leq t\} + \mathscr{P}_n\{V_n > v + \varepsilon\}.$$

Let

$$\delta(\varepsilon) = \sup_t \mathscr{P}_n\{t \leq U_n \leq t + \varepsilon\} = \sup_t [F(t + \varepsilon) - F(t-)],$$

so $\delta(\varepsilon) \downarrow 0$ as $\varepsilon \downarrow 0$. The absolute value of

$$\mathscr{P}_n\{U_n + V_n \leqq t\} - \mathscr{P}_n\{U_n + v \leqq t\}$$

is at most

$$\delta(\varepsilon) + \mathscr{P}_n\{|V_n - v| > \varepsilon\}. \qquad \bigstar$$

Let

$$S_n = \sum_{m=1}^{f(n)} T_{n,m}.$$

(8) Fact. S_n has mean $\sum_{m=1}^{f(n)} 1/q_{n,m}$ and variance $\sum_{m=1}^{f(n)} 1/q_{n,m}^2$.

PROOF. Use (5.31). $\qquad \bigstar$

Choosing f and q

The program is to define $f(n)$ and $\{q_{n,m}: m = 1, \ldots, f(n)\}$ inductively on n, and with them a sequence $t_n \downarrow 0$, such that

$$(9) \qquad\qquad P(t_n, a, d) \approx \tfrac{1}{2}d_n t_n \quad \text{for } d = b \text{ or } c;$$

$x_n \approx y_n$ means $x_n/y_n \to 1$. Relations (3) and (9) establish (c).

Fix a positive sequence ε_n with $\varepsilon_n \leqq 1$ and

$$(10) \qquad\qquad \varepsilon_n = o(d_n) \quad \text{for } d = b \text{ or } c.$$

Abbreviate

$$\theta_n = \sum_{m=1}^{f(n)} 1/q_{n,m}.$$

For $N \geqq 2$, I will require:

$$(11) \qquad\qquad 0 < t_N < \tfrac{1}{2}t_{N-1};$$

$$(12) \qquad\qquad \theta_N = \tfrac{1}{2}t_N;$$

$$(13) \qquad\qquad \mathscr{P}\{T_0 + S_n \leqq t_N\} < \varepsilon_N t_N \qquad \text{for } n = 1, \ldots, N-1;$$

$$(14) \qquad 1 - \varepsilon_N < \mathscr{P}\{T_0 + S_N \leqq t_n\}/\mathscr{P}\{T_0 + \theta_N \leqq t_n\} < 1 + \varepsilon_N$$

$$\text{for } n = 1, \ldots, N.$$

Let $f(1) = 1$ and $t_1 = 1$ and $q_{1,1} = 2$, so (12) holds even at $N = 1$. Let $N \geqq 2$. Suppose $f(n)$, t_n, and $q_{n,m}$ chosen for $n < N$, so that (11–12) holds. I will choose $f(N)$, t_N, and $q_{N,m}$. To begin with, (5.34) implies that for $n < N$,

$$\mathscr{P}\{T_0 + S_n \leqq t\} = o(t) \quad \text{as } t \to 0.$$

Choose t_N so (11) and (13) hold. Now choose $f(N)$ and $q_{N,m}$ so that

$$\sum_{m=1}^{f(N)} 1/q_{N,m} = t_N/2,$$

making (12) good; while

$$\sigma = \sum_{m=1}^{f(N)} 1/q_{N,m}^2$$

is so small that (14) holds. I can do this because (8) and Chebychev make $S_N \to \theta_N$ in probability as $\sigma \to 0$. But (11) and (12) make

$$\theta_N < t_N < t_{N-1} < \cdots < t_1,$$

so $\mathscr{P}\{T_0 + \theta_N \leq t_n\} > 0$ for $n = 1, \ldots, N$. Now (7) gets (14). This completes the construction.   ★

The rest of the proof

ARGUMENT FOR (9). I will continue with the notation established for (5.39). Let

$$\sigma_n = \tau_1 + \cdots + \tau_{f(n)}.$$

As usual, $d = b$ or c. Let

$$d(n, t) = \{\xi_1 = (d, n, 1) \text{ and } \tau_0 + \sigma_n \leq t\}.$$

As (4) implies, $\{X(t) = d\}$ differs by a π_a-null set from $\bigcup_{n=1}^{\infty} d(n, t)$. As (5,6) imply,

$$\pi_a\{d(n, t)\} = d_n \mathscr{P}\{T_0 + S_n \leq t\}.$$

So

(15) $$P(t, a, d) = \sum_{n=1}^{\infty} d_n \mathscr{P}\{T_0 + S_n \leq t\}.$$

Use (14) with $N = n$:

$$\mathscr{P}\{T_0 + S_n \leq t_n\} \approx \mathscr{P}\{T_0 + \theta_n \leq t_n\}.$$

Abbreviate

$$e(x) = 1 - e^{-x}.$$

Use (12) with $N = n$:

$$\mathscr{P}\{T_0 + \theta_n \leq t_n\} = e(t_n - \theta_n) = e(\tfrac{1}{2}t_n) \approx \tfrac{1}{2}t_n.$$

So

(16) $$d_n \mathscr{P}\{T_0 + S_n \leq t_n\} \approx \tfrac{1}{2}d_n t_n \quad \text{as } n \to \infty.$$

Suppose $N > n$. Use (14) and $\varepsilon_N \leq 1$ to check this estimate.

$$\mathscr{P}\{T_0 + S_N \leq t_n\} \leq (1 + \varepsilon_N)\mathscr{P}\{T_0 + \theta_N \leq t_n\}$$
$$\leq 2\mathscr{P}\{T_0 \leq t_n\}$$
$$= 2e(t_n)$$
$$\leq 2t_n.$$

Use (2):

(17) $$\sum_{N=n+1}^{\infty} d_N \mathscr{P}\{T_0 + S_N \leq t_n\} = o(d_n t_n) \quad \text{as } n \to \infty.$$

Let $v = 1, \ldots, n - 1$. Then (13) with n for N and v for n makes

$$\mathscr{P}\{T_0 + S_v \leq t_n\} \leq \varepsilon_n t_n.$$

Since $\Sigma_v d_v \leq 1$, relation (10) makes

(18) $$\sum_{v=1}^{n-1} d_v \mathscr{P}\{T_0 + S_v \leq t_n\} = o(d_n t_n) \quad \text{as } n \to \infty.$$

Combine (15–18) to get (9).   ★

Modifications.

Let T_1 be an exponential random variable with parameter 1 on $(\Omega, \mathscr{F}, \mathscr{P})$. Suppose the $f(n) + 2$ variables

$$T_0, T_{n,1}, \ldots, T_{n,f(n)}, T_1$$

are mutually independent, for each n. Modify the chain so that on first reaching b or c, it stays there time T_1, then jumps back to a. On returning to a, make the chain restart afresh. In formal (5.39) terms, this amounts to redefining $\Gamma(d, \cdot)$ and $q(d)$ for $d = b$ or c:

$$\Gamma(d, a) = 1;$$

$$q(d) = 1.$$

As before, let $\mathscr{X}_0$ be the set where $\xi_0 = a$, and $\xi_1 = (d, n, 1)$ for some d and n, while $\xi_1 = (d, n, 1)$ implies:

$$\xi_m = (d, n, m) \quad \text{for } m = 1, \ldots, f(n);$$

$$\xi_m = d \qquad\qquad \text{for } m = f(n) + 1;$$

$$\xi_m = a \qquad\qquad \text{for } m = f(n) + 2;$$

and $\tau_m < \infty$ for all $m = 0, 1, \ldots$. Use (5.37) to check:

$$\pi_a\{\mathscr{X}_0\} = 1;$$

$$\pi_a\{\xi_1 = (d, n, 1)\} = d_n;$$

given $\{\xi_1 = (d, n, 1)\}$, the conditional π_a-distribution of

$$\tau_0, \tau_1, \ldots, \tau_{f(n)}, \tau_{f(n)+1}$$

coincides with the $\mathscr{P}$-distribution of

$$T_0, T_{n,1}, \ldots, T_{n,f(n)}, T_1.$$

The conclusions and proof of (1) apply to this modified chain, provided that $\{t_n\}$ satisfies (19) in addition to (11–14):

(19) $$\mathscr{P}\{T_0 + T_1 \leq t_n\} \leq \varepsilon_n t_n \quad \text{for } n \geq 2.$$

Here are some clues. Keep

$$\sigma_n = \tau_1 + \cdots + \tau_{f(n)}.$$

Let

$$d^+(n, t) = \{\xi_1 = (d, n, 1) \text{ and } \tau_0 + \sigma_n \leqq t < \tau_0 + \sigma_n + \tau_{f(n)+1}\}$$

$$d^*(n, t) = \{\xi_1 = (d, n, 1) \text{ and } \tau_0 + \tau_{f(n)+1} \leqq t\}.$$

On $\mathscr{X}_0$,

$$\{\bigcup_{n=1}^{\infty} d^+(n, t)\} \subset \{X(t) = d\} \subset \{\bigcup_{n=1}^{\infty} d^+(n, t)\} \cup \{\bigcup_{n=1}^{\infty} \bigcup_{d=b,c} d^*(n, t)\}.$$

So the new $P(t, a, d)$ is trapped in $[D^+(t), D^+(t) + D^*(t)]$, where

$$D^+(t) = \Sigma_{n=1}^{\infty} \pi_a\{d^+(n, t)\}$$

and

$$D^*(t) = \Sigma_{n=1}^{\infty} \Sigma_{d=b,c} \pi_a\{d^*(n, t)\}.$$

But

$$\pi_a\{d^+(n, t)\} = d_n \mathscr{P}\{T_0 + S_n \leqq t < T_0 + S_n + T_1\}$$

and

$$\pi_a\{d^*(n, t)\} = d_n \mathscr{P}\{T_0 + T_1 \leqq t\}.$$

Check

$$\{T_0 + S_n \leqq t\} \cap \{T_1 > t\} \subset \{T_0 + S_n \leqq t < T_0 + S_n + T_1\}$$
$$\subset \{T_0 + S_n \leqq t\};$$

so

$$\mathscr{P}\{T_0 + S_n \leqq t\} \cdot \mathscr{P}\{T_1 > t\} \leqq \mathscr{P}\{T_0 + S_n \leqq t < T_0 + S_n + T_1\}$$
$$\leqq \mathscr{P}\{T_0 + S_n \leqq t\};$$

and

$$\pi_a\{d^+(n, t)\} \approx d_n \mathscr{P}\{T_0 + S_n \leqq t\}$$

as $t \to 0$, uniformly in n. This means you can estimate $D^+(t)$ by the old $P(t, a, d)$. Furthermore, $\Sigma_{n,d} d_n = 1$. So

$$D^*(t_N) = \mathscr{P}\{T_0 + T_1 \leqq t_N\} = o(d_N t_N)$$

by (10, 19). This term is trash. The overall conclusion: as $N \to \infty$, the new $P(t_N, a, d)$ is asymptotic to the old $P(t_N, a, d)$.   ★

Continue with the modified chain. Given the order of visits $\xi_0, \xi_1, \ldots$, the holding times $\tau_0, \tau_1, \ldots$ are independent and exponential, so I once expected

$$\pi_a\{\tau_0 + \tau_1 + \tau_2 \leqq t\} = o(t^2) \quad \text{as } t \to 0.$$

Since b can be reached from a in two jumps but not in one, I also expected

$$P(t, a, b) \sim t^2 \quad \text{as } t \to 0.$$

Both expectations were illusory. Let

$$g_d(t) = \pi_a\{X(t) = d \text{ and } \tau_0 + \tau_1 + \tau_2 \leqq t\}.$$

Then

(20) $P(t, a, d) = g_d(t) + \pi_a\{X(t) = d \text{ and } \tau_0 + \tau_1 + \tau_2 > t\}.$

Remember $f(1) = 1$. Suppose $f(n) > 1$ for $n > 1$. Except for a π_a-null set,

$$\{X(t) = d \text{ and } \tau_0 + \tau_1 + \tau_2 > t\}$$
$$= \{\xi_1 = (d, 1, 1) \text{ and } \tau_0 + \tau_1 \leqq t < \tau_0 + \tau_1 + \tau_2\}.$$

So the second term in (20) is

$$d_1\mathscr{P}\{T_0 + T_{1,1} \leqq t < T_0 + T_{1,1} + T_1\} = d_1 t^2 + o(t^2) \quad \text{as } t \to 0.$$

Consequently,

$$t^{-2} P(t, a, d) = t^{-2} g_d(t) + d_1 + o(1) \quad \text{as } t \to 0.$$

Put $d = b$ or c and $t = t_n$ and divide. Remember (9):

$$\frac{b_n}{c_n} \approx \frac{P(t_n, a, b)}{P(t_n, a, c)} = \frac{o(1) + b_1 + t_n^{-2} \cdot g_b(t_n)}{o(1) + c_1 + t_n^{-2} \cdot g_c(t_n)}.$$

Now $\lim \sup b_n/c_n = \infty$ forces $\lim \sup_{n \to \infty} t_n^{-2} \cdot g_b(t_n) = \infty$.
In particular,

(21) $\lim \sup_{t \to 0} t^{-2} \pi_a\{\tau_0 + \tau_1 + \tau_2 \leqq t\} = \infty;$

and

(22) $\lim \sup_{t \to 0} t^{-2} P(t, a, b) = \infty;$

here b can be reached from a in two jumps, but not in one. This disposes of the two illusions. Moreover, (7.52) implies that $P(t, a, b)$ is continuously differentiable on $[0, \infty)$. And (5.39) implies $P'(0, a, b) = 0$. Consequently, (22) implies that $P(t, a, b)$ does not have a finite second derivative at 0.

2. A SEMIGROUP WITH AN INFINITE SECOND DERIVATIVE

In this section, I will prove the following theorem of Juskevic (1959).

(23) Theorem.   *There is a countable set I containing a and b, and a standard stochastic semigroup P on I, such that:*

(a) *all elements of $Q = P'(0)$ are finite;*

(b) *there is a Markov chain with stationary transitions P, all of whose sample functions are step functions;*

(c) *$P''(1, a, b) = \infty$.*

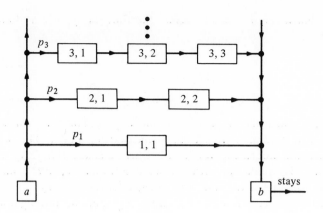

Figure 2.

The construction

OUTLINE. The states are a, b and (n, m) for $m = 1, \ldots, n$ and $n = 1, 2, \ldots$. Let p_n be positive, with $\Sigma_{n=1}^{\infty} p_n = 1$ and

(24) $\Sigma_{n=1}^{\infty} p_n n^{\frac{1}{2}} = \infty.$

Fix $\lambda > 0$. On a convenient probability triple $(\Omega, \mathscr{F}, \mathscr{P})$, let T_0 be exponential with parameter λ, and let $T_{n,m}$ be exponential with parameter n, for each state (n, m). Suppose the $n + 1$ variables

$$T_0, T_{n,1}, \ldots, T_{n,n}$$

are mutually independent for each n.

Construct an informal stochastic process as in Figure 2. The process starts in a, stays there time T_0, and then jumps to $(n, 1)$ with probability p_n. Having reached (n, m), the process stays there time $T_{n,m}$, and then jumps to $(n, m + 1)$ if $m < n$ or to b if $m = n$. Having reached b, the process stays there. By (5.39), this process is a Markov chain with stationary transitions P, which are standard and stochastic on I; and (a) holds. I will argue (c) soon.

Formal use of (5.39)

The elements of (5.39) are I, Γ, and q. The state space I has already been defined. Define the substochastic matrix Γ on I as follows, with $n = 1, 2, \ldots$:

$$\Gamma[a, (n, 1)] = p_n;$$

$$\Gamma[(n, m), (n, m + 1)] = 1 \quad \text{for } m = 1, \ldots, n - 1;$$

$$\Gamma[(n, n), b] = 1;$$

all other entries in Γ vanish. Define the function q on I as follows, with $n = 1, 2, \ldots$:

$$q(a) = \lambda$$

$$q(n, m) = n$$

$$q(b) = 0.$$

Now (5.39) constructs a process X and a probability π_i which makes X Markov with stationary transitions P and starting state i. The semigroup P is standard and stochastic on $I \cup \{\partial\}$, where ∂ is absorbing. As you will see in a minute, X cannot really reach ∂ starting from $i \in I$; so P is standard and stochastic when retracted to I. Use (5.39) to check (23a–b).

The visiting process in (5.39) was called $\{\xi_n\}$, and the holding time process, $\{\tau_n\}$. Let $\mathscr{X}_0$ be the set where $\xi_0 = a$, and $\xi_1 = (n, 1)$ for some n, while $\xi_1 = (n, 1)$ implies:

$$\tau_m < \infty \quad \text{and} \quad \xi_m = (n, m) \quad \text{for } m = 1, \ldots, n;$$

$$\tau_m = \infty \quad \text{and} \quad \xi_m = b \qquad \text{for } m = n + 1;$$

$$\tau_m = \infty \quad \text{and} \quad \xi_m = \partial \qquad \text{for } m > n + 1.$$

Use (5.37) to check (25–27):

(25) $$\pi_a(\mathscr{X}_0) = 1;$$

(26) $$\pi_a\{\xi_1 = (n, 1)\} = p_n;$$

(27) given $\{\xi_1 = (n, 1)\}$, the conditional π_a-distribution of

$$\tau_0, \tau_1, \ldots, \tau_n$$

coincides with the $\mathscr{P}$-distribution of

$$T_0, T_{n,1}, \ldots, T_{n,n}. \qquad\qquad ★$$

The rest of the proof

PROOF OF (c).   Relation (25) shows that except for a π_a-null set,

$$\{X(t) = b\} = \bigcup_{n=1}^{\infty} A_n(t),$$

where

$$A_n(t) = \{\xi_1 = (n, 1) \text{ and } \tau_0 + \tau_1 + \cdots + \tau_n \leqq t\}.$$

By (26, 27),

$$\pi_a\{A_n(t)\} = p_n F_n(t),$$

where

$$F_n(t) = \mathscr{P}\{T_0 + T_{n,1} + \cdots + T_{n,n} \leqq t\}.$$

Therefore,

$$P(t, a, b) = \Sigma_{n=1}^{\infty} p_n F_n(t).$$

As (5.34) implies, for all $t \geq 0$

(28) $$0 \leq F'_n(t) \leq \lambda.$$

By dominated convergence

(29) $$P'(t, a, b) = \Sigma_{n=1}^{\infty} p_n F'_n(t).$$

The density d_n of

$$S_n = T_{n,1} + \cdots + T_{n,n}$$

is the convolution of n exponential densities with parameter n:

$$d_n(t) = n^n t^{n-1} e^{-nt}/(n-1)! \quad \text{for } t \geq 0.$$

By Stirling,

(30) $$d_n(1) \approx (2\pi)^{-\frac{1}{2}} n^{\frac{1}{2}} \quad \text{as } n \to \infty.$$

Abbreviate $e(t)$ for $\lambda e^{-\lambda t}$. So T_0 has density $e(\cdot)$ on $[0, \infty)$. And the density F'_n of $T_0 + S_n$ is the convolution of $e(\cdot)$ and d_n. Namely,

$$F'_n(t) = \int_0^t d_n(s) \lambda e^{-\lambda(t-s)} ds \quad \text{for } t \geq 0.$$

Differentiate this:

$$F''_n(t) = \lambda d_n(t) - \lambda F'_n(t) \quad \text{for } t \geq 0.$$

Use (28):

(31) $$F''_n(t) \geq \lambda d_n(t) - \lambda^2 \geq -\lambda^2 \quad \text{for } t \geq 0.$$

In particular,

(32) $$h^{-1}[F'_n(1+h) - F'_n(1)] \geq -\lambda^2 \quad \text{for } -1 < h < 0 \text{ or } h > 0.$$

Let $-1 < h < 0$ or $0 < h < \infty$. Introduce the approximate second derivatives

$$s(h) = h^{-1}[P'(1+h, a, b) - P'(1, a, b)]$$

$$s_n(h) = h^{-1}[F'_n(1+h) - F'_n(1)].$$

By (29),

$$s(h) = \Sigma_{n=1}^{\infty} p_n s_n(h).$$

Fix c with $0 < c < (2\pi)^{-\frac{1}{2}}$. Using (30), find a positive integer N so large that

(33) $$\lambda d_n(1) - \lambda^2 \geq \lambda c n^{\frac{1}{2}} \quad \text{for } n \geq N.$$

For $1 \leq n < N$, inequality (32) gives

$$s_n(h) \geq -\lambda^2.$$

Of course, $\Sigma_{n=1}^{N-1} p_n \leq 1$. So

$$s(h) \geq -\lambda^2 + \Sigma_{n=N}^{\infty} p_n s_n(h).$$

For $n \geq N$,

$$\lim_{h \to 0} s_n(h) = F_n''(1) \qquad \text{(calculus)}$$

$$\geq \lambda d_n(1) - \lambda^2 \qquad (31)$$

$$\geq \lambda c n^{\frac{1}{2}} \qquad (33).$$

At this point, use Fatou:

$$\lim \inf_{h \to 0} s(h) \geq -\lambda^2 + \lambda c \Sigma_{n=N}^{\infty} p_n n^{\frac{1}{2}}$$

Now exploit assumption (24)   ★

Modifications

You can modify the construction so $P''(t, 0, 1) = \infty$ for all $t \in C$, a given countable subset of $[0, \infty)$. Count C off as $\{t_1, t_2, \ldots\}$. Suppose first that all t_v are positive. Let I consist of a, b, and (v, n, m) for $m = 1, \ldots, n$ and positive integer v and n. Rework the chain so that it jumps from a to $(v, n, 1)$ with positive probability $p_{v,n}$, where:

$$\Sigma_{v,n} p_{v,n} = 1; \quad \text{and} \quad \Sigma_n p_{v,n} n^{\frac{1}{2}} = \infty \quad \text{for each } v.$$

Make the chain jump from (v, n, m) to $(v, n, m + 1)$ when $m < n$, and to b when $m = n$. Make b absorbing. Let the holding time parameter for a be λ. Let the holding time parameter for (v, n, m) be n/t_v.

As before,

$$P(t, a, b) = \Sigma_{v,n} p_{v,n} F_{v,n}(t)$$

and

$$P'(t, a, b) = \Sigma_{v,n} p_{v,n} F_{v,n}'(t):$$

where $F_{v,n}$ is the distribution function of the sum of $n + 1$ independent, exponential random variables, of which the first has parameter λ and the other n have parameter n/t_v. The reason is like (28). I will work on t_1, the other t_v being symmetric. Let

$$s(h) = h^{-1} [P'(t_1 + h, a, b) - P'(t_1, a, b)].$$

Fix a large positive integer N. For $v > 1$ or $v = 1$ but $n < N$,

$$h^{-1} [F_{v,n}'(t_1 + h) - F_{v,n}'(t_1)] \geq -\lambda^2.$$

The reason is like (32). So

$$s(h) \geq -\lambda^2 + \Sigma_{n=N}^{\infty} p_{1,n} h^{-1} [F_{1,n}'(t_1 + h) - F_{1,n}'(t_1)].$$

Let $d_{n,\theta}$ be the density of the sum of n independent, exponential random variables, each having parameter n/θ. Thus,

$$d_{n,1} = d_n \quad \text{and} \quad d_{n,\theta}(t) = \frac{1}{\theta} d_n\left(\frac{t}{\theta}\right).$$

And $F'_{1,n}$ is the convolution of $e(\cdot)$ with d_{n,t_1}. Let $0 < c < (2\pi)^{-\frac{1}{2}}$. Arguing as for (31),

$$(\lambda c/t_1)n^{\frac{1}{2}} < \lim \inf_{h\to 0} h^{-1} [F'_{1,n}(t_1 + h) - F'_{1,n}(t_1)]$$

for large enough n. For large N, Fatou implies

$$-\lambda^2 + \frac{\lambda c}{t_1}\Sigma_{n=N}^\infty \, p_{1,n} \, n^{\frac{1}{2}} \leq \lim \inf_{h\to 0} s(h).$$

This completes the discussion for $C \subset (0, \infty)$.

Now suppose $t_1 = 0$, and all other $t_v > 0$. In this case, let I consist of a, b, and $(1, n)$ for positive integer n, and (v, n, m) for $m = 1, \ldots, n$ and $n = 1, 2, \ldots$ and $v = 2, 3, \ldots$. From a, let the chain jump to $(1, n)$ with probability $p_{1,n}$, and let the chain jump to $(v, n, 1)$ with probability $p_{v,n}$. Suppose the $p_{v,n}$ are positive and sum to 1, while

$$\Sigma_n \, p_{1,n} \, n = \infty \quad \text{and} \quad \Sigma_n \, p_{v,n} \, n^{\frac{1}{2}} = \infty \quad \text{for } v = 2, 3, \ldots.$$

From $(1, n)$ or from (v, n, n), let the chain jump to b, and let b be absorbing. From (v, n, m) with $m < n$, let the chain jump to $(v, n, m + 1)$. Let the holding time parameter for a be λ. Let the holding time parameter for $(1, n)$ be n. Let the holding time parameter for (v, n, m) be n/t_v.

The argument for $t_v > 0$ is essentially the same as before. Here is the program for $t_1 = 0$. Use the same formulas for $P(t, a, b)$ and $P'(t, a, b)$; but now $F_{1,n}$ is the distribution function of the sum of two independent, exponential random variables—the first having parameter λ and the second having parameter n. Define $s(h)$ the same way, for $h > 0$. For the usual reasons,

$$s(h) \geqq -\lambda^2 + \Sigma_{n=N}^\infty \, p_{1,n} \, h^{-1} [F'_{1,n}(h) - F'_{1,n}(0)];$$

and (5.34) implies

$$F'_{1,n}(0) = 0.$$

Now

$$F'_{1,n}(h) = \int_0^h n e^{-ns} \, \lambda e^{-\lambda(h-s)} \, ds,$$

so

$$\lim_{h \downarrow 0} h^{-1} F'_{1,n}(h) = \lambda n.$$

The rest is the same.   ★

If C is dense in $[0, \infty)$, it follows automatically that

$$\limsup_{h \downarrow 0} h^{-1} [P'(t + h, 0, 1) - P'(t, 0, 1)] = \infty$$

for a residual set of t in $[0, \infty)$, using the

(34) Fact. If $s_v(t)$ is a continuous function of t for $v = 1, 2, \ldots$, then $\{t : \sup_v s_v(t) = \infty\}$ is a G_δ.

According to (Chung, 1960, p. 268), the function $P'(t, a, b)$ is absolutely continuous, so the situation cannot get much worse.

3. LARGE OSCILLATIONS IN $P(t, 1, 1)$

The main results (35–36) of this section are taken from (Blackwell and Freedman, 1968). They should be compared with (1.1-4) of *ACM*. Let $I_n = \{1, 2, \ldots, n\}$. Let P_n be a generic standard stochastic semigroup on I_n.

(35) Theorem. *For any $\delta > 0$, there is a P_n with*

$$P_n(t, 1, 1) < \delta \quad for \; \delta \leqq t \leqq 1 - \delta \quad and \quad 1 + \delta \leqq t \leqq 2 - \delta$$

while

$$P_n(1, 1, 1) > 1/e.$$

In particular,

$$t \to P(t, 1, 1), \quad t \to \frac{1}{t} \int_0^t P(s, 1, 1) \, ds, \quad t \to \frac{1 - P(t, 1, 1)}{t}$$

are not monotone functions.

I remind you that

$$t \to \frac{1}{t} \int_0^t f(s) \, ds$$

is nonincreasing iff

$$f(t) \leqq \frac{1}{t} \int_0^t f(s) \, ds.$$

A more elegant nonmonotone P can be found in Section 4.

(36) Theorem. *For any $K < \frac{1}{2}$, for any small positive ε, there is a P_n with*

$$P_n(\tfrac{1}{2}, 1, 1) < 1 - \varepsilon - K\varepsilon^2$$

and

$$P_n(1, 1, 1) > 1 - \varepsilon.$$

I will prove (35) and (36) later. Here is some preliminary material, which will also be useful in *ACM*. Let q and c be positive real numbers. On a convenient probability triple $(\Omega, \mathscr{F}, \mathscr{P})$, let $T_0, T_1, \ldots$ be independent and exponential with common parameter q. Define a stochastic process Z as in Figure 3.

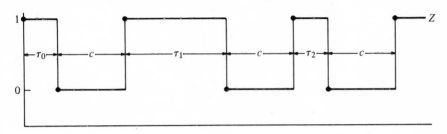

Figure 3.

$$Z(t) = 1 \quad \text{for} \qquad\qquad\qquad 0 \leq t < T_0$$
$$\text{and} \qquad\qquad T_0 + c \leq t < T_0 + c + T_1$$
$$\text{and} \quad T_0 + c + T_1 + c \leq t < T_0 + c + T_1 + c + T_2$$
$$\cdot$$
$$\cdot$$
$$\cdot$$

$$Z(t) = 0 \quad \text{for remaining } t.$$

Let

(37) $$f(t) = f(q, c; t) = \mathscr{P}\{Z(t) = 1\}.$$

The process Z is not Markov. But there are standard stochastic semigroups P_n on $\{1, 2, \ldots, n\}$ such that $P_n(t, 1, 1) \to f(t)$ uniformly on bounded t-sets. I will argue this later (43). Clearly,

(38) $$f(t) = e^{-qt} \quad \text{for } 0 \leq t \leq c.$$

by conditioning on T_0,

(39) $$f(t) = e^{-qt} + \int_0^{t-c} q e^{-q(t-c-s)} f(s) \, ds \quad \text{for } t \geq c.$$

LONG ARGUMENT FOR (39). Define a new process Z^* as follows:

$$Z^*(t) = Z(T_0 + c + t) \quad \text{for } 0 \leq t < \infty.$$

Now Z^* can be constructed from $T_1, T_2, \ldots$ just the way Z was constructed from $T_0, T_1, \ldots$; so Z^* is distributed like Z and is independent of T_0. For $t \geq c$,

$$\{Z(t) = 1\} = \{T_0 > t\} \cup \{T_0 \leq t - c \text{ and } Z^*(t - c - T_0) = 1\}.$$

So

$$f(t) = e^{-qt} + \mathscr{P}\{T_0 \leq t - c \text{ and } Z^*(t - c - T_0) = 1\}.$$

Fubini says

$$\mathscr{P}\{T_0 \leqq t - c \text{ and } Z^*(t - c - T_0) = 1\}$$
$$= \int_0^{t-c} \mathscr{P}\{Z^*(t - c - s) = 1\} \, qe^{-qs} \, ds$$
$$= \int_0^{t-c} f(t - c - s) \, qe^{-qs} \, ds$$
$$= \int_0^{t-c} qe^{-q(t-c-s)} f(s) \, ds.$$

This finishes (39).

Substitute (38) into (39):

(40) $f(t) = e^{-qt} + q(t - c)e^{-q(t-c)}$ for $c \leqq t \leqq 2c$.

Let $n \geqq 2$. Define a matrix Q_n on $I_n = \{1, \ldots, n\}$ as follows:

(41a) $Q_n(1, 1) = -q$ and $Q_n(1, 2) = q$

           and $Q_n(1, j) = 0$ for $j \neq 1, 2$;

if $1 \leqq i < n$,

(41b) $Q_n(i, i) = -(n - 1)/c$ and $Q_n(i, i + 1) = (n - 1)/c$

           and $Q_n(i, j) = 0$ for $j \neq i, i + 1$;

(41c) $Q_n(n, n) = -(n - 1)/c$ and $Q_n(n, 1) = (n - 1)/c$

           and $Q_n(n, j) = 0$ for $j \neq n, 1$.

Check that Q_n satisfies (5.28). Using (5.29),

(42) let P_n be the unique standard stochastic semigroup on I_n with $P_n'(0) = Q_n$.

(43) Proposition. *Fix positive, finite numbers q and c. Define f by (37) and P_n by (41, 42). Then*

$$\lim_{n \to \infty} P_n(t, 1, 1) = f(t)$$

uniformly in bounded t.

INFORMAL PROOF. Consider a Markov chain with stationary transitions P_n and starting state 1. The process moves cyclically $1 \to 2 \to \cdots \to n \to 1$. The holding times are unconditionally independent and exponentially distributed; the holding time in 1 has parameter q; the other holding time parameters are $(n - 1)/c$. There are $n - 1$ visits to other states intervening between successive visits to 1. So the gaps between the 1-intervals are independent and identically distributed, with mean c and variance $c^2/(n - 1)$. For large n, the first 10^{10} gaps are nearly c, so $P_n(t, 1, 1)$ is nearly $f(t)$ for all moderate t.   ★

FORMAL PROOF. Use (5.45). Let S be the set of right continuous I_n-valued step functions. Let X be the coordinate process on S. Let $\{\xi_m\}$ be the visiting process, and let $\{\tau_m\}$ be the holding time process. The probability $\pi = (P_n)_1$ on S makes X Markov with transitions P_n and starting state 1. Let $r(m)$ be one plus the remainder when m is divided by n, so

$$r(m) \in I_n \quad \text{and} \quad r(m) \equiv m + 1 \quad \text{modulo } n.$$

Let S_0 be the set where for all $m = 0, 1, \ldots$

$$0 < \tau_m < \infty \quad \text{and} \quad \xi_m = r(m).$$

Then

(44) $$\pi\{S_0\} = 1.$$

And with respect to π,

(45) $\tau_0, \tau_1, \ldots$ are unconditionally independent and exponentially distributed, the parameter for τ_m being q when m is a multiple of n, and $(n - 1)/c$ for other m.

Let $\theta_0, \theta_1, \ldots$ be the successive holding times of X in 1, and let $\gamma_0, \gamma_1, \ldots$ be the successive gaps between the 1-intervals of X. Formally, on S_0 let

$$\theta_m = \tau_{mn}$$
$$\gamma_m = \Sigma_\nu \{\tau_\nu : mn < \nu < (m + 1)n\}$$

for $m = 0, 1, \ldots$. With respect to π, I say:

(46) $\theta_0, \theta_1, \ldots$ are independent and exponentially distributed, with common parameter q;

(47) $\gamma_0, \gamma_1, \ldots$ are independent and identically distributed, with mean c and variance $c^2/(n - 1)$.

The reason for (46) is (45); the reason for (47) is (45) and (5.31). The θ's and the γ's are π-independent, but this doesn't affect the rest of the argument. Use (47) and Chebychev: for each $m = 0, 1, \ldots$.

(48) $\gamma_0 + \cdots + \gamma_m$ converges to $(m + 1)c$ in π-probability as $n \to \infty$.

WARNING. $S, X, \xi, \tau, \theta, \gamma$, and π all depend on n The π-distribution of γ depends on n. But the π-distribution of θ does not depend on n.

On some convenient probability triple $(\Omega, \mathscr{F}, \mathscr{P})$, let $T_0, T_1, \ldots$ be independent and exponentially distributed, with parameter q. So the π-distribution of $(\theta_0, \theta_1, \ldots)$ coincides by (46) with the $\mathscr{P}$-distribution of $(T_0, T_1, \ldots)$.

For $m = 0, 1, \ldots$, let $A_m(t)$ be the event that

$$\theta_0 + \gamma_0 + \cdots + \theta_m + \gamma_m \leq t < \theta_0 + \gamma_0 + \cdots + \theta_m + \gamma_m + \theta_{m+1}$$

and let $B_m(t)$ be the event that

$$T_0 + \cdots + T_m + (m + 1)c \leqq t < T_0 + \cdots + T_m + T_{m+1} + (m + 1)c.$$

WARNING. $A_m(t)$ depends on n; $B_m(t)$ doesn't.

Use (44): except for a π-null set,

$$\{X(t) = 1\} = \{\theta_0 > t\} \cup \{\mathsf{U}_{m=0}^{\infty} A_m(t)\};$$

so

$$P_n(t, 1, 1) = \pi\{\theta_0 > t\} + \Sigma_{m=0}^{\infty} \pi\{A_m(t)\}.$$

Use (7) and (48): for each $m = 0, 1, \ldots$

$$\pi\{A_m(t)\} \to \mathscr{P}\{B_m(t)\} \quad \text{uniformly in } t \text{ as } n \to \infty.$$

Fix t^* with $0 < t^* < \infty$, and confine t to $[0, t^*]$. Then

$$\pi\{A_m(t)\} \leqq \pi\{\theta_0 + \cdots + \theta_m \leqq t^*\} = \mathscr{P}\{T_0 + \cdots + T_m \leqq t^*\},$$

which is summable in m. And

$$\mathscr{P}\{B_m(t)\} = 0 \quad \text{when } (m + 1)c > T.$$

You can safely conclude

(49) $\lim_{n \to \infty} P_n(t, 1, 1) = \mathscr{P}\{T_0 > t\} + \Sigma_{m=0}^{\infty} \mathscr{P}\{B_m(t)\}$

uniformly in $t \leqq t^*$. Remember the definition (37) of f; by inspection, the right side of (49) is $f(t)$.   ★

PROOF OF (35). Define f by (37). Let $c < 1$ increase to 1, and let $q = 1/(1 - c)$. Then f tends to 0 uniformly on $[\delta, 1 - \delta]$ by (38), and on $[1 + \delta, 2 - \delta]$ by (40), while $f(1)$ decreases to $1/e$ by (40). That is, you can find q and c so $f = f(q, c: \cdot)$ satisfies the two inequalities of (35). Now use (43) to approximate f by $P_n(\cdot, 1, 1)$.   ★

PROOF OF (36). Define f by (37). Fix $c = \frac{1}{2}$, and let q decrease to 0. Abbreviate

$$g(q) = f(q, \tfrac{1}{2}: \tfrac{1}{2}) \quad \text{and} \quad h(q) = f(q, \tfrac{1}{2}: 1).$$

Then

$$g(q) = e^{-q/2} \qquad\qquad\qquad \text{by (38)}$$
$$= 1 - \tfrac{1}{2}q + \tfrac{1}{8}q^2 + 0(q^3);$$
$$h(q) = e^{-q} + \tfrac{1}{2}qe^{-q/2} \qquad\qquad \text{by (40)}$$
$$= 1 - \tfrac{1}{2}q + \tfrac{1}{4}q^2 + 0(q^3).$$

Consequently,

$$1 - h(q) = \tfrac{1}{2}q + 0(q^2);$$

$$h(q) - g(q) = \tfrac{1}{8}q^2 + 0(q^3)$$

$$= \tfrac{1}{2}[1 - h(q)]^2 + 0[1 - h(q)]^3.$$

Fix K with $0 < K < \tfrac{1}{2}$. Choose $q^* > 0$ but so small that on $(0, q^*)$:

$$h \text{ is strictly decreasing;}$$

$$h - g > K(1 - h)^2.$$

Let $0 < \varepsilon < 1 - h(q^*)$. Choose q' with $0 < q' < q^*$, so

$$h(q') = 1 - \varepsilon.$$

Then

$$g(q') < h(q') - K[1 - h(q')]^2 = 1 - \varepsilon - K\varepsilon^2.$$

By continuity, there is a positive q less than q', but very close, with

$$h(q) > 1 - \varepsilon$$

$$g(q) < 1 - \varepsilon - K\varepsilon^2.$$

That is, you can find small positive q and $c = \tfrac{1}{2}$ so that $f = f(q, c: \cdot)$ satisfies the two inequalities of (36). Now use (43) to approximate f by $P_n(\cdot, 1, 1)$.   ★

4. AN EXAMPLE OF SPEAKMAN

My object in this section is to give Speakman's (1967) example of two standard stochastic semigroups P and $\hat{P}$ on $I = \{1, 2, 3\}$, and $P = \hat{P}$ for some but not all times. Let

$$Q = \begin{pmatrix} -1 & 1 & 0 \\ 0 & -1 & 1 \\ 1 & 0 & -1 \end{pmatrix} \quad \text{and} \quad \hat{Q} = \begin{pmatrix} -1 & \tfrac{1}{2} & \tfrac{1}{2} \\ \tfrac{1}{2} & -1 & \tfrac{1}{2} \\ \tfrac{1}{2} & \tfrac{1}{2} & -1 \end{pmatrix}.$$

By (5.29), there are unique standard stochastic semigroups P and $\hat{P}$ on I with $P'(0) = Q$ and $\hat{P}'(0) = \hat{Q}$. In particular, $P \neq \hat{P}$.

(50) Theorem. $P(nc) = \hat{P}(nc)$ for $n = 0, 1, \ldots,$ where $c = 4\pi 3^{-\frac{1}{2}}$. However, $P \neq \hat{P}$.

PROOF.   It is even possible to write down P and $\hat{P}$ explicitly. Here is P.

$$P(t, 1, 1) = P(t, 2, 2) = P(t, 3, 3) = \frac{1}{3} + \frac{2}{3} e^{-3t/2} \cos\left(\frac{3^{\frac{1}{2}}t}{2}\right)$$

$$P(t, 1, 2) = P(t, 2, 3) = P(t, 3, 1) = \frac{1}{3} + \frac{2}{3} e^{-3t/2} \cos\left(\frac{3^{\frac{1}{2}}t}{2} - \frac{2\pi}{3}\right)$$

$$P(t, 1, 3) = P(t, 2, 1) = P(t, 3, 2) = \frac{1}{3} + \frac{2}{3} e^{-3t/2} \cos\left(\frac{3^{\frac{1}{2}}t}{2} + \frac{2\pi}{3}\right).$$

Here is $\hat{P}$.

$$\hat{P}(t, i, i) = \tfrac{1}{3} + \tfrac{2}{3}e^{-3t/2} \quad \text{for } i = 1, 2, 3$$

$$\hat{P}(t, i, j) = \tfrac{1}{3} - \tfrac{1}{3}e^{-3t/2} \quad \text{for } i \neq j \text{ in } I.$$

I will check P; you should check $\hat{P}$. By (5.29), the matrix Q generates a standard stochastic semigroup; by (5.26), the semigroup is the unique solution to the forward or backward system. Now $P(0)$ is the identity matrix, and P is differentiable; all I owe you is

$$P' = QP.$$

Ostensibly, there are 9 things to check. But P and Q are invariant under the cyclic permutation $1 \to 2 \to 3 \to 1$, so I only have to check row 1.

Abbreviate $\theta = 3^{\frac{1}{2}}t/2$ and $\lambda = 2\pi/3$. Remember from high school:

$$\cos \lambda = -\frac{1}{2} \quad \text{and} \quad \sin \lambda = \frac{3^{\frac{1}{2}}}{2}$$

$$\cos (\theta \pm \lambda) = -\frac{1}{2} \cos \theta \mp \frac{3^{\frac{1}{2}}}{2} \sin \theta.$$

In particular, rows 1 and 2 of P sum to 1; so row 1 of P' as well as row 1 of QP sum to 0, and I only have to check the $(1, 1)$ and $(1, 2)$ positions. I will do $(1, 1)$, and leave $(1, 2)$ to you.

Position $(1, 1)$ on the left is $P'(t, 1, 1)$, namely

$$-e^{-3t/2} \cos \theta - 3^{-\frac{1}{2}} e^{-3t/2} \sin \theta.$$

Position $(1, 1)$ on the right is $-P(t, 1, 1) + P(t, 2, 1)$, namely

$$-\tfrac{2}{3}e^{-3t/2} \cos \theta + \tfrac{2}{3}e^{-3t/2} \cos (\theta + \lambda).$$

Multiply both expressions by $\tfrac{3}{2}e^{3t/2}$ and add $\cos \theta$: the left side becomes

$$-\frac{1}{2} \cos \theta - \frac{3^{\frac{1}{2}}}{2} \sin \theta$$

and the right side becomes

$$\cos (\theta + \lambda).$$

I have already claimed the last two expressions are equal.

The same trigonometry, coupled with $\cos (2n\pi) = 1$, shows that $P(nc) = \hat{P}(nc)$.   ★

(51) Remark. $P(t, 1, 1)$ is not a monotone function of t.

5. THE EMBEDDED JUMP PROCESS IS NOT MARKOV

Consider a Markov chain with countable state space, stationary standard transitions, and continuous time parameter. Suppose all states are stable. As shown in (7.33), the embedded jump process is a discrete time Markov chain with stationary transitions, up to the first infinity. My object in this section is to show, by example, that the embedded jump process need not have this property between the first infinity and the second infinity, even when there is a second infinity. The possibility of this phenomenon was suggested by (Hunt, 1960); a related phenomenon is exhibited in Section 2.13 of *B & D*.

Let I be the integers. Let $\frac{1}{2} < p < 1$. A discrete time Markov chain is a *p-walk* iff it has state space I, and moves from j to $j + 1$ with probability p, and from j to $j - 1$ with probability $1 - p$. You should check that

$$\binom{n}{j} p^j (1 - p)^{n-j}$$

is maximized for j satisfying

$$np - (1 - p) < j \leqq np + p,$$

and is then of order $n^{-\frac{1}{2}}$: for example, use (Feller, 1968, Theorem on p. 158), followed by Stirling. In particular,

$$(52) \qquad \max_j \binom{n}{j} p^j (1 - p)^{n-j} \to 0 \quad \text{as } n \to \infty.$$

Let $\{T_n : n = \cdots -1, 0, 1, \ldots\}$ be a stochastic process on the probability triple $(\Omega, \mathscr{F}, \mathscr{P})$. Can $\{T_n\}$ be a p-walk? Suppose it were. Then

$$\mathscr{P}\{T_0 = k\} = \Sigma_j \, \mathscr{P}\{T_0 = k \mid T_{-n} = k - j\} \cdot \mathscr{P}\{T_{-n} = k - j\}$$

$$\leqq \max_j \mathscr{P}\{T_0 = k \mid T_{-n} = k - j\}.$$

But

$$\mathscr{P}\{T_0 = k \mid T_{-n} = k - j\} = \binom{n}{j} p^j (1 - p)^{n-j}.$$

Let $n \to \infty$ and use (52) to get $\mathscr{P}\{T_0 = k\} = 0$ for all k. Sum out k to get $1 = 0$.

(53)   No probability triple supports a p-walk with time parameter running through all the integers.

WARNING.   It's easy to manufacture a process $\{T_n:\text{integer } n\}$ such that $T_{n+1} - T_n$ are independent, 1 with probability p, and -1 with probability $1 - p$. For instance, start a p-walk at the value 0 with time going forward from 0; and start an independent $(1 - p)$-walk at the value 0 with time going backward from 0. Such a process has independent increments all right, but isn't Markov.

Endow I with the discrete topology, and let $\bar{I} = I \cup \{\varphi\}$ be its one point compactification. Here is the main result of this section.

(54) Theorem.   *There is a probability triple $(\mathscr{X}_0, \mathscr{F}, \pi_i)$, and a stochastic process $\{X(t): 0 \leqq t < \infty\}$ on $(\mathscr{X}_0, \mathscr{F}, \pi_i)$, such that;*

(a) *for each $x \in \mathscr{X}_0$, the function $X(\cdot, x)$ is $\bar{I}$-valued, continuous from the right, and has a limit from the left, at all times;*

(b) *$X(t, x) = \varphi$ iff $t = \varphi_n(x)$ for $n = 1, 2, \ldots$, where*

$$0 < \varphi_1(x) < \varphi_2(x) < \cdots < \varphi_n(x) \to \infty \quad as \quad n \to \infty;$$

at these times, $X(\cdot, x)$ is continuous;

(c) *Between $\varphi_n(x)$ and $\varphi_{n+1}(x)$, the function $X(\cdot, x)$ visits each state at least once;*

(d) *On $[0, \varphi_1)$, the sequence of states visited by X is a p-walk starting from i;*

(e) *X is a Markov chain with stationary standard transitions.*

Let $\zeta = \{\ldots, \zeta_{-1}, \zeta_0, \zeta_1, \ldots\}$ be $\mathscr{F}$-measurable, I-valued functions giving the order of visits paid by X on (φ_1, φ_2). Here is one way to formalize this idea. Suppose ϕ_0 is a function, with $\varphi_1 < \phi_0 < \varphi_2$, and X discontinuous at ϕ_0. For $n = 1, 2, \ldots$, let ϕ_n be the time of the nth discontinuity in X after ϕ_0, and let ϕ_{-n} be the time of the nth discontinuity in X before ϕ_0. Then ζ gives the order of visits paid by X on (φ_1, φ_2) iff there exists a ϕ_0 with $\varphi_1 < \phi_0 < \varphi_2$, and X discontinuous at ϕ_0, and

$$\zeta_n = X(\phi_n) \quad \text{for } n = \ldots, -1, 0, 1, \ldots.$$

The paradox embodied in (54) is

(55) Proposition.   *ζ cannot be Markov with stationary transitions.*

PROOF.   Suppose it were. I will use the strong Markov property separately on ζ and on X, to show that ζ is a p-walk. This contradicts (53). Fix $i = 0$. Let σ be the least n with $\zeta_n = 0$; this makes sense by (54c). Let τ be the least $t > \varphi_1$ with $X(t) = 0$. Then $\tau < \varphi_2$ by (54c). Now $\zeta_{\sigma+}$ is the sequence of states visited by $X(\tau + \cdot)$ up to its first infinity. By strong Markov (7.38), the shifted process $X(\tau + \cdot)$ is distributed like X. So $\zeta_{\sigma+}$ is a p-walk by (54d). But $\zeta_{\sigma+}$ has the same transitions as ζ, by discrete-time strong Markov. ★

Here are some preliminaries to (54). The segments of the visiting process in X up to the first infinity and between the successive infinities will be independent. The visiting process between the nth and $n + 1$st infinity will be a *strongly approximate p-walk*. I prefer not to define these objects now. But I will point to one. I warn you that the rest of this section is ridiculously hard.

A strongly approximate p-walk

Let $(\Omega, \mathcal{F}, \mathcal{P})$ be a convenient probability triple. On it, let $\{S_m(n) : n = 0, 1, \ldots\}$ be independent p-walks starting from m, for non-positive integer m. I am entitled to assume $S_m(n + 1) = S_m(n) \pm 1$ for all m and n. In view of (1.95, 96), I can also assume $S_m(n) \to \infty$ as $n \to \infty$ for all m. Let $\sigma_0 = 0$. For $m = -1, -2, \ldots$, let τ_m be the least n such that $S_m(n) = m + 1$, and let $\sigma_m = \tau_{-1} + \cdots + \tau_m$. Define a new process S as in Figure 4.

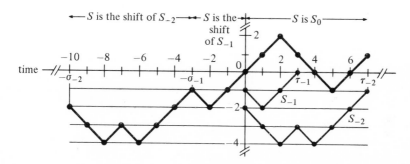

Figure 4.

$$
\begin{aligned}
(56) \qquad S(n) &= S_0(n) && \text{for } n = 0, 1, \ldots \\
&= S_m(\sigma_m + n) && \text{for } n = -\sigma_m, \ldots, -\sigma_{m+1} \\
& && \text{and } m = -1, -2, \ldots.
\end{aligned}
$$

Formally, S is defined twice at $-\sigma_m$; but the definitions agree and make $S = m$. If $n < -\sigma_m$, then $S(n) < m$. On the time interval $[-\sigma_m, -\sigma_{m+1}]$, the process S moves from first hitting m to first hitting $m + 1$; the sequence of moves performed by S on this interval coincides with the sequence of moves performed by S_m until S_m first hits $m + 1$. Clearly,

(57) $S(n + 1) = S(n) \pm 1$ for all n,   and $\lim_{n \to \pm \infty} S(n) = \pm \infty$.

NOTE. S is a strongly approximate p-walk.

Aside. You get the same effect by starting a p-walk at the value 0 with time going forward from 0, and an independent $(1 - p)$-walk at the value 0 with time going backward from 0, provided you condition the second walk on never returning to 0.

(58) Lemma.   *Let $j \in I$. Let λ be the least n with $S(n) = j$. Then*

$$\{S(\lambda + n) : n = 0, 1, \ldots\}$$

is a p-walk starting from j, and is independent of

$$\{S(\lambda - n) : n = 0, 1, \ldots\}.$$

PROOF.   You should do the easier case $j \geq 0$. I will argue the case $j < 0$. Then $\lambda = -\sigma_j$. You should check that $\{S(\lambda - n) : n \geq 0\}$ is measurable on $\{S_{-1}, S_{-2}, \ldots\}$. I will show that $\{S(\lambda + n) : n \geq 0\}$ is measurable on $\{S_j, S_{j+1}, \ldots, S_0\}$. This will prove the independence. I also have to argue that $S(\lambda + \cdot)$ is distributed like S_j.

To begin with,

$$S(\lambda + \cdot) = S_j(\cdot) \qquad \text{on } [0, \tau_j)$$
$$S(\lambda + \tau_j + \cdot) = S_{j+1}(\cdot) \qquad \text{on } [0, \tau_{j+1})$$

$$\cdot$$
$$\cdot$$
$$\cdot$$

$$S(\lambda + \tau_j + \cdots + \tau_{-2} + \cdot) = S_{-1}(\cdot) \qquad \text{on } [0, \tau_{-1})$$
$$S(\lambda + \tau_j + \cdots + \tau_{-1} + \cdot) = S_0(\cdot) \qquad \text{on } [0, \infty);$$

where τ_k by definition is the least n with $S_k(n) = k + 1$. Introduce the corresponding times $\theta_j, \ldots, \theta_{-1}$ for S_j: namely, $\theta_j = \tau_j$ is the least n with $S_j(n) = j + 1$; and θ_k is the least n with

$$S_j(\theta_j + \cdots + \theta_{k-1} + n) = k + 1.$$

Thus, $\theta_j + \cdots + \theta_k$ is the least n with $S_j(n) = k + 1$. Use successive doses of strong Markov on S_j to check the next assertion. The joint $\mathscr{P}$-distribution of the $|j| + 1$ fragments

$$S_j(\cdot) \quad \text{on } [0, \tau_j)$$
$$S_{j+1}(\cdot) \quad \text{on } [0, \tau_{j+1})$$

$$\cdot$$
$$\cdot$$
$$\cdot$$

$$S_{-1}(\cdot) \quad \text{on } [0, \tau_{-1})$$
$$S_0(\cdot) \quad \text{on } [0, \infty)$$

coincides with the joint $\mathscr{P}$-distribution of the $|j| + 1$ fragments

$$S_j(\cdot) \quad \text{on } [0, \theta_j)$$
$$S_j(\theta_j + \cdot) \quad \text{on } [0, \theta_{j+1})$$
$$\vdots$$
$$S_j(\theta_j + \cdots + \theta_{-2} + \cdot) \quad \text{on } [0, \theta_{-1})$$
$$S_j(\theta_j + \cdots + \theta_{-1} + \cdot) \quad \text{on } [0, \infty).$$

But $S(\lambda + \cdot)$ is obtained by laying the first lot of fragments together, end to end; and $S_j(\cdot)$ is obtained in a similar way from the second lot.   ★

It will be helpful to generalize (58). Let $N = 0, 1, \ldots$. Let Δ be the set where $S(n) = j$ for $N + 1$ or more n's. On Δ, let λ be the n such that $S(n) = j$ for the $N + 1$st time.

(59) Lemma.   *Given* Δ, *the process* $\{S(\lambda + n): n = 0, 1, \ldots\}$ *is a p-walk starting from* j, *independent of* $\{S(\lambda - n): n = 0, 1, \ldots\}$.

PROOF.   Use (58) and strong Markov (1.22).   ★

This ends the *p*-walk story.

The construction of π_i

Let C be the set of pairs (m, n) with $m = 0, 1, \ldots$, and $n = 0, 1, \ldots$ when $m = 0$; while n is any integer when $m > 0$. Write $(m', n') < (m, n)$ iff $m' < m$, or $m' = m$ but $n' < n$. The intervals of constancy in X will be indexed by C. Let Ω be the set of functions ω from C to I. Let

$$\xi(c)(\omega) = \omega(c) \quad \text{for} \quad c \in C \text{ and } \omega \in \Omega.$$

This ξ will be the visiting process in X. Give Ω the smallest σ-field over which all $\xi(c)$ are measurable.

(60) Definition.   *Let* Ω_0 *be the subset of* Ω *where:*

$$\xi(m, n + 1) = \xi(m, n) \pm 1 \quad \text{for all } (m, n) \in C;$$

$$\lim_{n \to \infty} \xi(0, n) = \infty;$$

$$\lim_{n \to \pm\infty} \xi(m, n) = \pm\infty \qquad \text{for all } m = 1, 2, \ldots .$$

(61) Definition. *For each $i \in I$, let Γ_i be the probability on Ω for which:*

$\xi(m, \cdot)$ *are independent for $m = 0, 1, \ldots$;*

$\xi(0, \cdot)$ *is a p-walk starting from i;*

$\xi(m, \cdot)$ *is distributed like the S of (56) for $m = 1, 2, \ldots$.*

Using (57),

(62) $\Gamma_i\{\Omega_0\} = 1.$

For $i \in I$, let $0 < q(i) < \infty$, with

(63) $\Sigma_{i \in I} \, 1/q(i) < \infty.$

The holding time parameter in i will be $q(i)$. Let W be the set of functions w from C to $(0, \infty)$. Let

$$\tau(c)(w) = w(c) \quad \text{for } c \in C \text{ and } w \in W.$$

The length of interval c will be $\tau(c)$. Give W the smallest σ-field over which all $\tau(c)$ are measurable.

Let $\mathscr{X} = \Omega \times W$, in the product σ-field. Let

(64) $\xi(c)(\omega, w) = \omega(c) \quad \text{and} \quad \tau(c)(\omega, w) = w(c)$

          for $c \in C$ and $\omega \in \Omega$ and $w \in W$.

For $A \subset \mathscr{X}$, let $A(\omega)$ be the ω-section of A:

(65) $A(\omega) = \{w : w \in W \text{ and } (\omega, w) \in A\}.$

(66) Definition. (a) *For $\omega \in \Omega$, let $\eta_{q(\omega)}$ be the probability on W which makes the variables $\tau(c)$ independent and exponentially distributed, the parameter for $\tau(c)$ being $q[\omega(c)]$.*

(b) *For $i \in I$, let π_i be this probability on $\mathscr{X}$:*

$$\pi_i\{A\} = \int_\Omega \eta_{q(\omega)}\{A(\omega)\} \, \Gamma_i(d\omega):$$

where Γ_i was defined in (61).

INFORMAL NOTE. The π_i-distribution of ξ is Γ_i. The π_i-conditional distribution of τ given $\xi = \omega$ is $\eta_{q(\omega)}$.

The main line of argument starts at (80).

Some lemmas

To state (67), let F be a finite subset of C. Let f be a function from F to I. Let W_F be the set of functions from F to $(0, \infty)$, with the product σ-field.

Let η_f be the probability on W_F which makes the coordinates independent and exponentially distributed, the parameter for coordinate $c \in F$ being $f(c)$. Let τ_F map $\mathscr{X}$ into W_F:

$$(\tau_F x)(c) = \tau(c)(x) \quad \text{for } c \in F \text{ and } x \in \mathscr{X}.$$

(67) Lemma. *For any measurable subset B of W_F,*

$$\pi_i\{\xi(c) = f(c) \text{ for } c \in F, \text{ and } \tau_F \in B\} = \Gamma_i\{\xi(c) = f(c) \text{ for } c \in F\} \cdot \eta_f\{B\}.$$

PROOF.   Use definition (66).   ★

To state (68), let $\zeta_m = \{\xi(m, \cdot), \tau(m, \cdot)\}$.

(68) Lemma. *With respect to π_i, the processes $\zeta_0, \zeta_1, \ldots$ are independent. For $m > 0$, the π_i-distribution of ζ_m depends neither on i nor on m.*

PROOF.   Fix positive integers M and N. Let $i(n) \in I$ and $0 \leq t(n) < \infty$ for $n = 0, \ldots, N$. Let $i(m, n) \in I$ and $0 \leq t(m, n) < \infty$ for $m = 1, \ldots, M$ and $n = -N, \ldots, N$. Let

$$s(0) = \Sigma_{n=0}^N q[i(n)]t(n)$$
$$s(m) = \Sigma_{n=-N}^N q[i(m, n)]t(m, n) \quad \text{for } m = 1, \ldots, M$$
$$s = s(0) + s(1) + \cdots + s(M).$$

Let

$$D_0 = \{\xi(0, n) = i(n) \text{ for } n = 0, \ldots, N\}$$

$$D_m = \{\xi(m, n) = i(m, n) \text{ for } n = -N, \ldots, N\} \quad \text{for } m = 1, \ldots, M$$

$$E_0 = \{\tau(0, n) > t(n) \text{ for } n = 0, \ldots, N\}$$

$$E_m = \{\tau(m, n) > t(m, n) \text{ for } n = -N, \ldots, N\} \quad \text{for } m = 1, \ldots, M.$$

Let

$$F = D_0 \cap D_1 \cap \cdots \cap D_M \cap E_0 \cap E_1 \cap \cdots \cap E_M.$$

Then

$$\pi_i\{F\} = \Gamma_i\{D_0 \cap D_1 \cap \cdots \cap D_M\} \cdot e^{-s} \quad \text{by (67)}$$

$$= \Gamma_i\{D_0\} \cdot \Gamma_i\{D_1\} \cdots \Gamma_i\{D_M\} \cdot e^{-s} \quad \text{by definition (61).}$$

By (67),

(69) $$\pi_i\{D_m \cap E_m\} = \Gamma_i\{D_m\} \cdot e^{-s(m)} \quad \text{for } m = 0, \ldots, M.$$

So

$$\pi_i\{F\} = \mathrm{II}_{m=0}^M \Gamma_i\{D_m \cap E_m\},$$

proving the first assertion through (10.16). The second assertion also follows from (69). Suppose $i(m, \cdot)$ and $t(m, \cdot)$ do not depend on $m > 0$. Then $s(m)$ does not depend on $m > 0$. And $\Gamma_i\{D_m\}$ depends neither on i nor on $m > 0$, by definition (61).   ★

The set Ω_1, the index λ, and the σ-field $\mathscr{A}$

Fix $M = 1, 2, \ldots$ and $N = 0, 1, \ldots$ and $j \in I$.

(70) Definition.   **(a)** *Let Ω_1 be the subset of Ω_0, as defined in (60), where $\xi(M, \cdot)$ visits j at least $N + 1$ times.*

(b) *On Ω_1, let λ be the index at which $\xi(M, \cdot)$ visits j for the $N + 1$st time. So*

(71) $$\xi(M, \lambda) = j,$$

and there are N integers n with $n < \lambda$ and $\xi(M, n) = j$.

(c) *Let $\mathscr{A}$ be the σ-field in Ω_1 generated by $\xi(m, \cdot)$ with $m < M$ and by $\xi(M, \lambda - n)$ with $n \geq 0$.*

(d) *Let $\lambda(\omega, w) = \lambda(\omega)$ for $\omega \in \Omega_1$ and $w \in W$.*

NOTATION. Ω_1, λ, and $\mathscr{A}$ all depend on M and N.

For (72) and the proof of (73), let $\mathscr{A}_0$ be the σ-field in Ω generated by $\xi(m, \cdot)$ with $m < M$. Let $\mathscr{A}_1$ be the σ-field in Ω_1 generated by $\xi(M, \lambda - n)$ with $n \geq 0$. Check

(72) $\mathscr{A}$ is the σ-field in Ω_1 generated by sets $J \cap K$ with $J \in \mathscr{A}_0$ and $K \in \mathscr{A}_1$.

NOTATION $K \in \mathscr{A}_1$ forces $K \subset \Omega_1$.

To state (73), let n_B be a positive integer. Let $i_B(n) \in I$ for $n = 0, \ldots, n_B$. Let

$$D = \{\omega : \omega \in \Omega_1 \text{ and } \omega[M, \lambda(\omega) + n] = i_B(n) \text{ for } n = 0, \ldots, n_B\}$$
$$D^* = \{\omega : \omega \in \Omega \text{ and } \omega(0, n) = i_B(n) \text{ for } n = 0, \ldots, n_B\}.$$

(73) Lemma.   *Let h be any nonnegative, $\mathscr{A}$-measurable function on Ω_1. Then*

$$\int_D h \, d\Gamma_i = \Gamma_j\{D^*\} \int_{\Omega_1} h \, d\Gamma_i.$$

PROOF.   Let $h = 1_{J \cap K}$, with $J \in \mathscr{A}_0$ and $K \in \mathscr{A}_1$. Now Ω_1, D, and K are in the σ-field spanned by $\xi(M, \cdot)$, which is Γ_i-independent of $\mathscr{A}_0$, from definition (61) of Γ_i. So

$$\Gamma_i\{J \cap K\} = \Gamma_i\{J\} \cdot \Gamma_i\{K\}$$
$$\Gamma_i\{J \cap K \cap D\} = \Gamma_i\{J\} \cdot \Gamma_i\{K \cap D\}.$$

Use (59) and (61):

$$\Gamma_i\{K \cap D\} = \Gamma_i\{K\} \cdot \Gamma_j\{D^*\}.$$

That is,

$$\Gamma_i\{J \cap K \cap D\} = \Gamma_i\{J \cap K\} \cdot \Gamma_j\{D^*\};$$

and (73) holds for this h. By (72) and (10.16), the result holds for $h = 1_A$ with $A \in \mathscr{A}$. It then holds for simple $\mathscr{A}$-functions by linearity, and non-negative $\mathscr{A}$-functions by monotone approximation.   ★

To state (74–79), let C^* be the set of all pairs $(m, n) \in C$ with $m < M$, together with the nonnegative integers. Let W^* be the set of all functions from C^* to $(0, \infty)$, with the product σ-field. Let n_A be a nonnegative integer. Let C_A be a finite subset of C, with $m < M$ for all $(m, n) \in C_A$. Suppose $0 < t_A(c) < \infty$ for $c \in C_A$ or $c = 1, \ldots, n_A$. Let W_A be the set of $w \in W^*$ such that

(74) $w(c) > t_A(c)$ for $c \in C_A$, and

$w(n) > t_A(n)$ for $n = 1, \ldots, n_A$, and

$\Sigma \{w(c) : c \in C^* \text{ but } c \neq 0\} \leqq t < \Sigma \{w(c) : c \in C^*\}.$

Check:

(75) W_A is a measurable subset of W^*.

Define a mapping τ_ω from W to W^*:

(76) $\tau_\omega(w)(m, n) = w(m, n)$ for $(m, n) \in C$ with $m < M$

$\tau_\omega(w)(n) = w[M, \lambda(\omega) - n]$ for $n = 0, 1, \ldots.$

Check:

(77) τ_ω is measurable

(78)   the $\eta_{q(\omega)}$-distribution of τ_ω is $\mathscr{A}$-measurable, as ω varies over Ω_1.

Conclude from (75, 77–78):

(79)   the function $\omega \to \eta_{q(\omega)}\{\tau_\omega \in W_A\}$ is $\mathscr{A}$-measurable on Ω_1.

The construction of $\mathscr{X}_0$ and X

(80) Definition.   *On W:*

   (a) *Let $\sigma(c)$ be the sum of $\tau(d)$ over $d \in C$ with $d < c$.*

   (b) *Let $r(m)$ be the sum of $\tau(m, n)$ over n with $(m, n) \in C$.*

   (c) *Let $\sigma(c)(\omega, w) = \sigma(c)(w)$ and $r(m)(\omega, w) = r(m)(w)$ for $\omega \in \Omega$ and $w \in W$.*

   (d) *Let W_0 be the subset of W where $\sigma(c) < \infty$ for all c, but $\sup_c \sigma(c) = \infty$.*

(e) *Let* $\mathscr{X}_0 = \Omega_0 \times W_0$, *where* Ω_0 *was defined in* (60). *Give* $\mathscr{X}_0$ *the product σ-field.*

(f) *On $\mathscr{X}_0$, let*

$$X(t) = \xi(m, n) \quad \text{when } \sigma(m, n) \leqq t < \sigma(m, n + 1)$$
$$= \varphi \qquad\quad \text{when } \sigma(m, n) \leqq t < \sigma(m, n + 1) \quad \text{for no } (m, n).$$

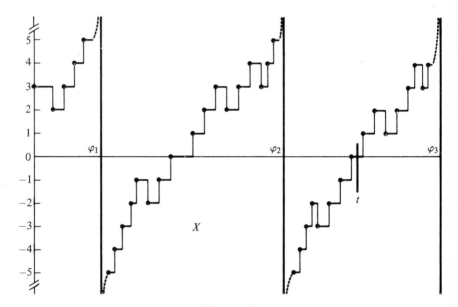

Figure 5.

Definition (80f) is illustrated by Figure 5. You should check

(81) X is jointly measurable.

With some thinking, you can get

(82) X has properties (54a–c). The map $(m, n) \to [\sigma(m, n), \sigma(m, n + 1))$ is order-preserving from C onto the intervals of constancy in X. The value of X on interval c is $\xi(c)$, and the length of interval c is $\tau(c)$. Furthermore,

$$\varphi_n = r(0) + \cdots + r(n - 1) \quad \text{for } n = 1, 2, \ldots;$$

where r is defined in (80b), and φ_n is defined in (54b).

(83) Lemma. $\pi_i\{\mathscr{X}_0\} = 1.$

PROOF.   You have to remember definitions (66) of π_i and (80) of $\mathcal{X}_0$. I will refer to r, defined in (80b). I will argue that

$$(84) \qquad \int_{\Omega_0 \times W} r(m) \, d\pi_i < \infty \qquad \text{for } m = 0, 1, \ldots.$$

Relations (62, 84) and (10.10b) make

$$\pi_i\{\Omega_0 \times W \text{ and } r(m) < \infty\} = 1 \quad \text{for } m = 0, 1, \ldots.$$

Countable additivity makes

$$\pi_i\{\Omega_0 \times W \text{ and } r(m) < \infty \text{ for } m = 0, 1, \ldots\} = 1.$$

By thinking,

$$\{r(m) < \infty \text{ for } m = 0, 1, \ldots\} = \{\sigma(c) < \infty \text{ for } c \in C\}.$$

Next,

$$\sup_c \sigma(c) = \Sigma_{m=0}^\infty r(m).$$

Lemma (68) makes $r_0, r_1, \ldots$ independent; the r_m being identically distributed for $m > 0$. Since r_1 is positive,

$$\pi_i\{\Omega_0 \times W \text{ and } \Sigma_{m=0}^\infty r(m) = \infty\} = 1.$$

So (83) reduces to (84).

To argue (84), let $V(j, m)(\omega)$ be the number of n with $\omega(m, n) = j$. Remember the definition (61) of Γ_i. Use (58) and (1.98):

$$(85a) \qquad \int_{\Omega_0} V(j, m) \, d\Gamma_i = \frac{1}{2p-1} \qquad \text{for } m > 0$$

$$(85b) \qquad \int_{\Omega_0} V(j, 0) \, d\Gamma_i = \frac{1}{2p-1} \qquad \text{for } j \geqq i$$

$$(85c) \qquad \int_{\Omega_0} V(j, 0) \, d\Gamma_i = \left(\frac{1-p}{p}\right)^{i-j} \cdot \frac{1}{2p-1} \quad \text{for } j < i.$$

By (85),

$$(86) \qquad \int_{\Omega_0} V(j, m) \, d\Gamma_i \leqq \frac{1}{2p-1} \qquad \text{for all } i, j, m.$$

Use (5.31) and definition (55) of η:

(87)   for each ω, the function $w \to w(m, n)$ on W has $\eta_{q(\omega)}$-expectation equal to $1/q[\omega(m, n)]$.

Let us compute together, keeping j in I, and m fixed, and (m, n) in C.

$$\int_{\Omega_0 \times W} r(m)\, d\pi_i = \int_{\Omega_0} \int_W r(m)(\omega, w)\, \eta_{q(\omega)}(dw)\, \Gamma_i(d\omega) \qquad \text{by (66)}$$

$$= \int_{\Omega_0} \sum_n \int_W \tau(m, n)(\omega, w)\, \eta_{q(\omega)}(dw)\, \Gamma_i(d\omega) \quad \begin{array}{l}\text{by monotone} \\ \text{convergence}\end{array}$$

$$= \int_{\Omega_0} \sum_n \int_W w(m, n)\, \eta_{q(\omega)}\, \Gamma_i(d\omega) \qquad \text{by (64)}$$

$$= \int_{\Omega_0} \sum_n 1/q[\omega(m, n)]\, \Gamma_i(d\omega) \qquad \text{by (87)}$$

$$= \int_{\Omega_0} \sum_j V(j, m)(\omega)/q(j)\, \Gamma_i(d\omega) \qquad \text{by rearranging}$$

$$= \sum_j \frac{1}{q(j)} \int_{\Omega_0} V(j, m)(\omega)\, \Gamma_i(d\omega) \qquad \begin{array}{l}\text{by monotone} \\ \text{convergence}\end{array}$$

$$\leqq \sum_j \frac{1}{q(j)} \frac{1}{(2p - 1)} \qquad \text{by (86)}$$

$$< \infty \qquad \text{by (63).} \qquad \bigstar$$

This completes the construction: $(\mathscr{X}_0, \pi_i)$ is a bona fide probability triple by (83), and X is an $\bar{I}$-valued process on this triple by (81). Properties (54a–c) have already been claimed (82). At this point, you could check property (54d), and

(88) $\pi_i\{X(0) = i\} = 1$ for all $i \in I$.

(89) Lemma. $\pi_i\{X(t) \in I\} = 1$.

PROOF.   Using (54b),

$$\text{Lebesgue } \{t : X(t, x) = \varphi\} = 0.$$

Let E be the set of t such that

$$\pi_j\{X(t) = \varphi\} > 0 \quad \text{for some } j.$$

Use (81) and Fubini to deduce

(90) $\text{Lebesgue } E = 0.$

Fix $i \in I$. Let S map Ω into itself:

$$(S\omega)(0, n) = \omega(0, n + 1)$$
$$(S\omega)(m, n) = \omega(m, n) \quad \text{for } m > 0.$$

Using (60, 61), check that S maps Ω_0 into itself, and

(91)   the Γ_i-distribution of S is $p\Gamma_{i+1} + (1 - p)\Gamma_{i-1}$.

Let T map W into itself:

$$(Tw)(0, n) = w(0, n + 1)$$
$$(Tw)(m, n) = w(m, n) \quad \text{for } m > 0.$$

Using definition (80), check that T maps W_0 into itself. Remember definition (66a) of η. Relative to $\eta_{q(\omega)}$:

(92) T is independent of $\tau(0, 0)$ and has distribution $\eta_{q(S\omega)}$;
             $\tau(0, 0)$ is exponential with parameter $q[\omega(0, 0)]$.

Let U map $\mathscr{X}$ into itself:

$$U(\omega, w) = (S\omega, Tw).$$

Using (80), check that U maps $\mathscr{X}_0$ into itself, and

$$X(t, x) = X[t - \tau(0, 0)(x), U(x)] \quad \text{when } \tau(0, 0)(x) \leqq t.$$

INFORMAL NOTE. U shifts X by $\tau(0, 0)$.
So

(93) $\{X(0, \cdot) = i \text{ and } X(t, \cdot) = \varphi\}$
       $= \{X(0, \cdot) = i \text{ and } \tau(0, 0) \leqq t \text{ and } X[t - \tau(0, 0), U] = \varphi\}$.

Abbreviate $q = q(i)$ and $\pi = p\pi_{i+1} + (1 - p)\pi_{i-1}$. Combine (91, 92) to see that relative to π_i,

(94) U is independent of $\tau(0, 0)$

       U has distribution π

       $\tau(0, 0)$ is exponential with parameter q.

Fubini up (93, 94):

$$\pi_i\{X(t) = \varphi\} = q \int_0^t e^{-qs} \pi\{X(t - s) = \varphi\} \, ds.$$

But $\pi\{X(t - s) = \varphi\} = 0$ for Lebesgue almost all $s \in [0, t]$ by (90).   ★

Let $\mathscr{F}$ be the product σ-field on $\mathscr{X}_0$. Let $\mathscr{F}(t)$ be the σ-field spanned by $X(s)$ for $s \leqq t$. Let $s, t \geqq 0$. Let $i, j, k \in I$. Let $A \in \mathscr{F}(t)$ with $A \subset \{X(t) = j\}$. I will eventually succeed in arguing

(95) Markov property. $\pi_i\{A \text{ and } X(t + s) = k\} = \pi_i\{A\} \cdot \pi_j\{X(s) = k\}$.

Relations (88, 89, 95) and lemma (5.4) make X Markov with stationary transitions and starting state i, relative to π_i. These transitions have to be standard by (54a). So (54d) reduces to (95); but the proof of (95) is hairy.

WARNING.   Theorem (6.108) does not apply, because condition (6.66) fails.

The special A

(96) Definition.   *Remember the definition (80) of $\mathscr{X}_0$ and X. Let $\mathscr{X}(0)$ be the subset of $\mathscr{X}_0$ where*

$$0 \leqq t < \varphi_1 \quad and \quad X(t) = j.$$

Let $M = 1, 2, \ldots$ and $N = 0, 1, \ldots$. Let $\mathscr{X}(M, N)$ be the subset of $\mathscr{X}_0$ where:

$\varphi_M < t < \varphi_{M+1} \quad and \quad X(t) = j$; and

the number of j-intervals in X after φ_M but before the one surrounding t is N.

Then $\mathscr{X}(0)$ and $\mathscr{X}(M, N)$ are in $\mathscr{F}(t)$. These sets are pairwise disjoint as M and N vary; their union is $\{X(t) = j\}$. You should prove (95) when $A \subset \mathscr{X}(0)$: it's similar to (5.39). The proof I will give works for these A's, if you treat the notation gently; but it's silly.

(97)   I only have to prove (95) for $A \subset \mathscr{X}(M, N)$.

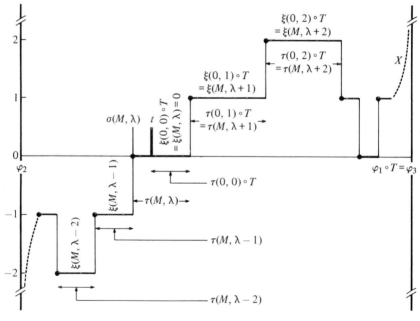

Figure 6.

So fix positive integer M and nonnegative integer N. Define Ω_1, λ, and $\mathscr{A}$ by (70), with this choice of M and N. Review (80) and look at Figure 6. Use

(71, 82) to make sure that

(98) $\quad \mathscr{X}(M, N) = \{\Omega_1 \times W_0 \text{ and } \sigma(M, \lambda) \leq t < \sigma(M, \lambda + 1)\}.$

On $\Omega_1 \times W_0$, check:

(99) $\quad\quad\quad \varphi_M = \Sigma_{(m,n)} \{\tau(m, n) : m < M\};$

(100) $\quad \sigma(M, \lambda - n) = \varphi_M + \Sigma_{\nu=n+1}^\infty \tau(M, \lambda - \nu);$

(101) $\quad\quad X(s) = \xi(M, \lambda - n) \quad \text{for } \sigma(M, \lambda - n) \leq s < \sigma(M, \lambda - n + 1).$

(102) Lemma. *Let $m < M$. Then $\xi(m, n)$ and $\tau(m, n)$ can be measurably computed from $\{X(s): 0 \leq s < \varphi_M\}$, at least on $\mathscr{X}_0$.*

PROOF. Number the intervals of constancy in $[0, \varphi_1]$ from left to right so that interval number 0 has left endpoint 0. Then X is $\xi(0, n)$ on interval n, which has length $\tau(0, n)$, for $n = 0, 1, \ldots$. Let $1 \leq m < M$. Let ϕ be the least $t > \varphi_m$ with $X(t) = 0$. Number the intervals of constancy on $(\varphi_m, \varphi_{m+1})$ from left to right so that interval number 0 has left endpoint ϕ. Then X is $\xi(m, n)$ on interval n, which has length $\tau(m, n)$, for integer n. ★

(103) Let $\mathscr{B}$ be the σ-field in $\mathscr{X}(M, N)$ spanned by $\xi(m, \cdot)$ and $\tau(m, \cdot)$ with $m < M$ and domain cut down to $\mathscr{X}(M, N)$. Let $\mathscr{C}$ be the σ-field in $\mathscr{X}(M, N)$ spanned by $\xi(M, \lambda - n)$ and $\tau(M, \lambda - n)$ with $n = 1, 2, \ldots$ and domain cut down to $\mathscr{X}(M, N)$.

(104) Lemma. *$\mathscr{X}(M, N) \cap \mathscr{F}(t)$ is spanned by $\mathscr{B}$ and $\mathscr{C}$.*

PROOF. $\varphi_M < t$ on $\mathscr{X}(M, N)$. So $\mathscr{B} \subset \mathscr{X}(M, N) \cap \mathscr{F}(t)$ by (102). Next, $\mathscr{C} \subset \mathscr{X}(M, N) \cap \mathscr{F}(t)$, because the nth interval of constancy in X before the one at time t is a visit to $\xi(M, \lambda - n)$ of length $\tau(M, \lambda - n)$: use (98–101). I now have to compute $\{X(s): 0 \leq s \leq t\}$ on $\mathscr{X}(M, N)$ from $\mathscr{B}$ and $\mathscr{C}$. To begin, you can compute $\{X(s): 0 \leq s \leq \varphi_M\}$ on $\mathscr{X}(M, N)$ from $\mathscr{B}$, using definition (80): and φ_M retracted to $\mathscr{X}(M, N)$ is $\mathscr{B}$-measurable by (99). So $\sigma(M, \lambda - n)$ retracted to $\mathscr{X}(M, N)$ is $\mathscr{B} \vee \mathscr{C}$-measurable by (100), for $n = 0, 1, \ldots$. You can now compute the fragment $\{X(s): \varphi_M < s < \sigma(M, \lambda)\}$ on $\mathscr{X}(M, N)$ from $\mathscr{B} \vee \mathscr{C}$, using (101). Finally,

$$X(s) = j \quad \text{for} \quad \sigma(M, \lambda) \leq s \leq t \quad \text{on } \mathscr{X}(M, N);$$

my authority is (71, 98, 101). But I peek at Figure 6. ★

WARNING. λ retracted to $\mathscr{X}(M, N)$ is not $\mathscr{F}(t)$-measurable.

(105) Definition. *Review (70a, b). Call a set A special iff there is a finite subset C_A of C, with $m < M$ for all $(m, n) \in C_A$, a nonnegative integer n_A,* *a function i_A from $\{1, \ldots, n_A\} \cup C_A$ to I, and a function t_A from $\{1, \ldots, n_A\} \cup C_A$ to $(0, \infty)$, such that A is the subset of $\Omega_1 \times W_0$ where:*

$$\xi(c) = i_A(c) \quad and \quad \tau(c) > t_A(c) \quad for \ all \ c \in C_A;$$
$$\xi(M, \lambda - n) = i_A(n) \quad and \quad \tau(M, \lambda - n) > t_A(n) \quad for \ all \quad n = 1, \ldots, n_A;$$
$$\sigma(M, \lambda) \leqq t < \sigma(M, \lambda + 1).$$

I claim

(106)   I only have to prove (95) for special A.

Indeed, the special A are subsets of $\mathcal{X}(M, N)$ by (98). They span $\mathcal{X}(M, N) \cap \mathcal{F}(t)$ by (104). Two different special A are disjoint or nested, by inspection. And $\mathcal{X}(M, N)$ is special. So (106) follows from (97) and (10.16).

The mapping T

Remember definition (96) of $\mathcal{X}(M, N)$. Define a mapping T of $\mathcal{X}(M, N)$ into $\mathcal{X}$, as in Figure 6 on page 286:

(107a) $\xi(0, n) \circ T = \xi(M, \lambda + n)$ for $n \geqq 0$;

(107b) $\xi(m, n) \circ T = \xi(M + m, n)$ for $m > 0$;

(107c) $\tau(0, 0) \circ T = \sigma(M, \lambda + 1) - t$;

(107d) $\tau(0, n) \circ T = \tau(M, \lambda + n)$ for $n > 0$;

(107e) $\tau(m, n) \circ T = \tau(M + m, n)$ for $m > 0$.

Review definition (80) of $\mathcal{X}_0$ and X. Check that T maps into $\mathcal{X}_0$, and

(108) $X(t + s) = X(s) \circ T$ on $\mathcal{X}(M, N)$.

Relation (108) is a straightforward but tedious project, which I leave to you. Consider the assertion

(109) $\pi_i\{A \text{ and } T \in B\} = \pi_i\{A\} \cdot \pi_j\{B\}.$

I claim

(110)   it is enough to prove (109) for special A and all measurable subsets B of $\{\xi(0, 0) = j\}$.

To see this, put $B = \{X(0) = j \text{ and } X(s) = k\}$ in (109). Then use (108) to get (95) for the special A. Then use (106).

The special B

(111) **Definition.**   *A set B is special iff there is a finite subset C_B of C, with $m > 0$ for all $(m, n) \in C_B$, a nonnegative integer n_B, a function i_B from*

$\{0, \ldots, n_B\} \cup C_B$ to I, and a function t_B from $\{0, \ldots, n_B\} \cup C_B$ to $(0, \infty)$, such that

$$i_B(0) = j, \quad \text{and}$$

$$B = B_1 \cap B_2, \quad where$$

$$B_1 = \{\xi(0, n) = i_B(n) \text{ and } \tau(0, n) > t_B(n) \text{ for } n = 0, \ldots, n_B\}, \quad while$$

$$B_2 = \{\xi(c) = i_B(c) \text{ and } \tau(c) > t_B(c) \text{ for } c \in C_B\}.$$

I claim

(112)   it is enough to prove (109) for special A and special B.

Indeed, the special B span the full σ-field on $\{\xi(0, 0) = j\}$. Two different special B are disjoint or nested, by inspection. And $\{\xi(0, 0) = j\}$ is special. So (112) follows from (110) and (10.16).

The ultraspecial B

(113)   Call B *ultraspecial* iff B is special in the sense of (111), and C_B is empty: so $B = B_1$, as defined in (111).

I claim

(114)   it is enough to prove (109) for special A and ultraspecial B.

Fix a special B, in the sense of (111). Remember C_B, n_B, i_B, t_B, B_1 and B_2 from (111). Remember

(115) $m > 0 \quad$ for $(m, n) \in C_B$.

Let D_1 be the subset of $\mathscr{X}(M, N)$ where $\xi(M, \lambda + n) = i_B(n)$ for $n = 0, \ldots, n_B$ and $\sigma(M, \lambda + 1) > t + t_B(0)$ and $\tau(M, \lambda + n) > t_B(n)$ for $n = 1, \ldots, n_B$.

Let D_2 be the subset of $\mathscr{X}$ where $\xi(M + m, n) = i_B(m, n)$ and $\tau(M + m, n) > t_B(m, n)$ for all $(m, n) \in C_B$.

Check

(116a) $T^{-1}B_1 = D_1 \quad$ and $\quad T^{-1}B = D_1 \cap D_2$.

Get an A from (105). I claim:

(116b) $\pi_i\{A \cap D_1 \cap D_2\} = \pi_i\{A \cap D_1\} \cdot \pi_i\{D_2\}$

(116c) $\pi_i\{D_2\} = \pi_j\{B_2\}$

(116d) $\pi_j\{B_1 \cap B_2\} = \pi_j\{B_1\} \cdot \pi_j\{B_2\}.$

Remember $\zeta_m = \{\xi(m, \cdot), \tau(m, \cdot)\}$. In order, $\mathscr{X}(M, N)$, D_1, and $A \cap D_1$ are all measurable on $\zeta_0, \ldots, \zeta_M$: use (98) for the first move. Next, D_2 is measurable on $(\zeta_{M+1}, \zeta_{M+2}, \ldots)$ by (115). So (68) proves (116b). Relation (116c) follows from (115) and (68). Finally, B_1 is measurable on ζ_0; and (115) makes B_2 measurable on $(\zeta_1, \zeta_2, \ldots)$. So (68) proves (116d). Suppose (109) for ultraspecial B. Compute:

$$
\begin{aligned}
\pi_i\{A \cap T^{-1}B\} &= \pi_i\{A \cap D_1 \cap D_2\} && \text{by (116a)} \\
&= \pi_i\{A \cap D_1\} \cdot \pi_i\{D_2\} && \text{by (116b)} \\
&= \pi_i\{A \cap D_1\} \cdot \pi_j\{B_2\} && \text{by (116c)} \\
&= \pi_i\{A \cap T^{-1}B_1\} \cdot \pi_j\{B_2\} && \text{by (116a)} \\
&= \pi_i\{A\} \cdot \pi_j\{B_1\} \cdot \pi_j\{B_2\} && \text{by (109 on } B_1) \\
&= \pi_i\{A\} \cdot \pi_j\{B_1 \cap B_2\} && \text{by (116d)} \\
&= \pi_i\{A\} \cdot \pi_j\{B\} && \text{by (111).}
\end{aligned}
$$

This settles (114). I wish I could reward you for coming this far, but the worst lies ahead.

The proof of (109) for special A and ultraspecial B

Review definitions (70, 80). For the rest of the argument, $\tau(m, n)$ and $\sigma(m, n)$ have domain W. Fix one special A in the sense of (105), and one ultraspecial B in the sense of (113). Remember C_A, n_A, i_A and t_A from (105). Remember $n_B, i_B,$ and t_B from (113, 111). Introduce the following subsets of Ω: read (117–119) with the list.

$A_1 = \{\omega : \omega \in \Omega_1 \text{ and } \omega(c) = i_A(c) \text{ for all } c \in C_A, \text{ and}$
$\qquad\qquad\qquad \omega[M, \lambda(\omega) - n] = i_A(n) \text{ for all } n = 1, \ldots, n_A\}.$

$D = \{\omega : \omega \in \Omega_1 \text{ and } \omega[M, \lambda(\omega) + n] = i_B(n) \text{ for all } n = 0, \ldots, n_B\}.$

$D^* = \{\omega : \omega \in \Omega \text{ and } \omega(0, n) = i_B(n) \text{ for all } n = 0, \ldots, n_B\}.$

Introduce the subset H of W:

$\qquad H = \{w : w \in W \text{ and } w(0, n) > t_B(n) \text{ for all } n = 0, \ldots, n_B\}.$

For each $\omega \in \Omega_1$, introduce the following subsets of W.

$A_\omega^- = \{w : w \in W \text{ and } w(c) > t_A(c) \text{ for all } c \in C_A, \text{ and}$
$\qquad\qquad\qquad w[M, \lambda(\omega) - n] > t_A(n) \text{ for all } n = 1, \ldots, n_A\}.$
$A_\omega^+ = \{w : w \in W \text{ and } \sigma[M, \lambda(\omega)](w) \leqq t < \sigma[M, \lambda(\omega) + 1](w)\}.$
$A_\omega = A_\omega^- \cap A_\omega^+.$
$E_\omega = \{w : w \in W \text{ and } \sigma[M, \lambda(\omega) + 1](w) > t + t_B(0)\}.$
$F_\omega = \{w : w \in W \text{ and } w[M, \lambda(\omega) + n](w) > t_B(n) \text{ for all } n = 1, \ldots, n_B\}.$

Check

(117) $A = \{(\omega, w): \omega \in A_1 \text{ and } w \in A_\omega \cap W_0\}$

(118) $B = D^* \times H.$

Remember $i_B(0) = j$; use (117–118) and definition (107) to check

(119) $A \cap T^{-1}B = \{(\omega, w): \omega \in A_1 \cap D \text{ and } w \in A_\omega \cap E_\omega \cap F_\omega \cap W_0\}.$

From (119) and definition (66b):

(120) $\pi_i\{A \cap T^{-1}B\} = \int_{A_1 \cap D} \eta_{q(\omega)}\{A_\omega \cap E_\omega \cap F_\omega \cap W_0\}\, \Gamma_i(d\omega).$

As (83) and definition (66b) imply,

(121) $\Gamma_i\{\omega: \eta_{q(\omega)}(W_0) = 1\} = 1.$

By (120, 121):

(122) $\pi_i\{A \cap T^{-1}B\} = \int_{A_1 \cap D} \eta_{q(\omega)}\{A_\omega \cap E_\omega \cap F_\omega\}\, \Gamma_i(d\omega).$

Fix $\omega \in A_1 \cap D$. Let $\mathscr{A}_\omega$ be the σ-field in W spanned by $\tau(m, n)$ with $(m, n) \leqq (M, \lambda(\omega))$. Relative to $\eta_{q(\omega)}$,

$$\mathscr{A}_\omega \quad \text{and} \quad \tau[M, \lambda(\omega) + n] \quad \text{for } n = 1, \ldots, n_B$$

are mutually independent, the nth variable being exponential with parameter $q[i_B(n)]$. This follows from definition (66a). But $A_\omega \cap E_\omega \in \mathscr{A}_\omega$. So

(123) $\eta_{q(\omega)}\{A_\omega \cap E_\omega \cap F_\omega\} = e^{-v}\,\eta_{q(\omega)}\{A_\omega \cap E_\omega\}, \quad \text{where}$

$$v = \sum_{n=1}^{n_B} t_B(n)q[i_B(n)].$$

Let Σ_ω be the σ-field in W spanned by $\tau(m, n)$ with $(m, n) < (M, \lambda(\omega))$. Relative to $\eta_{q(\omega)}$:

Σ_ω and $\tau[M, \lambda(\omega)]$ are independent;

$\tau[M, \lambda(\omega)]$ is exponential with parameter $q(j)$.

This follows from (71) and definition (66a). But A_ω^- and $\sigma[M, \lambda(\omega)]$ are Σ_ω-measurable. And

$$\sigma[M, \lambda(\omega) + 1] = \sigma[M, \lambda(\omega)] + \tau[M, \lambda(\omega)].$$

So (5.30) makes

(124) $\eta_{q(\omega)}\{A_\omega \cap E_\omega\} = e^{-u}\,\eta_{q(\omega)}\{A_\omega\}, \quad \text{where } u = q(j)t_B(0).$

Combine (122–124):

(125) $\pi_i\{A \cap T^{-1}B\} = e^{-u-v}\int_D 1_{A_1}\,\eta_{q(\omega)}\{A_\omega\}\, \Gamma_i(d\omega).$

Review definition (70c) of $\mathscr{A}$, definition (74) of W_A, and (76) of τ_ω. Check

$$A_\omega = \{\tau_\omega \in W_A\}.$$

Conclude from (79) that

$$\omega \to \eta_{q(\omega)}\{A_\omega\}$$

is $\mathscr{A}$-measurable on Ω_1. Check $A_1 \in \mathscr{A}$. Use (73, 125):

$$\pi_i\{A \cap T^{-1}B\} = \left[\int_{A_1} \eta_{q(\omega)}\{A_\omega\}\Gamma_i(d\omega)\right] \cdot [e^{-u-v}\,\Gamma_j\{D^*\}].$$

Use (117, 121) and definition (66b):

$$\int_{A_1} \eta_{q(\omega)}\{A_\omega\}\,\Gamma_i(d\omega) = \int_{A_1} \eta_{q(\omega)}\{A_\omega \cap W_0\}\,\Gamma_i(d\omega) = \pi_i\{A\}.$$

Use (118) and (67):

$$e^{-u-v}\,\Gamma_j\{D^*\} = \pi_j\{B\}.$$

This settles (109) for special A and ultraspecial B.   ★ ★ ★ ★ ★

6. ISOLATED INFINITIES

The ideas of Section 5 can be used to construct the most general Markov chain with all states stable and isolated infinities. In this section, an *infinity* is a time t such that: in any open interval around t, the sample function makes infinitely many jumps. I will sketch the program. To begin with, let Γ be a substochastic matrix on the countably infinite set I. A *strongly approximate* Γ-chain on the probability triple $(\Omega, \mathscr{F}, \mathscr{P})$ is a partially defined, I-valued process $\xi(n)$, which has the strong Markov property for hitting times. More exactly, let J be a finite subset of I. Let Ω_J be the set where $\xi(n) \in J$ for some n; on Ω_J, let τ_J be the least such n: assume there is a least. Given Ω_J, the process $\xi(\tau_J + \cdot)$ is required to be Markov with stationary transitions Γ. This does not make ξ Markov. Incidentally, the time parameter n runs over a random subinterval of the integers; the most interesting case is where n runs over all the integers, so the $\xi(n)$ are defined everywhere.

Let X be a Markov chain with stationary, standard transitions P, and regular sample functions.

Suppose the infinities of X are isolated, almost surely;

and occur at times $\varphi_1 < \varphi_2 < \cdots$. Let $\varphi_0 = 0$, and $\varphi_m = \infty$ if there are fewer than m infinities.

NOTE. You can show that the set of paths with isolated infinities is measurable, so its probability can in principle be computed from P and the starting state.

As usual, let $Q = P'(0)$ and $q(i) = -Q(i,j) < \infty$, and

$$\Gamma(i,j) = Q(i,j)/q(i) \quad \text{for } i \neq j \text{ and } q(i) > 0$$
$$= 0 \qquad\qquad \text{elsewhere.}$$

Given $\{\varphi_m < \infty\}$, the order of visits $\xi_m(\cdot)$ paid by X on $(\varphi_m, \varphi_{m+1})$ is a strongly approximate Γ-chain. This is more or less obvious from (7.33) and strong Markov (7.38).

A strongly approximate Γ-chain ξ has an exit boundary B_∞, consisting of extreme, excessive functions; and an entrance boundary $B_{-\infty}$, consisting of extreme, subinvariant measures. As time n increases, $\xi(n)$ converges almost surely to a random point $\xi(\infty) \in B_\infty$; as time n decreases, $\xi(n)$ converges almost surely to a random point $\xi(-\infty) \in B_{-\infty}$. Up to null sets, the point $\xi(\infty)$ generates the σ-field of invariant subsets of the far future, and the point $\xi(-\infty)$ generates the σ-field of invariant subsets of the remote past. Most of this is in Chapter 4.

On $\{\varphi_m < \infty\}$, let $\xi_m(-\infty) \in B_{-\infty}$ be the limit of $\xi_m(n)$ as n decreases, and let $\xi_m(\infty) \in B_\infty$ be the limit of $\xi_m(n)$ as n increases. Now P is specified by q, Γ, and a kernel $K(h, d\mu)$, which is a probability on $B_{-\infty}$ for each $h \in B_\infty$. At the infinity φ_{m+1}, the process X crosses from the exit point $\xi_m(\infty)$ to the entrance point $\xi_{m+1}(-\infty)$ using the kernel K. That is, the distribution of $\xi_{m+1}(-\infty)$ given $\varphi_{m+1} < \infty$ and given $X(t)$ for $t \leq \varphi_{m+1}$ is

$$K[\xi_m(\infty), d\mu].$$

Given $\varphi_{m+1} < \infty$ and $X(t)$ for $t \leq \varphi_{m+1}$ and $\xi_{m+1}(-\infty)$, the visiting process $\xi_{m+1}(\cdot)$ on $(\varphi_{m+1}, \varphi_{m+2})$ is a strongly approximate Γ-chain starting from $\xi_{m+1}(-\infty)$. As usual, given the visiting process, the holding times are conditionally independent and exponentially distributed, holding times in j having parameter $q(j)$.

In particular, the conditional distribution of $X(\varphi_{m+1} + \cdot)$ given $\varphi_{m+1} < \infty$ and $X(t)$ for $t \leq \varphi_{m+1}$ depends only on $\xi_m(\infty)$. On $\varphi_{m+1} < \infty$, the pre-φ_{m+1} sigma field is spanned up to null sets by $X(t)$ for $t \leq \varphi_{m+1}$ and $\xi_{m+1}(-\infty)$; the conditional distribution of $X(\varphi_{m+1} + \cdot)$ given $\varphi_{m+1} < \infty$ and the pre-φ_{m+1} sigma field depends only on $\xi_{m+1}(-\infty)$.

By reversing the procedure, you can construct the general chain with stable states and isolated infinities, from the holding time parameters q, the jump matrix Γ, and the crossing kernel K. It's like Section 5. For details in a similar construction, see (Chung, 1963, 1966).

The construction of a strongly approximate Γ-chain starting from $\mu \in B_{-\infty}$ is not trivial. For simplicity, suppose all $i \in I$ are transient relative to Γ. Let $\{\xi(n)\}$ be a strongly approximate Γ-chain starting from $\mu \in B_{-\infty}$, on the probability triple $(\Omega, \mathscr{F}, \mathscr{P})$. More or less by definition, $\mu(j)$ is the mean number of n with $\xi(n) = j$. As before, let Ω_J be the set where $\xi(n) \in J$ for some n, and let τ_J be the least such n. Let $\mu_{(J)}$ be the distribution of $\xi(\tau_J)$:

$$\mu_{(J)}(j) = \mathscr{P}\{\Omega_J \text{ and } \xi(\tau_J) = j\} \quad \text{for } j \in J.$$

It turns out that the main problem in constructing $\{\xi(n)\}$ is the computation of $\mu_{(J)}$: because $\mu_{(J)}$ determines the distribution of $\xi(\tau_J + \cdot)$. One method is sketched in (Hunt, 1960). Here is another.

Let $G(i, j)$ be the mean number of visits to j by a true Γ-chain starting from i. By the strong Markov property,

$$\Sigma_{i \in J}\, \mu_{(J)}(i)\, G(i, j) = \mu(j) \quad \text{for } j \in J.$$

This system of linear equations uniquely determines $\mu_{(J)}$: I will write down the inversion matrix. Let $\Gamma_{(J)}$ be the transition matrix of a true Γ-chain watched only when in J, as discussed in Section 2.5. Let Δ_J be the identity matrix on J:

$$\begin{aligned} \Delta_J(i, j) &= 1 \quad \text{for } i = j \text{ in } J \\ &= 0 \quad \text{for } i \neq j \text{ in } J. \end{aligned}$$

Define G_J and μ_J as follows:

$$\begin{aligned} G_J(i, j) &= G(i, j) \quad \text{for } i \text{ and } j \text{ in } J \\ \mu_J(j) &= \mu(j) \quad \text{for } j \text{ in } J. \end{aligned}$$

The system of equations for $\mu_{(J)}$ can now be rewritten in matrix notation:

$$\mu_{(J)} \cdot G_J = \mu_J.$$

I claim

$$G_J = \Sigma_{n=0}^{\infty} (\Gamma_{(J)})^n:$$

erasing non-J times doesn't affect the number of visits to $j \in J$. The sum converges beautifully, and

$$G_J^{-1} = \Delta_J - \Gamma_{(J)}:$$

boldly multiply both sides of the equation for G_J by $(\Delta_J - \Gamma_{(J)})$. Therefore,

$$\mu_{(J)} = \mu_J \cdot (\Delta_J - \Gamma_{(J)}). \qquad \qquad \bigstar$$

NOTE. Δ_J, G_J, and μ_J are respectively Δ, G, and μ retracted to J. But $\Gamma_{(J)}$ and $\mu_{(J)}$ are the restrictions of Γ and μ to J in a much subtler sense.

This is all I want to say about the isolated infinities case. There are two drawbacks to the theory. First, it is hard to tell from P when the infinities are isolated. Second, there are extreme, invariant μ which do not give the expected number of visits by a strongly approximate Γ-chain on any finite measure space; and you can't tell the players without a program.

This kind of construction probably fails when the set of infinities is countable but complicated. It certainly fails when the set of infinities is uncountable. See Section 2.13 of *B & D* for an example.

7. THE SET OF INFINITIES IS BAD

Consider a stochastic process $\{X(t): 0 \leq t < \infty\}$ on a probability triple $(\Omega, \mathscr{F}, \mathscr{P})$. Suppose the process is

(126) a Markov chain with countable state space I, stationary standard transitions, and all states stable.

Give I the discrete topology, and compactify by adjoining one state, φ. Suppose further:

(127) for each $\omega \in \Omega$ and $0 \leq t < \infty$, the $I \cup \{\varphi\}$-valued function $t \to X(t, \omega)$ is continuous from the right, has a limit from the left, and is continuous when it takes the value φ;

(128) it is possible to jump from any $i \in I$ to any $j \in I$;

(129) each $i \in I$ is recurrent.

Let $S_\varphi(\omega) = \{t : X(t, \omega) = \varphi\}$. How bad is $S_\varphi(\omega)$? In view of (127) and (7.17),

(130) $S_\varphi(\omega)$ is closed and has Lebesgue measure 0 for $\mathscr{P}$-almost all ω.

The object of this section is to indicate, by construction, that (130) is best possible. More precisely, fix a compact subset S of the real line, having Lebesgue measure 0. Call a subset S' of the real line *similar* to S iff there is a strictly increasing function f of the real line onto itself, which carries S onto S', such that f and f^{-1} are absolutely continuous.

(131) Theorem.   *There is a stochastic process $\{X(t): 0 \leq t < \infty\}$ on a triple $(\Omega, \mathscr{F}, \mathscr{P})$ satisfying (126) through (129), such that for all $\omega \in \Omega$, the set $S_\varphi(\omega)$ includes a countably infinite number of disjoint sets similar to S.*

A similar phenomenon appears in Section 2.13 of *B & D*. I provide an outline of the construction. We're both too tired for a formal argument.

OUTLINE OF THE CONSTRUCTION.   Suppose $S \subset [0, 1]$, and $0 \in S$, and $1 \in S$. Suppose further that 1 is a limit point of S, the other case being similar and easier. Let A be the set of maximal open subintervals of $(-\infty, 1]$ complementary to S. The state space I consists of all pairs (a, n) with $a \in A$ and n an integer. The construction has parameters $p(i)$ and $q(i)$ for $i \in I$, arbitrary subject to

$$0 < q(i) < \infty \quad \text{and} \quad \Sigma_{i \in I} 1/q(i) < \infty$$
$$0 < p(i) < 1 \quad \text{and} \quad \Pi_{i \in I} p(i) > 0.$$

These parameters enter in the following way. The X process will visit various states i; on reaching i, the process remains in i for a length of time which is exponential with parameter $q(i)$. The holding time is independent of everything else. On leaving $i = (a, n)$, the process jumps to $(a, n + 1)$ with overwhelming probability, namely $p(i)$. It jumps to each other state in I with positive probability summing to $1 - p(i)$. These other probabilities also constitute parameters for the construction, but they are not important, and can be fixed in any way subject to the constraints given.

The local behavior of the process is now fixed. To explain the global behavior, say $a < b$ for $a \in A$ and $b \in A$ iff a is to the left of b as a subset of the line; say $(a, m) < (b, n)$ for $(a, m) \in I$ and $(b, n) \in I$ iff $a < b$, or $a = b$ and $m < n$. Fix t and ω with $X(t, \omega) = \varphi$. The global behavior of X is determined by the requirement that either case 1 or case 2 holds.

Case 1.   There is an $\varepsilon > 0$, an $a \in A$, and an integer L, such that: as u increases through $(t - \varepsilon, t)$, the function $X(u, \omega)$ runs in order through precisely the states $(a, n): n \geq L$ in order. Then there is a $\delta > 0$, an interval $c \in A$ with $c > a$, and an integer K, such that: as u increases through $(t, t + \delta)$, skipping times u' with $X(u', \omega) = \varphi$, the function $X(u, \omega)$ runs in order through precisely the states: (b, n) with $a < b < c$, all n; and $b = c$, but $n \leq K$.

Case 2.   There is an $\varepsilon > 0$ and an interval $a \in A$ such that: as u increases through $(t - \varepsilon, t)$, skipping times u' with $X(u', \omega) = \varphi$, the function $X(u, \omega)$ runs in order through precisely the states (b, n) with $b > a$, all n. Then there is a $\delta > 0$ and an integer K such that: as u increases through $(t, t + \delta)$, the function $X(u, \omega)$ runs in order through precisely the states (l, n) with $n \leq K$ and $l = (-\infty, 0)$.   ★

OUTLINE OF THE PROOF.   Whenever case 2 occurs, there is positive probability $\Pi p(i)$ that the chain proceeds to move through its states in order. Whenever this occurs, the corresponding section of $S_\varphi(\omega)$ is similar to S. By Borel-Cantelli, infinitely many disjoint sections of $S_\varphi(\omega)$ are similar to S, as required.   ★

9

THE GENERAL CASE

1. AN EXAMPLE OF BLACKWELL

My object in this section is to present Blackwell's (1958) example of a standard stochastic semigroup, all of whose states are instantaneous. For other examples of this phenomenon, see Sections 3.3 of *ACM* and Section 2.12 of *B & D*. To begin with, consider the matrix

$$Q = \begin{pmatrix} -\lambda & \lambda \\ \mu & -\mu \end{pmatrix}$$

on $\{0, 1\}$, with λ and μ nonnegative, $\lambda + \mu$ positive. There is exactly one standard stochastic semigroup P on $\{0, 1\}$ with $P'(0) = Q$, namely:

(1)

$$P(t, 0, 0) = \frac{\mu}{\mu + \lambda} + \frac{\lambda}{\mu + \lambda} e^{-(\mu+\lambda)t}$$

$$P(t, 0, 1) = 1 - P(t, 0, 0)$$

$$P(t, 1, 1) = \frac{\lambda}{\mu + \lambda} + \frac{\mu}{\mu + \lambda} e^{-(\mu+\lambda)t}$$

$$P(t, 1, 0) = 1 - P(t, 1, 1).$$

One way to see this is to use (5.29): define P by (1); check P is continuous, $P(0)$ is the identity, $P'(0) = Q$, and $P(t + s) = P(t) \cdot P(s)$. Dull computations in the last step can be avoided by thinking: it is enough to do $\mu + \lambda = 1$ by re-scaling time; since $P(u)$ is 2×2 and stochastic when u is t or s or $t + s$,

---

I want to thank Mike Orkin for checking the final draft of this chapter.

it is enough to check that $P(t + s) = P(t) \cdot P(s)$ on the diagonal; by interchanging μ and λ, so 0 and 1, it is enough to check the $(0, 0)$ position. This is easy.

Parenthetically, (1) implies the following. Let M be a stochastic matrix on $\{0, 1\}$. There is a standard stochastic semigroup P on $\{0, 1\}$ with $P(1) = M$ iff trace $M > 1$. The corresponding question for $\{0, 1, 2\}$, let alone for an infinite state space, is open. For a recent discussion, see (Kingman, 1962).

Now let

$$Q_n = \begin{pmatrix} -\lambda_n & \lambda_n \\ \mu_n & -\mu_n \end{pmatrix}$$

with λ_n and μ_n positive, and let P_n be the standard stochastic semigroup on $\{0, 1\}$ with $P_n'(0) = Q_n$. Let I be the countable set of infinite sequences $i = (i_1, i_2, i_3, \ldots)$ of 0's and 1's, with only finitely many 1's. Let $N(i)$ be the least N such that $n \geq N$ implies $i_n = 0$. Suppose $\Pi_n \mu_n/(\mu_n + \lambda_n) > 0$, that is,

$$(2) \qquad \sum_n \frac{\lambda_n}{\mu_n + \lambda_n} < \infty.$$

For $t \geq 0$, define the matrix $P(t)$ on I as

$$(3) \qquad P(t, i, j) = \Pi_n P_n(t, i_n, j_n).$$

Let $\{X_n(t): 0 \leq t < \infty\}$ be 0-1 valued stochastic processes with right continuous step functions for sample functions, on a convenient measurable space $(\Omega, \mathscr{F})$. Let $X(t)$ be the sequence $(X_1(t), X_2(t), \ldots)$. For each $i \in I$, let $\mathscr{P}_i$ be a probability on $\mathscr{F}$ for which $X_1, X_2, \ldots$ are independent, and X_n is Markov with stationary transitions P_n, starting from i_n. This construction is possible by (5.45) and the existence of product measure. I say

$$\mathscr{P}_i\{X(t) \in I\} = 1 \quad \text{for all } t \geq 0.$$

Indeed, for $n \geq N(i)$,

$$\mathscr{P}_i\{X_n(t) = 1\} = P_n(t, 0, 1) \leq \lambda_n/(\mu_n + \lambda_n)$$

by (1), so $\sum_{n=1}^{\infty} \mathscr{P}_i\{X_n(t) = 1\} < \infty$ by (2). Use Borel-Cantelli.

I will now check that X is a Markov chain, with stationary transitions P, which are stochastic on I. Indeed, suppose $0 < t_1 < \cdots < t_N < \infty$ and $j(1), \ldots, j(N)$ are in I. Let $t_0 = 0$ and $j(0) = i$. Let $j(n, m)$ be the mth component of $j(n)$. Then

$$\mathscr{P}_i\{X(t_n) = j(n) \text{ for } n = 1, \ldots, N\}$$
$$= \mathscr{P}_i\{X_m(t_n) = j(n, m) \text{ for } n = 1, \ldots, N \text{ and } m = 1, 2, \ldots\}$$
$$= \Pi_{m=1}^{\infty} \mathscr{P}_i\{X_m(t_n) = j(n, m) \text{ for } n = 1, \ldots, N\}$$
$$= \Pi_{m=1}^{\infty} \Pi_{n=0}^{N-1} P_m[t_{n+1} - t_n, j(n, m), j(n + 1, m)]$$
$$= \Pi_{n=0}^{N-1} \Pi_{m=1}^{\infty} P_m[t_{n+1} - t_n, j(n, m), j(n + 1, m)]$$
$$= \Pi_{n=0}^{N-1} P[t_{n+1} - t_n, j(n), j(n + 1)].$$

Use (5.4) to clinch the argument.

I will now verify that P is standard. Let $N \geqq N(i)$. Then

$$P(t, i, i) = \mathscr{P}_i\{X(t) = i\}$$
$$= [\Pi_{n=1}^{N-1} P_n(t, i_n, i_n)] \cdot [\Pi_{n=N}^{\infty} P_n(t, 0, 0)].$$

The first factor tends to 1 as $t \to 0$. Using (1), the second factor is at least $\Pi_{n=N}^{\infty} \mu_n/(\mu_n + \lambda_n)$, which is nearly 1 for large N by (2).

Finally, suppose

(4) $\Sigma_n \lambda_n = \infty.$

I claim each $i \in I$ is instantaneous. Indeed, fix $t > 0$ and consider

$$\mathscr{P}_i\{X(r) = i \text{ for all binary rational } r \text{ with } 0 \leqq r \leqq t\}.$$

This number is at most

$$\mathscr{P}_i\{X_n(r) = 0 \text{ for all binary rational } r \text{ with } 0 \leqq r \leqq t \text{ and } n \geqq N(i)\},$$

which by independence is the product as n runs from $N(i)$ to ∞ of

$$\mathscr{P}_i\{X_n(r) = 0 \text{ for all binary rational } r \text{ with } 0 \leqq r \leqq t\} = e^{-\lambda_n t},$$

where the last equality comes from (5.48) or (7.4). Using (4),

$$\mathscr{P}_i\{X(r) = i \text{ for all binary rational } r \text{ with } 0 \leqq r \leqq t\} = 0.$$

Now (7.4) forces $P'(0, i, i) = -\infty$.   ★

2. QUASIREGULAR SAMPLE FUNCTIONS

For the rest of this chapter, let I be a finite or countably infinite set in the discrete topology. Let $\bar{I} = I$ for finite I. Let $\bar{I} = I \cup \{\varphi\}$ be the one-point compactification of I for infinite I. Call φ the *infinite* or *fictitious* or *adjoined* state, as opposed to the *finite* or *real* states $i \in I$. Let P be a standard stochastic semigroup on I, with $Q = P'(0)$ and $q(i) = -Q(i, i)$. Do not assume

$q(i) < \infty$. The main point of this section is to construct a Markov chain with stationary transitions P, all of whose sample functions have this smoothness property at all nonnegative t:

if $f(t) = \varphi$, then $f(r)$ converges to φ as binary rational r decreases to t;

if $f(t) \in I$, then $f(r)$ has precisely one limit point in I, namely $f(t)$, as binary rational r decreases to t.

This result is in (Chung, 1960, II.7). The key lemma (9) is due to Doob (1942).

Downcrossings

For any finite sequence $\mathbf{s} = (s(1), s(2), \ldots, s(N))$ of real numbers, and pair $u < v$ of real numbers, the number of *downcrossings* $\beta(u, v, \mathbf{s})$ of $[u, v]$ by $\mathbf{s}$ is the largest positive integer d such that: there exist integers

$$1 \leq n_1 < n_2 < \cdots < n_{2d} \leq N$$

with

$$s(n_1) \geq v, \; s(n_2) \leq u, \ldots, s(n_{2d-1}) \geq v, \; s(n_{2d}) \leq u.$$

If no such d exists, the number of downcrossings is 0. If $\mathbf{s}$ and $\mathbf{t}$ are finite sequences, and $\mathbf{s}$ is a subsequence of $\mathbf{t}$, then

$$\beta(u, v, \mathbf{s}) \leq \beta(u, v, \mathbf{t}).$$

Of course, $\beta(u, v, \mathbf{s})$ depends on the enumeration of $\mathbf{s}$.

Functions with right and left limits

Let R be the set of nonnegative binary rationals. Let F be the set of functions f from R to $[0, 1]$, with the product σ-field, namely the smallest σ-field over which all Y_r are measurable, where $Y_r(f) = f(r)$ for $r \in R$ and $f \in F$. Let $0 < M < \infty$. Let M^* be the set of $f \in F$ such that: for all real t with $0 \leq t < M$, as $r \in R$ decreases to t, the generalized sequence $f(r)$ converges; and for all real t with $0 < t \leq M$, as $r \in R$ increases to t, the generalized sequence $f(r)$ converges. For $f \in F$, let $\beta_n(u, v, f)$ be the number of downcrossings of $[u, v]$ by

$$\{f(m/2^n) : m = 0, \ldots, \langle M2^n \rangle\}:$$

where $\langle x \rangle$ is the greatest integer no more than x. Verify that $\beta_n(u, v, \cdot)$ is measurable, and nondecreasing in n.

(5) Lemma. *M^* is the set of f such that $\lim_n \beta_n(u, v, f) < \infty$ for all rational u and v with $u < v$. In particular, M^* is measurable.*

PROOF. Suppose $f \notin M^*$. Suppose that for some $t \in [0, M)$ and sequence $r_m \in R \cap [0, M)$ with $r_m \downarrow t$,

$$a = \lim \inf f(r_m) < \lim \sup f(r_m) = b.$$

The increasing case is similar. Choose rational u, v with

$$a < u < v < b.$$

For large N, the number of downcrossings D of $[u, v]$ by $f(r_1), \ldots, f(r_N)$ is large. The number of downcrossings of $[u, v]$ by $f(r_N), \ldots, f(r_1)$ is at least $D - 1$. If n is so large that $2^n r_1, \ldots, 2^n r_N$ are all integers,

$$\beta_n(u, v, f) \geqq D - 1.$$

So

$$\lim_{n \to \infty} \beta_n(u, v, f) = \infty.$$

Conversely, suppose $f \in M^*$. Fix $u < v$. Let $0 < \varepsilon < \frac{1}{2}(v - u)$. Abbreviate

$$f(t+) = \lim \{f(r) : r \in R \text{ and } r \downarrow t\}$$
$$f(t-) = \lim \{f(r) : r \in R \text{ and } r \uparrow t\}.$$

For any $t \in [0, M]$, there is a $\delta = \delta(t) > 0$, such that:

if $r \in R \cap [0, M] \cap (t, t + \delta)$, then $|f(r) - f(t+)| < \varepsilon$;

if $r \in R \cap [0, M] \cap (t - \delta, t)$, then $|f(r) - f(t-)| < \varepsilon.$

The first condition is vacuous for $t = M$, and the second is vacuous for $t = 0$. In particular, $|f(r) - f(s)| < 2\varepsilon < v - u$ if r and s are in $R \cap [0, M]$ and: either r and s are in $(t, t + \delta)$, or r and s are in $(t - \delta, t)$. Let $I(t)$ be the open interval $(t - \delta(t), t + \delta(t))$. By compactness, there are finitely many points $t_1, \ldots, t_N$ in $[0, M]$ such that the union of $I(t_n)$ for $n = 1, \ldots, N$ covers $[0, M]$. I claim that for all v,

$$\beta_v(u, v, f) \leqq 3N/2.$$

Indeed, suppose $0 \leqq r_1 < r_2 < \cdots < r_{2d} \leqq M$ are in R, and

$$f(r_1) \geqq v, f(r_2) \leqq u, \ldots, f(r_{2d-1}) \geqq v, f(r_{2d}) \leqq u,$$

as in Figure 1. I say that $I(t_n)$ contains at most three r_m's. For suppose $I(t_n)$ contains $r_m, r_{m+1}, r_{m+2}, r_{m+3}$, as in Figure 1. Then t_n is either to the right of r_{m+1} or to the left of r_{m+2}. In either case, there is a forbidden oscillation. So there are at most $3N$ points r_m. That is, $2d \leqq 3N$.   ★

Pedro Fernandez eliminated the unnecessary part of an earlier proof.

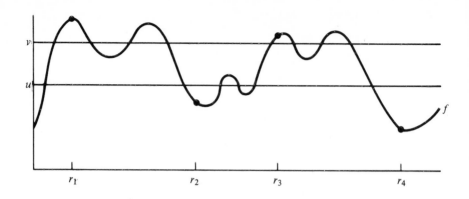

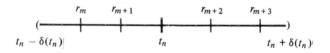

Figure 1.

Quasiconvergence

(6) Definition. *Let A be a set directed by $>$ and let $i_a \in I$ for each $a \in A$. That is, $\{i_a\}$ is a* generalized sequence. *Say i_a* quasiconverges *to $j \in I$, or*

$$q\text{-}\lim i_a = j,$$

iff: for any finite subset D of $I\backslash\{j\}$, there is some $a(D) \in A$ such that $a > a(D)$ implies $i_a \notin D$; and for any $a \in A$ there is some $b > a$ such that $i_b = j$. Say i_a quasiconverges *to φ, or*

$$q\text{-}\lim i_a = \varphi,$$

iff: for any finite subset D of I, there is some $a(D) \in A$ such that $a > s(D)$ implies $i_a \notin D$.

    The directed sets of main interest are: the nonnegative integers; the nonnegative binary rationals less than a given real number; the nonnegative real numbers less than a given real number; the binary rationals greater than a given nonnegative real number; the real numbers greater than a given nonnegative real number. In the first three cases, $a > b$ means a is greater than b. In the last two cases, $a > b$ means a is less than b. Here is the usual notation for these five quasilimits; t is the given real number and R is the set of nonnegative binary rationals.

$$q\text{-lim } i_n.$$

$$q\text{-lim } \{i_r : r \in R \text{ and } r \uparrow t\}.$$

$$q\text{-lim } \{i_s : s \uparrow t\} = q\text{-lim}_{s \uparrow t} i_s.$$

$$q\text{-lim } \{i_r : r \in R \text{ and } r \downarrow t\}.$$

$$q\text{-lim } \{i_s : s \downarrow t\} = q\text{-lim}_{s \downarrow t} i_s.$$

Quasiconvergence is not topological. In fact, a typical sequence which quasiconverges to $j \in I$ has subsequences quasiconverging to φ. Try

$$1, 2, 1, 3, 1, 4, \ldots$$

whose q-lim is 1. On the brighter side, if $q\text{-lim } i_a = j \in I$, and a^* is coterminous generalized subsequence, then $q\text{-lim } i_{a*}$ exists, and is either j or φ.

Coterminous means: for any $a \in A$, there is an $a^* > a$.
Conversely, if $q\text{-lim } i_a$ exists, and $i_a = j \in I$ for arbitrarily remote a, then $q\text{-lim } i_a = j$.

Quasilimits cohere with convergence in probability, a fact that will be in use later. Let $\{X_n\}$ be a sequence of I-valued random variables on $(\Omega, \mathscr{F}, \mathscr{P})$. Suppose

$$X_q = q\text{-lim}_n X_n$$

exists $\mathscr{P}$-almost surely. And suppose X_n converges in $\mathscr{P}$-probability to an I-valued limit X_p. Then

$$\mathscr{P}\{X_p = X_q\} = 1.$$

To see this, choose a subsequence n^* such that

$$X_{n*} = X_p \quad \text{for infinitely many } n$$

with $\mathscr{P}$-probability 1.

If $\mathscr{P}\{X_p = \varphi\} > 0$, the only safe assertion is

$$\mathscr{P}\{X_p = X_q \mid X_p \in I\} = 1.$$

The situation is similar for generalized sequences, provided the directing set A has a coterminous countable subset A^*.

EXAMPLE.   Suppose X_n is 0 with probability $1/n$, and n with probability $(n-1)/n$. Suppose the X_n are independent. Then

$$X_n \rightarrow \varphi \quad \text{in probability}$$

$$q\text{-lim}_n X_n = 0 \quad \text{with probability 1.}$$

Let R be the set of binary rationals in $[0, \infty)$ as before. The next definition is key.

(7) Definition.   *A function f from R to I is* quasiregular *iff:*

$$q\text{-lim } \{f(r):r \in R \text{ and } r \downarrow t\}$$

exists for all nonnegative real t, and equals f(t) for all t ∈ R; while

$$q\text{-lim } \{f(r):r \in R \text{ and } r \uparrow t\}$$

exists for all positive real t, and equals f(t) for all positive t ∈ R.
   A function f from $[0, \infty)$ *to* $\bar{I}$ *is* quasiregular *iff:* $f(r) \in I$ *for all* $r \in R$; *and f retracted to R is quasiregular; and*

$$q\text{-lim } \{f(r):r \in R \text{ and } r \downarrow t\} = f(t)$$

for all nonnegative real t.

   WARNING.   *t* runs over all of $[0, \infty)$.

   Suppose *f* is quasiregular from $[0, \infty)$ to $\bar{I}$. I claim *f* is quasicontinuous from the right: that is, $f(t) = q\text{-lim } \{f(s):s \downarrow t\}$. Begin the check by supposing $f(t) = i \in I$. By definition, $f(s) = i$ for *s* arbitrarily close to *t* on the right. Conversely, suppose $f(s_n) = i \in I$ for some sequence $s_n \downarrow t$. Without loss, make the s_n strictly decreasing, and use the definition to find binary rational r_n with $s_n \leqq r_n < s_{n-1}$ and $f(r_n) = i$. But $r_n \downarrow t$, so the definition forces $f(t) = i$. Similarly, *f* has a quasilimit from the left at all positive times, and is quasicontinuous at all binary rational *r*.

The process on R

   Let Ω be the set of all functions ω from *R* to *I*. Let $\{X(r):r \in R\}$ be the coordinate process on Ω, that is, $X(r)(\omega) = \omega(r)$ for $r \in R$ and $\omega \in \Omega$. Endow *I* with the σ-field of all its subsets, and Ω with the product σ-field, that is, the smallest σ-field over which each $X(r)$ is measurable. For each $i \in I$, let P_i be the probability on Ω for which $\{X(r):r \in R\}$ is Markov with stationary transitions *P* and $X(0) = i$. Namely, for $0 = r_0 < r_1 < \cdots < r_n$ in *R* and $i_0 = i, i_1, \ldots, i_n$ in *I*,

$$P_i\{X(r_m) = i_m \text{ for } m = 0, \ldots, n\} = \Pi_{m=0}^{n-1} P(r_{m+1} - r_m, i_m, i_{m+1}).$$

By convention, an empty product is 1.
   For $0 < L < \infty$, let L^* be the set of $\omega \in \Omega$ such that: for all real *t* with $0 \leqq t < L$, as $r \in R$ decreases to *t*, the generalized sequence $\omega(r)$ quasi-converges; and for all real *t* with $0 < t \leqq L$, as $r \in R$ increases to *t*, the generalized sequence $\omega(r)$ quasiconverges.

(8) Lemma.   *The set L^* is measurable.*

PROOF. If $i_0, i_1, \ldots, i_m$ is an I-sequence, there is a *change* at index $\nu < m$ iff $i_\nu \neq i_{\nu+1}$. Let j and k be different states. Consider the state sequence

$$\{\omega(m2^{-n}): m = 0, \ldots, \langle L2^n\rangle\}.$$

Delete those terms which are neither j nor k. Count the number of changes in this reduced sequence, and call it $\beta_n(j, k, \omega)$. Check that $\beta_n(j, k, \cdot)$ is measurable, and nondecreasing with n. You can show that L^* is the set of $\omega \in \Omega$ with:

$$\lim_{n \to \infty} \beta_n(j, k, \omega) < \infty$$

for all pairs of different states j and k. The argument is similar to the one in (5). For the second part, fix $j \neq k$ in I and $\omega \in L^*$. For each $t \in [0, L]$, there is a $\delta = \delta(t) > 0$ such that:

there do not exist r, s in $R \cap [0, L] \cap (t, t + \delta)$ with $\omega(r) = j$ and $\omega(s) = k$;

there do not exist r, s, in $R \cap [0, L] \cap (t - \delta, t)$ with $\omega(r) = j$ and $\omega(s) = k$. ★

(9) Lemma.   *The set* $\bigcap_{L>0} L^*$ *is measurable and has P_i-probability 1.*

PROOF.   Clearly,

$$\bigcap \{L^*: L \text{ is a positive real number}\} = \bigcap \{L^*: L \text{ is a positive integer}\}.$$

It is therefore enough to prove $P_i\{L^*\} = 1$ for each $L > 0$. Without real loss of generality, fix $L = 1$. For $r \in R \cap [0, 1]$ and $k \in I$, define a real-valued function $f_{r,k}$ of pairs $s \in R \cap [0, 1]$ and $\omega \in \Omega$ by

$$f_{r,k}(s, \omega) = P(r - s, \omega(s), k) \quad \text{for } s \leq r,$$
$$= f_{r,k}(r, \omega) \qquad \text{for } s > r.$$

Let $F_{r,k}$ be the set of all $\omega \in \Omega$ such that:

for all $t \in (0, 1]$, the generalized sequence $f_{r,k}(s, \omega)$ converges as $s \in R$ increases to t; and,

for all $t \in [0, 1)$, the generalized sequence $f_{r,k}(s, \omega)$ converges as $s \in R$ decreases to t.

For $0 \leq s \leq r$ and $s \in R$,

$$f_{r,k}(s, \cdot) = P_i\{X(r) = k \mid X(u) \text{ for } u \in R \text{ and } u \leq s\}.$$

Therefore, $\{f_{r,k}(s, \cdot): 0 \leq s \leq r \text{ and } s \in R\}$ is a martingale relative to P_i. Consequently,

$$\{f_{r,k}(s, \cdot): 0 \leq s \leq 1 \text{ and } s \in R\}$$

is a martingale relative to P_i. Let $u < v$ be rational, and let $\beta_n(u, v, \omega)$ be the number of downcrossings of $[u, v]$ by

$$\{f_{r,k}(m/2^n, \omega): m = 0, \ldots, 2^n\}.$$

As (5) implies, $F_{r,k}$ is the set of ω with $\lim_n \beta_n(u, v, \omega) < \infty$ for all rational pairs $u < v$. By the downcrossings inequality (10.33),

$$\int_\Omega \beta_n(u, v, \omega)\, P_i(d\omega)$$

is bounded in n for each pair u, v. But $\beta_n(u, v, \omega)$ is nondecreasing with n. Therefore,

$$\int_\Omega \lim_n \beta(u, v, \omega)\, P_i(d\omega) < \infty.$$

By (10.10b),

$$P_i\{\lim_n \beta_n(u, v, \cdot) < \infty\} = 1$$

for each pair u, v. So $P_i\{F_{r,k}\} = 1$, and $\bigcap_{r,k} F_{r,k}$ has P_i-probability 1.

Let $\omega \in \bigcap_{r,k} F_{r,k}$. To see that $\omega \in 1^*$, suppose $s_n \in R$ and $\hat{s}_n \in R$ both increase to $t \in (0, 1]$, while $\omega(s_n) = j \in I$ and $\omega(\hat{s}_n) = \hat{\jmath} \in I$. The decreasing case is similar. Let $r \in R \cap [0, 1]$, and $r \geq t$. Let $k \in I$. Then

$$f_{r,k}(\hat{s}_n, \omega) = P(r - \hat{s}_n, \hat{\jmath}, k) \to P(r - t, \hat{\jmath}, k)$$
$$f_{r,k}(s_n, \omega) = P(t - s_n, j, k) \to P(r - t, j, k)$$

by (5.9). Since $\omega \in F_{r,k}$, the two subsequential limits must be equal:

$$P(r - t, \hat{\jmath}, k) = P(r - t, j, k).$$

Let $r \downarrow t$ and use (5.9) again to get

$$P(0, \hat{\jmath}, k) = P(0, j, k).$$

The left side is 1 iff $\hat{\jmath} = k$, and the right side is 1 iff $j = k$, so $\hat{\jmath} = j$.   ★

(10) Lemma. (a)   *Let $s \in R$. As $r \in R$ converges to s, the generalized sequence $X(r)$ converges in P_i-probability to $X(s)$.*

(b) *Let t be real. As $r \in R$ converges to t, the generalized sequence $X(r)$ converges in P_i-probability to an I-valued random variable $\hat{X}(t)$.*

(c) *If $t \in R$, then $\hat{X}(t) = X(t)$ with P_i-probability 1.*

PROOF.   *Claim (b)*.   It is enough to check that $P_i\{X(r) = X(s)\} \to 1$ as $r, s \in R$ converge to t. If $0 \leq r \leq s$,

$$P_i\{X(r) = X(s)\} = \Sigma_j\, P(r, i, j) P(s - r, j, j).$$

Now $P(s - r, j, j) \to 1$ for each j. Use Fatou.

Claim (a) is similar.
Claim (c) is immediate from (a) and (b).   ★

WARNING.   The limiting random variable $\hat{X}(t)$ is well defined only a.e.

(11) Lemma.   *Let t be positive and real. Choose a version of $\hat{X}(t)$, as defined in (10b). For P_i-almost all ω, for any $\varepsilon > 0$, there are r and s in R with*

$$t - \varepsilon < r < t < s < t + \varepsilon \quad and \quad \omega(r) = \hat{X}(t, \omega) = \omega(s).$$

PROOF.   Choose a sequence $r_n \in R$ with $r_n \uparrow t$. By (10), $X(r_n) \to \hat{X}(t)$ in P_i-probability. Now choose a subsequence n' with $P_i\{X(r_{n'}) \to \hat{X}(t)\} = 1$. Similarly for the right.   ★

(12) Definition.   *Let $\Omega_q = \{\omega : \omega \in \Omega \text{ and } \omega \text{ is quasiregular}\}$.*

Quasiregularity was defined in (7).

(13) Lemma.   *The set Ω_q is measurable and $P_i\{\Omega_q\} = 1$.*

PROOF.   For $r \in R$, let $G(r)$ be the set of $\omega \in \Omega$ for which: there are $s \in R$ with $s > r$ but arbitrarily close having $\omega(s) = \omega(r)$; and if $r > 0$, there are $s \in R$ with $s < r$ but arbitrarily close having $\omega(s) = \omega(r)$. Clearly, $G(r)$ is measurable; and $P_i\{G(r)\} = 1$ by (10c, 11). Remember (8–9). Check

$$\Omega_q = (\bigcap\nolimits_{L>0} L^*) \cap (\bigcap\nolimits_{r \in R} G(r)).$$ ★

The process on $[0, \infty)$

(14) Definition.   *For $\omega \in \Omega_q$, let $X(t, \omega) = q\text{-}\lim X(r, \omega)$ as $r \in R$ decreases to t.*

Check $X(r, \omega) = \omega(r)$ for $\omega \in \Omega_q$ and $r \in R$.

(15) Lemma.   *The function $(t, \omega) \to X(t, \omega)$ is measurable.*

PROOF.   The set of pairs (t, ω) with $0 \leq t < \infty$ and $\omega \in \Omega_q$ and $X(t, \omega) = i$ is

$$\bigcap\nolimits_{n=1}^{\infty} \bigcup\nolimits_{r \in R} A(n, r),$$

where $A(n, r)$ is the set of pairs (t, ω) with $0 \leq t < \infty$ and $\omega \in \Omega_q$ and $t < r < t + \dfrac{1}{n}$ and $\omega(r) = i$.   ★

(16) Theorem.   *The process $\{X(t) : 0 \leq t < \infty\}$, defined by (14) on the probability triple (Ω_q, P_i), is a Markov chain with stationary transitions P, starting state i, and all sample functions quasiregular in the sense of (7).*

NOTE.   The σ-field on Ω_q is the product σ-field on Ω relativized to Ω_q; and P_i is retracted to this smaller σ-field.

PROOF. Fix $t \geq 0$. As $r \in R$ decreases to t, I claim $X(r)$ converges to $X(t)$ in P_i-probability; you derive (16) from this claim and (13). To argue the claim, fix a version of $\hat{X}(t)$, as defined in (10b). Now $X(r)$ converges to $\hat{X}(t)$ in P_i-probability, by (10b); so it is enough to show

$$P_i\{X(t) = \hat{X}(t)\} = 1.$$

By (11), for P_i-almost all $\omega \in \Omega_q$, there are $r \in R$ greater than but arbitrarily close to t, with $X(r, \omega) = \hat{X}(t, \omega)$. But $X(r, \omega)$ quasiconverges to $X(t, \omega)$ as r decreases to t, by definition (14); this identifies $X(t, \omega)$ with $\hat{X}(t, \omega)$. ★

Abstract processes

For (17), let $(\mathcal{X}, \mathcal{F}, \mathcal{P})$ be an abstract probability triple, and $\{Y(t) : 0 \leq t < \infty\}$ be a Markov chain on $(\mathcal{X}, \mathcal{F}, \mathcal{P})$, with stationary transitions P.

(17) Theorem. Let $\mathcal{X}_q$ be the set of $x \in \mathcal{X}$ such that $Y(\cdot, x)$ retracted to R is I-valued and in Ω_q. Then $\mathcal{X}_q \in \mathcal{F}$ and $\mathcal{P}\{\mathcal{X}_q\} = 1$. On $\mathcal{X}_q$, let $Y^*(t, x) = q$-lim $Y(r, x)$ as $r \in R$ decreases to t. Then $Y^*(\cdot, x)$ is quasiregular and $\mathcal{P}\{Y(t) = Y^*(t)\} = 1$.

PROOF. Use (13) to get $\mathcal{X}_q \in \mathcal{F}$ and $\mathcal{P}\{\mathcal{X}_q\} = 1$. Plainly, $Y^*(\cdot, x)$ is quasiregular for $x \in \mathcal{X}_q$. Suppose $\mathcal{P}\{Y(0) = i\} = 1$. The joint $\mathcal{P}$-distribution of $\{Y(t)$ and $Y(r) : r \in R\}$ coincides with the joint P_i-distribution of $\{X(t)$ and $X(r) : r \in R\}$, by (16). So,

$$\mathcal{P}\{Y(t) = Y^*(t)\} = P_i\{X(t) = q\text{-lim } X(r) \text{ as } r \in R \text{ decreases to } t\} = 1,$$

by definition (14). See (5.45–48) or (7.15) in case of trouble. ★

3. THE SETS OF CONSTANCY

Let $A(i, s)$ be the set of $\omega \in \Omega$ with $\omega(r) = i$ when $r \in R$ and $r \leq s$.

(18) Lemma. $A(i, s)$ is measurable, and

$$P_i\{A(i, s)\} = e^{-q(i)s}.$$

Here $e^{-\infty} = 0$, and $\infty \cdot s = \infty$ for $s > 0$, while $\infty \cdot 0 = 0$.

PROOF. This repeats (7.4). ★

For $j \in I$, let

$$R_j(\omega) = \{r : r \in R \text{ and } \omega(r) = j\}.$$

For $j \in I$ and $\omega \in \Omega_q$, let

$$S_j(\omega) = \{t : 0 \leq t < \infty \text{ and } X(t, \omega) = j\}.$$

The process X was defined in (14). These sets are called the *level sets* or *sets of constancy*.

(19) Definition. *Let $j \in I$. Then $\Omega(j)$ is the set of $\omega \in \Omega_q$ such that: for all $t \geq 0$,*

$$X(t, \omega) = j \quad implies \quad \lim_{s \downarrow t} X(s, \omega) = j;$$

and for all $t > 0$,

$$q\text{-}\lim_{s \uparrow t} X(s, \omega) = j \quad implies \quad \lim_{s \uparrow t} X(s, \omega) = j.$$

The set Ω_q was defined in (12).

(20) Theorem. *Let $j \in I$, and let $\omega \in \Omega_q$. Then $\omega \in \Omega(j)$ iff the set $S_j(\omega)$ is either empty or a finite or countably infinite union of intervals $[a_n, b_n)$ with $a_1 < b_1 < a_2 < \cdots$. Of course, a and b depend on ω, and are not binary rational. If there are infinitely many intervals, $a_n \to \infty$.*

PROOF. The "if" part is easy. For "only if," let $\omega \in \Omega(j)$. Suppose $t \in S_j(\omega)$. Then $[t, t + \varepsilon) \subset S_j(\omega)$ for some $\varepsilon > 0$. Of course, $S_j(\omega)$ is closed from the right, by quasiregularity. Consequently, $S_j(\omega)$ is a finite or countably infinite union of maximal disjoint nonempty intervals $[a_n, b_n)$. Suppose there are an infinite number. By way of contradiction, suppose $a_{n'} \uparrow c < \infty$ for a subsequence n'. Then $X(t, \omega)$ quasiconverges, so converges, to j as t increases to c. So $a_m < c \leq b_m$ for some m, contradicting the disjointness. Similarly for the right.   ★

(21) Theorem. *The set $\Omega(j)$ is measurable. And $P_i\{\Omega(j)\} = 1$ if $j \in I$ is stable.*

PROOF. First, $\Omega(j)$ is the set of $\omega \in \Omega_q$ such that the indicator function of $R_j(\omega)$, as a function on R, is continuous on R, and has limits from left and right at all positive $t \notin R$: use (19, 20). So $\Omega(j)$ is measurable by (5). Check that the indicator function of $R_j(\omega)$, as a function on R, is continuous for P_i-almost all ω: use the argument for (7.6). For such an ω, any $r \in R_j(\omega)$ is interior to a maximal j-interval of ω, as in the paragraph before (7.8). The argument for (7.8) shows that for P_i-almost all ω, for any n, only a finite number of these intervals meet $[0, n]$. This disposes of the second claim. ★

Recall that j is instantaneous iff $q(j) = \infty$.

(22) Lemma. *Suppose $j \in I$ is instantaneous. The set of $\omega \in \Omega$ for which $R_j(\omega)$ includes no proper interval of R is measurable and has P_i-probability 1.*

A *proper* interval has a nonempty interior.

PROOF. For r and s in R with $s > 0$, let $B(r, s)$ be the set of $\omega \in \Omega$ such that $\omega(t) = j$ for all $t \in R \cap [r, r + s]$. Then $B(r, s)$ is measurable. I say $P_i\{B(r, s)\} = 0$. Indeed, consider this mapping T_r of Ω into Ω:

$$(T_r\omega)(s) = \omega(r + s) \quad \text{for } s \in R.$$

Use (10.16) on the definition of P_i and P_j, to see that for all measurable A:

$$P_i\{X(r) = j \text{ and } T_r \in A\} = P(r, i, j)P_j\{A\}.$$

Put $A(j, s)$ for A: then

$$B(r, s) = \{T_r \in A(j, s)\} \subset \{X(r) = j\},$$

so (18) makes

$$P_i\{B(r, s)\} = P(r, i, j)P_j\{A(j, s)\} = 0.$$

But $\bigcup_{r,s} B(r, s)$ is the complement of the set described in the lemma.   ★

(23) Theorem. *Suppose* $j \in I$ *is instantaneous: then the set of* $\omega \in \Omega_q$ *satisfying* (a) *is measurable and has* P_i-*probability* 1. *Properties* (b) *and* (c) *hold for all* $\omega \in \Omega_q$ *and all* $j \in I$.

(a) $S_j(\omega)$ *is nowhere dense.*

(b) *Each point of* $S_j(\omega)$ *is a limit from the right of* $S_j(\omega)$.

(c) $S_j(\omega)$ *is closed from the right.*

PROOF.   You should check (b) and (c). I will then get (a) from (22). In fact, for $\omega \in \Omega_q$, property (a) coincides with the property described in (22). To see this, suppose $\omega \in \Omega_q$ and $R_j(\omega) \supset [a, b] \cap R$ for $a < b$ in R. Then $S_j(\omega) \supset [a, b)$ by (c). Conversely, suppose $\omega \in \Omega_q$ and $S_j(\omega)$ is dense in (a, b) with $a < b$. Choose a pair of binary rationals c, d with $a < c < d < b$. Then $S_j(\omega) \supset [c, d]$ by (c), so $R_j(\omega) \supset [c, d] \cap R$.   ★

(24) Remarks.   (a) Suppose $\omega \in \Omega_q$. Then $[0, \infty) \backslash S_j(\omega)$ is the finite or countably infinite union of intervals $[a, b)$ whose closures $[a, b]$ are pairwise disjoint. This follows from properties (23b–c).

For (b-d), suppose $\omega \in \Omega_q$ satisfies (23a).

(b) $[0, \infty) \backslash S_j(\omega)$ is dense in $[0, \infty)$, and is therefore a countably infinite union of maximal intervals.

(c) That is, $S_j(\omega)$ looks like the Cantor set, except that the left endpoints of the complementary intervals have been removed from the set. And $S_j(\omega)$ has positive Lebesgue measure, as will be seen in (28, 32).

(d) If $t \in S_j(\omega)$, there is a sequence $r_n \in R$ with $r_n \downarrow t$ and $X(r_n, \omega) \neq j$, so $X(r_n, \omega) \to \varphi$.

(25) Theorem.   *For* $\omega \in \Omega_q$ *and* $i \neq j$ *in* I, *the set* $\overline{S_i(\omega)} \cap \overline{S_j(\omega)}$ *is finite in finite intervals.*

Here, $\bar{A}$ is the closure of A.

PROOF.   Use compactness.   ★

This set is studied again in Section 2.2 of *ACM*.

(26) Theorem.   *The set of $\omega \in \Omega_q$ for which $S_\varphi(\omega)$ has Lebesgue measure 0 is measurable and has P_i-probability 1.*

PROOF.   Fubini on (15, 16).   ★

NOTE.   Suppose all $j \in I$ are instantaneous. For almost all ω, the set $S_\varphi(\omega)$ is the complement of a set of the first category. Consequently, any nonempty interval meets $S_\varphi(\omega)$ in uncountably many points. For a discussion of category, see (Kuratowski, 1958, Sections 10 and 30).

(27) Definition.   *Call a Borel subset B of $[0, \infty)$ metrically perfect iff: for any nonempty open interval (a, b), the set $B \cap (a, b)$ is either empty or of positive Lebesgue measure. Let Ω_m be the set of $\omega \in \Omega_q$, as defined in (12), such that: for all $j \in I$, the set $S_j(\omega)$ is metrically perfect. This is no restriction for stable j and $\omega \in \Omega(j)$, as defined in (19).*

(28) Lemma.   *The set Ω_m is measurable and $P_i\{\Omega_m\} = 1$.*

PROOF.   For $a < r < b$ and $r \in R$ and $j \in I$, let $A(j, r, a, b)$ be the set of $\omega \in \Omega_q$ such that: either $\omega(r) \neq j$, or

$$\text{Lebesgue } \{S_j(\omega) \cap (a, b)\} > 0.$$

Any proper interval contains a proper subinterval with rational endpoints. Moreover, $S_j(\omega) \cap (a, b)$ is nonempty iff $R_j(\omega) \cap (a, b)$ is nonempty. Consequently, Ω_m is the intersection of $A(j, r, a, b)$ for $j \in I$ and $r \in R$ and rational a, b with $a < r < b$. So Ω_m is measurable, and it is enough to prove

$$P_i\{A(j, r, a, b)\} = 1.$$

Suppose $P(r, i, j) > 0$, for otherwise there is little to do. Let $\varepsilon > 0$. Let

$$L(\varepsilon, \omega) = \frac{1}{\varepsilon} \text{ Lebesgue } \{S_j(\omega) \cap (r, r + \varepsilon)\}.$$

Use (15) and Fubini in the first line, and (16) in the last:

$$\int_{\{X(r)=j\}} L(\varepsilon) \, dP_i = \frac{1}{\varepsilon} \int_r^{r+\varepsilon} P_i\{X(r) = j \text{ and } X(t) = j\} \, dt$$

$$= \frac{1}{\varepsilon} \int_0^\varepsilon P_i\{X(r) = j \text{ and } X(r + t) = j\} \, dt$$

$$= P(r, i, j) \frac{1}{\varepsilon} \int_0^\varepsilon P(t, j, j) \, dt.$$

Now $0 \leqq L(\varepsilon) \leqq 1$, and the preceding computation shows

$$\int_{\{X(r)=j\}} L(\varepsilon) \, dP_i \to P_i\{X(r) = j\} \quad \text{as } \varepsilon \to 0.$$

For any $c < 1$, Chebychev on $1 - L(\varepsilon)$ makes

$$P_i\{L(\varepsilon) > c \mid X(r) = j\} \to 1 \quad \text{as } \varepsilon \to 0.$$

So

$$P_i\{A(j, r, a, b) \mid X(r) = j\} = 1.$$

By definition,

$$P_i\{A(j, r, a, b) \mid X(r) \neq j\} = 1. \qquad \bigstar$$

The Markov property

The next main result on the sets of constancy is (32). To prove it, I need the Markov property (31). Lemmas (29–30) are preliminary to (31).

(29) Lemma. *Let $t_n \to t$. Then:*

  (a) *The sequence $X(t_n)$ tends to $X(t)$ in P_i-probability;*

  (b) $P_i\{q\text{-lim } X(t_n) = x(t)\} = 1.$

PROOF. The argument for (10) proves the first claim. For the second, suppose without much loss of generality that $t_n < t$ for infinitely many n; and $t_n > t$ for infinitely many n; and $t_n = t$ for no n. Consider first the set L of n with $t_n < t$. Use (a) to find a subsequence $n' \in L$ with

$$P_i\{X(t_{n'}) \to X(t)\} = 1.$$

As $n \to \infty$ through L, the sequence $X(t_n, \omega)$ has at most one finite limit, by quasiregularity. So, for P_i-almost all ω, it has exactly one, namely $X(t, \omega)$. Similarly for the right. $\qquad \bigstar$

Let R^* be a countable dense subset of $[0, \infty)$. Say a function f from $[0, \infty)$ to $\bar{I}$ is *quasiregular relative to* R^* iff for all $t \geq 0$,

$$f(t) = q\text{-lim } f(r) \quad \text{as } r \in R^* \text{ decreases to } t,$$

and for all $t > 0$,

$$q\text{-lim } f(r) \quad \text{as } r \in R^* \text{ increases to } t$$

exists, and f is finite and quasicontinuous when retracted to R^*.

(30) Lemma. *The set of $\omega \in \Omega_q$ such that $X(\cdot, \omega)$ is quasiregular relative to R^* is measurable and has P_i-probability 1.*

PROOF. Let G be the set of $\omega \in \Omega_q$ such that $X(\cdot, \omega)$ is finite and quasicontinuous when retracted to R^*, and satisfies

$$q\text{-lim } X(r^*, \omega) = X(r, \omega) \quad \text{as } r^* \in R^* \text{ decreases to } r.$$

Now G is measurable, and (29) implies $P_i\{G\} = 1$. If $X(\cdot, \omega)$ is quasiregular relative to R^*, then $\omega \in G$. Suppose $\omega \in G$. I say $X(\cdot, \omega)$ is quasiregular relative to R^*. Fix $t \geq 0$, and let $r^* \in R^*$ decrease to t. Since $X(\cdot, \omega)$ is quasiregular, $X(r^*, \omega)$ quasiconverges; if the limit is $j \in I$, then $X(t, \omega) = j$. On the other hand, if $X(t, \omega) = j \in I$, there is a sequence $r_n \in R$ decreasing to t with $X(r_n, \omega) = j$. Find $r_n^* \in R^*$ to the right of r_n but close to it, with $X(r_n^*, \omega) = j$, and make $X(r^*, \omega)$ quasiconverge to j. I haven't handled the value φ explicitly. But $X(t, \omega) \neq q$-lim $\{X(r^*):r^* \in R^*$ and $r^* \downarrow t\}$ forces at least one side to be finite, so the infinite case follows from the finite case. The argument on the left is similar   ★

Let $t \geq 0$. Let W_t be the set of $\omega \in \Omega_q$ such that $u \to X(t + u, \omega)$ is quasiregular on $[0, \infty)$. For $\omega \in \Omega_q$, let $T_t\omega$ be the function $r \to X(t + r, \omega)$ from R to $\bar{I}$. Let $\mathscr{F}(t)$ be the σ-field spanned by $X(s)$ for $s \leq t$. Here is the *Markov Property*.

(31) Lemma. *Fix $t \geq 0$.*

(a) W_t *is measurable and has P_i-probability* 1.

(b) T_t *is a measurable mapping of W_t into Ω_q.*

(c) *If $\omega \in W_t$ and $u \geq 0$, then $X(t + u, \omega) = X(u, T_t\omega)$.*

(d) *Suppose $A \in \mathscr{F}(t)$ and $A \subset \{X(t) = j\}$. Suppose B is a measurable subset of Ω. Then*
$$P_i\{A \text{ and } T_t^{-1}B\} = P_i\{A\} \cdot P_j\{B\}$$

(e) *On $\{X(t) = j\} \cap W_t$, a regular conditional P_i-distribution for T_t given $\mathscr{F}(t)$ is P_j.*

(f) *Given $\{X(t) = j\}$, the shift T_t is conditionally P_i-independent of $\mathscr{F}(t)$, and its conditional P_i-distribution is P_j.*

(g) *Let $\mathscr{F}$ be the product σ-field in Ω relativized to Ω_q. Let F be a non-negative, measurable function on the cartesian product*
$$(\Omega_q, \mathscr{F}(t)) \times (\Omega_q, \mathscr{F}).$$
Then
$$\int_{W_t} F(\omega, T_t\omega) \, P_i(d\omega) = \int_{\Omega_q} F^*(\omega) \, P_i(d\omega),$$
where
$$F^*(\omega) = \int_{\Omega_t} F(\omega, \omega') \, P_{X(t,\omega)}(d\omega').$$

PROOF. *Claim (a)* is clear from (30). For R^* take the union of $R \cap [0, t]$ with $\{t + r : r \in R$ and $r > 0\}$. You have to worry separately that
$$X(t) = q\text{-lim } \{X(t + r) : r \in R \text{ and } r \downarrow 0\}.$$

Claim (b) is clear.

Claim (c) follows from this computation, for $\omega \in W_t$.
$$\begin{aligned}
X(u, T_t\omega) &= q\text{-lim } \{X(r, T_t\omega) : r \in R \text{ and } r \downarrow u\} \\
&= q\text{-lim } \{X(t + r, \omega) : r \in R \text{ and } r \downarrow u\} \\
&= X(t + u, \omega).
\end{aligned}$$

The first line is definition (14). The second line is the definition of T_t. The third line uses $\omega \in W_t$.

Claim (d). By inspection or (10.16), it is enough to do this for special A and B, of the form

$$A = \{X(s_n) = i_n \text{ for } n = 0, \ldots, N\}$$
$$B = \{X(u_m) = j_m \text{ for } m = 0, \ldots, M\};$$

where $0 = s_0 < \cdots < s_N = t$ and $i_0 = i, \ldots, i_N = j$ are in I; for the second line $0 \leq u_0 < \cdots < u_M < \infty$ and the j's are in I. Let

$$B^* = \{X(t + u_m) = j_m \text{ for } m = 0, \ldots, M\}.$$

Now compute.

$$\begin{aligned}
P_i\{A \text{ and } T_t^{-1}B\} &= P_i\{A \text{ and } W_t \text{ and } T_t^{-1}B\} &&\text{by (a)}\\
&= P_i\{A \text{ and } W_t \text{ and } B^*\} &&\text{by (c)}\\
&= P_i\{A \text{ and } B^*\} &&\text{by (a)}\\
&= P_i\{A\} \cdot P_j\{B\} &&\text{by (16).}
\end{aligned}$$

Claims (e, f) follow from *(d)*.

Claim (g). If $F(\omega, \omega') = 1_A(\omega) \cdot 1_B(\omega')$, this reduces to *(d)*. Now extend.   ★

For a general discussion of *(g)*, see (10.44).

WARNING. The condition $\omega \in \Omega_q$ does not imply $T_t\omega \in \Omega_q$. The conditions $\omega \in \Omega_q$ and $T_t\omega \in \Omega_q$ do not imply $X(s, T_t\omega) = X(t + s, \omega)$.

Metric density

The next result is due to Chung (1960, Theorem 3 on p. 146). It has content only for instantaneous j.

(32) Theorem. *Let $j \in I$. The set of $\omega \in \Omega_q$ with*

$$\lim_{\varepsilon \to 0} \varepsilon^{-1} \text{ Lebesgue } \{S_j(\omega) \cap [0, \varepsilon]\} = 1$$

is measurable and has P_j-probability 1.

PROOF. Confine ω to Ω_q. The function f_n whose value at (t, ω) is

$$n \text{ Lebesgue } \left\{s: t \leq s \leq t + \frac{1}{n} \text{ and } X(s, \omega) = j\right\}$$

is measurable in ω by (15) and continuous in t by inspection. Consequently, it is jointly measurable. Let G be the set of pairs (t, ω) where $\lim_{n\to\infty} f_n(t, \omega) = 1$. Then G is measurable. Let G_t be the t-section of G, namely, the set of $\omega \in \Omega_q$ with $(t, \omega) \in G$. Similarly, G_ω is the set of t with $(t, \omega) \in G$. For each $\omega \in \Omega_q$, the set G_ω differs by a Lebesgue-null set from $S_j(\omega)$, by the metric density theorem (10.59). By Fubini, G differs from the set of pairs (t, ω) with $X(t, \omega) = j$ by a Lebesgue $\times$ P_j-null set. Using Fubini again, there is a Lebesgue-null set N, and for each $t \notin N$ there is a P_j-null set $N(t)$, such that: if $t \notin N$ and $\omega \notin N(t)$ then $X(t, \omega) = j$ iff $\omega \in G_t$. Fix any $t \notin N$. Now

$$P_j\{X(t) = j\} = P(t, j, j) > 0$$

by (5.7), so $P_j\{G_t \mid X(t) = j\} = 1$. If $\omega \in W_t$, as defined for (31), then

$$f_n(t, \omega) = f_n(0, T_t\omega)$$

by (31c). So $\omega \in G_t$ iff $T_t\omega \in G_0$. Then (31d) implies

$$P_j\{G_0\} = P_j\{G_t \mid X(t) = j\} = 1.$$

To complete the argument, check that for measurable subsets B of $[0, \infty)$,

$$\lim_{n\to\infty} n \text{ Lebesgue } \{B \cap [0, 1/n]\} = \lim_{\varepsilon\to 0} \frac{1}{\varepsilon} \text{ Lebesgue } \{B \cap [0, \varepsilon]\},$$

because Lebesgue $\{B \cap [0, \varepsilon]\}$ is monotone in ε and $n/(n+1) \to 1$.   ★

An attractive conjecture is: for P_i-almost all ω, the set $S_j(\omega)$ has right metric density 1 at all its points.

4. THE STRONG MARKOV PROPERTY

You should review Section 7.4 before tackling this section, which is pretty technical. One reason is a breakdown in the proof of (7.35). I'll use its notation for the moment. Then

$$\lim \sup (A \cap B_n) \subset A \cap B$$

survives, by quasiregularity. But

$$A \cap B \subset \lim \inf (A \cap B_n)$$

collapses. To patch this up in (34), I need (33). Here is the permanent notation.

As in Section 7.4, let $\mathscr{F}(t)$ be the smallest σ-field over which all $X(s)$ with $0 \leq s \leq t$ are measurable: the coordinate process X on Ω_q has smooth sample functions and is Markov relative to P_i, with stationary standard transitions P and starting state i. A nonnegative random variable τ on Ω_q is a *Markov time* or just *Markov* iff for all t, the set $\{\tau < t\}$ is in $\mathscr{F}(t)$. Let $\mathscr{F}(\tau+)$, the *pre-τ sigma field*, be the collection of all measurable sets A with $A \cap \{\tau < t\}$ in $\mathscr{F}(t)$ for all t. Let $\Delta = \{\tau < \infty\}$, so $\Delta \subset \Omega_q$. Let τ_n be the least $m/2^n > \tau$, on Δ.

EXAMPLE. $X(\tau)$ is measurable relative to $\mathscr{F}(\tau+)$.

PROOF.   Let $j \in I$. You should check that

$$X(\tau) = j \quad \text{and} \quad \tau < t$$

iff for any positive rational r, there are rational u and v with

$$u \leq \tau < v < t \quad \text{and} \quad v - u < r \quad \text{and} \quad X(v) = j. \qquad \bigstar$$

(33) **Lemma.**   *Suppose τ is Markov. For each $t \geq 0$ and $j \in I$,*

$$\lim_{n \to \infty} P_i\{X(\tau + t) = j \text{ and } X(\tau_n + t) \neq j\} = 0.$$

PROOF.   I will only argue $t = 0$. Fix $\varepsilon > 0$. Then fix n so large that $t \leq 2^{-n}$ implies $1 - P(t, j, j) \leq \varepsilon$. Now

$$\{X(\tau) = j \text{ and } X(\tau_n) \neq j\} = \bigcup_{m=1}^{\infty} A(m),$$

where

$$A(m) = \{(m - 1)2^{-n} \leq \tau < m2^{-n} \text{ and } X(\tau) = j \text{ and } X(m2^{-n}) \neq j\}.$$

But $A(m) \subset \{B(m) \text{ and } X(m2^{-n}) \neq j\}$, where $B(m)$ is the event:

$$(m - 1)2^{-n} \leq \tau < m2^{-n},$$

and there is a binary rational r with

$$\tau < r < m2^{-n} \quad \text{and} \quad X(r) = j.$$

Furthermore, $B(m)$ is the increasing limit of $B(m, N)$, where $B(m, N)$ is the event:

$$(m - 1)2^{-n} \leq \tau < m2^{-n},$$

and there is a binary rational r with

$$\tau < r < m2^{-n} \quad \text{and} \quad X(r) = j$$

and $2^N r$ an integer. Finally, $B(m, N)$ is the disjoint union of $B(m, N, M)$, where $B(m, N, M)$ is the event:

$$(m - 1)2^{-n} \leq \tau < M2^{-N} < m2^{-n} \quad \text{and} \quad X(M2^{-N}) = j,$$

and M is the least such nonnegative integer.

Now $B(m, N, M) \in \mathscr{F}(M/2^N)$. The definition of P_i or the Markov property (31) imply:

$$P_i\{B(m, N, M) \text{ and } X(m2^{-n}) \neq j\}$$
$$= P_i\{B(m, N, M)\} \cdot \left[1 - P\left(\frac{m}{2^n} - \frac{M}{2^N}, j, j\right)\right]$$
$$\leqq \varepsilon P_i\{B(m, N, M)\}.$$

Sum on M:
$$P_i\{B(m, N) \text{ and } X(m2^{-n}) \neq j\} \leqq \varepsilon P_i\{B(m, N)\}.$$

Let $N \uparrow \infty$:
$$P_i\{B(m) \text{ and } X(m2^{-n}) \neq j\} \leqq \varepsilon P_i\{B(m)\}.$$

But the $B(m)$ are disjoint. Sum on m:
$$P_i\{X(\tau) = j \text{ and } X(\tau_n) \neq j\} \leqq \varepsilon. \qquad \bigstar$$

Let $Y(t) = X(\tau + t)$. Call Y the *post-τ process*. I say $(t, \omega) \to Y(t, \omega)$ is measurable. Indeed, $(t, \omega) \to \tau(\omega) + t$ is the sum of two measurable functions, and is therefore measurable. Consequently, $(t, \omega) \to (\tau(\omega) + t, \omega)$ is measurable. But $(t, \omega) \to X(t, \omega)$ is measurable by (15). The composition of the last two mappings is $(t, \omega) \to Y(t, \omega)$.

(34) Proposition. *Suppose τ is a Markov time. Let Y be the post-τ process. Let $0 \leqq s_0 < s_1 < \cdots < s_M$. Let $i_0, i_1, \ldots, i_M$ be in I. Let*
$$B = \{Y(s_m) = i_m \text{ for } m = 0, \ldots, M\}$$
and
$$\pi = \Pi_{m=0}^{M-1} P(s_{m+1} - s_m, i_m, i_{m+1}).$$
Let $A \in \mathscr{F}(\tau+)$ with $A \subset \Delta$. Then

$$(35) \qquad P_i\{A \cap B\} = P_i\{A \text{ and } Y(s_0) = i_0\} \cdot \pi.$$

PROOF. As in (33), let τ_n be the least $m/2^n > \tau$, on Δ. As in (7.35),

$$(36) \quad P_i\{A \text{ and } X(\tau_n + s_m) = i_m \text{ for } m = 0, \ldots, M\}$$
$$= P_i\{A \text{ and } X(\tau_n + s_0) = i_0\} \cdot \pi.$$

But
$$\{A \text{ and } X(\tau + s_0) = i_0\} \supset \limsup_n \{A \text{ and } X(\tau_n + s_0) = i_0\}$$

by quasiregularity. So
$$P_i\{A \text{ and } X(\tau + s_0) = i_0\} \geqq \limsup_n P_i\{A \text{ and } X(\tau_n + s_0) = i_0\}.$$

By (33),
$$P_i\{A \text{ and } X(\tau + s_0) = i_0\} \leqq \liminf_n P_i\{A \text{ and } X(\tau_n + s_0) = i_0\}.$$

Thus, the right side of (36) converges to the right side of (35). Similarly for the left sides.   ★

The next problem is controlling the sample functions of the post-τ process. It is convenient to prove the more general lemma (37) before cleaning the post-τ process in (39). Proposition (39) is a preliminary form of the strong Markov property (41), on the set $\{X(\tau) \in I\}$.

(37) Lemma. *Let Y be a jointly measurable, $\bar{I}$-valued process on a probability triple $(\Omega, \mathscr{F}, \mathscr{P})$. Suppose Y is a Markov chain with stationary transitions P and starting state $j \in I$. Suppose:*

    (a) *$Y(\cdot, \omega)$ is quasicontinuous from the right for all ω;*

    (b) *$\{t: Y(t, \omega) = k\}$ is metrically perfect for all $k \in I$ and all ω.*

Then the set of ω such that $Y(\cdot, \omega)$ is quasiregular has inner $\mathscr{P}$-probability 1.

PROOF.   Let Ω_R be the set of $\omega \in \Omega$ such that $Y(\cdot, \omega)$ retracted to R is quasiregular. As (17) shows, $\Omega_R \in \mathscr{F}$ and $\mathscr{P}\{\Omega_R\} = 1$. Let

$$Y^*(t, \omega) = q\text{-lim } Y(r, \omega) \quad \text{as } r \in R \text{ decreases to } t$$

for all $t \geq 0$ and $\omega \in \Omega_R$. In particular, the Y^* sample functions are quasi-regular. As (17) shows,

$$\mathscr{P}\{Y^*(t) = Y(t)\} = 1 \quad \text{for each } t \geq 0.$$

Let Ω_0 be the set of $\omega \in \Omega_R$ such that

$$\text{Lebesgue } \{t: Y^*(t, \omega) \neq Y(t, \omega)\} = 0.$$

As Fubini implies, $\mathscr{P}\{\Omega_0\} = 1$. The proof of (37) is accomplished by showing that for all $\omega \in \Omega_0$:

$$Y(t, \omega) = Y^*(t, \omega) \quad \text{for all } t \geq 0.$$

Indeed,

(38)   if $Y^*(t, \omega) = k \in I,$ then $Y(t, \omega) = k,$

because of (a). Conversely, suppose $Y(t, \omega) = k \in I$. Now (b) implies Lebesgue $D > 0$ for any $\delta > 0$, where

$$D = \{s: t < s < t + \delta \text{ and } Y(s, \omega) = k\}.$$

Because $\omega \in \Omega_0$, there is an $s \in D$ with $Y^*(s, \omega) = Y(s, \omega)$, that is, with $Y^*(s, \omega) = k$. Because $Y^*(\cdot, \omega)$ is quasiregular, $Y^*(t, \omega) = k$. I haven't handled the value φ explicitly. But $Y(t, \omega) \neq Y^*(t, \omega)$ implies that at least one is in I, so the infinite case follows from the finite case.   ★

The final polish on this argument is due to Pedro Fernandez.

Let Δ_q be the set of $\omega \in \Delta$ such that $Y(\cdot, \omega)$ is quasiregular: where Y is the post-τ process.

WARNING.   Select an ω such that $Y(\cdot, \omega)$ is quasiregular when retracted to R. Even though $Y(\cdot, \omega)$ is quasicontinuous from the right, it is still possible that

$$q\text{-}\lim_{r \downarrow t} Y(r, \omega) = \varphi \quad \text{while} \quad Y(t, \omega) \in I;$$

so $\omega \notin \Delta_q$. This hurts me more than you.

(39) Proposition.   *Let τ be Markov, and $\Delta = \{\tau < \infty\}$.*

(a) *Given Δ and $X(\tau) = j \in I$, the pre-τ sigma field $\mathscr{F}(\tau+)$ and the post-τ process Y are conditionally P_i-independent, Y being conditionally Markov with stationary transitions P and starting state j.*

(b) *Δ_q is measurable.*

(c) *$P_i\{\Delta_q \mid \Delta \text{ and } X(\tau) = j \in I\} = 1$.*

PROOF.[†]   *Claim (a).* Use (34).

Claim (b). Let Δ_R be the set of $\omega \in \Delta$ such that $Y(\cdot, \omega)$ retracted to R is quasiregular. As (17) shows, Δ_R is measurable. For $\omega \in \Delta_R$, let

$$Y^*(t, \omega) = q\text{-}\lim Y(r, \omega) \quad \text{as } r \in R \text{ decreases to } t.$$

Then $Y^*(\cdot, \omega)$ is quasiregular. Now $\omega \in \Delta_q$ iff $\omega \in \Delta_R$ and

$$Y(t, \omega) = Y^*(t, \omega) \quad \text{for all } t \geq 0.$$

If $\omega \in \Delta_R$, then automatically

$$Y(r, \omega) = Y^*(r, \omega) \quad \text{for all } r \in R.$$

Let Δ_Q be the set of $\omega \in \Delta_R$ such that for all $r \in R$, either $r < \tau(\omega)$ or

$$X(r, \omega) = Y^*(r - \tau(\omega), \omega).$$

Plainly, Δ_Q is measurable.

I say $\Delta_Q = \Delta_q$. To begin with, I will argue $\Delta_q \subset \Delta_Q$. Fix $\omega \in \Delta_q$. Then

$$Y^*(t, \omega) = Y(t, \omega) = X[\tau(\omega) + t, \omega]$$

for $t \geq 0$. If $r \geq \tau(\omega)$, put $t = r - \tau(\omega)$, and get $\omega \in \Delta_Q$. Next, I will argue $\Delta_Q \subset \Delta_q$. Suppose $\omega \in \Delta_Q$ and $t \geq 0$. Then $\omega \in \Delta_R$, and all I have to get is $Y(t, \omega) = Y^*(t, \omega)$. If $Y^*(t, \omega) = j \in I$, there are $r \in R$ close to t on the right with $Y^*(r, \omega) = Y(r, \omega) = j$; and $Y(\cdot, \omega)$ is quasicontinuous from the right, so $Y(t, \omega) = j$. Suppose $Y(t, \omega) = j \in I$. There is a sequence $r_n \in R$ with $r_n \downarrow \tau(\omega) + t$ and $X(r_n, \omega) = j$. Because $\omega \in \Delta_Q$,

$$Y^*(r_n - \tau(\omega), \omega) = X(r_n, \omega) = j.$$

[†] I know this is a bad one, but it's plain sailing afterwards.

But $r_n - \tau(\omega) \downarrow t$ and $Y^*(\cdot, \omega)$ is quasicontinuous from the right, so $Y^*(t, \omega) = j = Y(t, \omega)$. The case of infinite values follows by logic. This completes the argument that $\Delta_Q = \Delta_q$, and shows Δ_q is measurable.

Claim (c). Use (a–b) and (37). If $\omega \in \Omega_m$, as defined in (27), then condition (37b) holds; and $P_i\{\Omega_m\} = 1$ by (28).   ★

Let $\bar{\Omega}$ be the set of all functions from R to $\bar{I}$. Let $\bar{X}(r, \omega) = \omega(r)$ for $r \in R$ and $\omega \in \bar{\Omega}$. Give $\bar{\Omega}$ the smallest σ-field over which all $\bar{X}(r)$ are measurable. Here is the shift mapping S from Δ to $\bar{\Omega}$:

$$(40) \qquad \bar{X}(r, S\omega) = Y(r, \omega) = X[\tau(\omega) + r, \omega] \quad \text{for all } r \in R.$$

You should check that S is measurable.

Here is the strong Markov property on the set $\{X(\tau) \in I\}$.

(41) Theorem. *Suppose τ is Markov. Let $\Delta = \{\tau < \infty\}$ and let Y be the post-τ process. Remember that Δ_q is the set of $\omega \in \Delta$ such that $Y(\cdot, \omega)$ is quasiregular. Define the shift S by (40).*

    (a) $P_i\{\Delta_q \mid \Delta \text{ and } X(\tau) = j \in I\} = 1.$

    (b) *If $\omega \in \Delta_q$, then $Y = X \circ S$; that is,*

$$Y(t, \omega) = X(t, S\omega) \quad \text{for all } t \geqq 0.$$

    (c) *Suppose $A \in \mathscr{F}(\tau+)$ and $A \subset \{\Delta \text{ and } X(\tau) = j \in I\}$. Suppose B is a measurable subset of Ω. Then*

$$P_i\{A \text{ and } S \in B\} = P_i\{A\} \cdot P_j\{B\}.$$

    (d) *Given Δ and $X(\tau) = j \in I$, the pre-τ sigma field $\mathscr{F}(\tau+)$ is conditionally P_i-independent of the shift S, and the conditional P_i-distribution of S is P_j.*

    (e) *Let $\mathscr{F}$ be the product σ-field in Ω relativized to Ω_q. Let F be a nonnegative, measurable function on the cartesian product*

$$(\Omega_q, \mathscr{F}(\tau+)) \times (\Omega_q, \mathscr{F}).$$

Let $j \in I$, and

$$D = \{\Delta_q \text{ and } X(\tau) = j\}.$$

Then

$$\int_D F(\omega, S\omega) \, P_i(d\omega) = \int_D F^*(\omega) \, P_i(d\omega)$$

where

$$F^*(\omega) = \int_{\Omega_q} F(\omega, \omega') \, P_j(d\omega').$$

NOTE. Claims (b–e) make sense, if you visualize Ω as this subset of $\bar{\Omega}$:

$$\{\omega : \omega \in \bar{\Omega} \text{ and } \omega(r) \in I \text{ for all } r \in R\}.$$

Then S maps Δ_q into Ω.

PROOF. *Claim* (*a*). Use (39b, c).

Claim (*b*). Use the definitions.

Claim (*c*). Use (34) to handle the special B, of the form

$$\{X(s_m) = i_m \text{ for } m = 0, \ldots, M\},$$

with $0 \leqq s_0 < s_1 < \cdots < s_M$ and $i_0, i_1, \ldots, i_M$ in I. Then use (10.16).

Claim (*d*). Use (c).

Claim (*e*). When $F(\omega, \omega') = 1_A(\omega) \cdot 1_B(\omega')$ this reduces to claim (c). Now extend.   ★

For a general discussion of (e), see (10.44).

Remember $\bar{\Omega}$ is the set of all functions from R to $\bar{I}$; and $X(r, \omega) = \omega(r)$ for $\omega \in \bar{\Omega}$ and $r \in R$; and $\bar{\Omega}$ is endowed with the smallest σ-field over which all $\bar{\Omega}(r)$ are measurable.

Let f be a function from R to $\bar{I}$. Say f is *quasiregular on* $(0, \infty)$ iff:

(a) $f(r) \in I$ for all $r > 0$;

(b) q-$\lim f(r)$ exists as $r \in R$ decreases to t for all $t \geqq 0$, and is $f(t)$ for $t \in R$;

(c) q-$\lim f(r)$ exists as $r \in R$ increases to t for all $t > 0$, and is $f(t)$ for positive $t \in R$.

Let f be a function from $[0, \infty)$ to $\bar{I}$. Say f is *quasiregular on* $(0, \infty)$ iff: f retracted to R is quasiregular on $(0, \infty)$, and

$$f(t) = q\text{-}\lim f(r) \quad \text{as } r \in R \text{ decreases to } t$$

for all $t \geqq 0$. Let $\bar{\Omega}_q$ be the set of $\omega \in \bar{\Omega}$ which are quasiregular on $(0, \infty)$.

CRITERION. Let $\omega \in \bar{\Omega}$. Then $\omega \in \bar{\Omega}_q$ iff $\omega(r + \cdot) \in \Omega_q$ for all positive $r \in R$, and

$$\omega(0) = q\text{-}\lim \{\omega(s) : s \in R \text{ and } s \downarrow 0\}.$$

As (8) implies, $\bar{\Omega}_q$ is measurable. For $\omega \in \bar{\Omega}_q$, let

(42) $\bar{X}(t, \omega) = q\text{-}\lim \bar{X}(r, \omega)$ as $r \in R$ decreases to t.

Introduce the class **P** of probabilities μ on $\bar{\Omega}$, having the properties:

(43a) $\mu\{\bar{X}(r) \in I\} = 1$ for all positive $r \in R$;

and

(43b) $\mu\{\bar{X}(r_n) = i_n \text{ for } n = 0, \ldots, N\}$

$$= \mu\{\bar{X}(r_0) = i_0\} \cdot \Pi_{n=0}^{N-1} P(r_{n+1} - r_n, i_n, i_{n+1})$$

for all nonnegative integers N, and $0 \leqq r_0 < r_1 < \cdots < r_N$ in R, and $i_0, \ldots, i_N$ in I. By convention, an empty product is 1.

CRITERION. $\mu \in \mathbf{P}$ iff for all positive $r \in R$,

$$\mu\{\overline{X}(r) \in I\} = 1,$$

and the μ-distribution of $\{\overline{X}(r + s) : s \in R\}$ is

$$\Sigma_{j \in I}\, \mu\{\overline{X}(r) = j\} \cdot P_j.$$

This makes sense, because $\{s \to \overline{X}(r + s, \omega) : s \in R\}$ is an element of Ω for μ-almost all $\omega \in \overline{\Omega}$. And P_j acts on Ω.

The results of this chapter extend to all $\mu \in \mathbf{P}$ in a fairly obvious way, with $\overline{X}$ replacing X. In particular, $\mu \in \mathbf{P}$ concentrates on $\overline{\Omega}_q$ by (13). And (16) shows that $\mu \in \mathbf{P}$ iff relative to μ, the process $\{\overline{X}(t) : 0 \leq t < \infty\}$ is finitary and Markov on $(0, \infty)$ with stationary transitions P, as defined in (7.18).

(44) Proposition   *Suppose τ is Markov. Let $\Delta = \{\tau < \infty\}$, and let Y be the post-τ process. Let $\overline{\Delta}_q$ be the set of $\omega \in \Delta$ such that $Y(\cdot, \omega)$ is quasiregular on $(0, \infty)$.*

(a) *With respect to $P_i\{\cdot \mid \Delta\}$, the post-τ process Y is a Markov chain on $(0, \infty)$, with stationary transitions P.*

(b) $P_i\{Y(t) \in I \mid \Delta\} = 1$ *for $t > 0$, so Y is finitary.*

(c) $\overline{\Delta}_q$ *is measurable.*

(d) $P_i\{\overline{\Delta}_q \mid \Delta\} = 1.$

PROOF.   *Claim (a)*   follows from (34): put $A = \Delta$.
Claim (b)   follows from (a), (26), and (7.20).
Claims (c, d).   Let R^+ be the set of positive $r \in R$. Let G_r be the set of $\omega \in \Delta$ such that $Y(r + \cdot, \omega)$ is quasiregular. Let

$$Y(0+, \omega) = q\text{-}\lim\, \{Y(r, \omega) : r \in R \text{ and } r \downarrow 0\}$$

$$H = \{\omega : \omega \in \Delta \text{ and } Y(0, \omega) = Y(0+, \omega)\}.$$

Then

$$\overline{\Delta}_q = H \cap (\textstyle\bigcap_{r \in R^+} G_r).$$

Clearly, H is measurable. Because $Y(\cdot, \omega)$ is quasicontinuous from the right, $Y(0+, \omega) \in I$ implies $Y(0, \omega) = Y(0+, \omega)$. Proposition (39c) implies

$$P_i\{Y(0, \omega) \in I \text{ and } Y(0+, \omega) \neq Y(0, \omega) \text{ and } \Delta\} = 0.$$

If two elements of $\overline{I}$ are unequal, not both are φ; so

$$P_i\{H \mid \Delta\} = 1.$$

Fix $r \in R^+$. Then $\tau + r$ is Markov, so G_r is measurable by (39b) on $\tau + r$. And

$$P_i\{X(\tau + r) \in I \mid \Delta\} = 1$$

by (b), so I can use (39c) on $\tau + r$ to get

$$P_i\{G_r \,|\, \Delta\} = 1. \qquad\qquad \star$$

Here is the *strong Markov property*.

(45) Theorem. *Suppose τ is Markov. Let $\Delta = \{\tau < \infty\}$, and let Y be the post-τ process. Let $\bar{\Delta}_q$ be the set of $\omega \in \Delta$ such that $Y(\cdot, \omega)$ is quasiregular on $(0, \infty)$. Define the shift S by (40).*

   (a) *$\bar{\Delta}_q$ is measurable and $P_i\{\bar{\Delta}_q \,|\, \Delta\} = 1$.*

   (b) *If $\omega \in \bar{\Delta}_q$, then $Y = \bar{X} \circ S$; that is,*

$$Y(t, \omega) = \bar{X}(t, S\omega) \quad \text{for all } t \geqq 0.$$

On Δ, let $Q(\cdot, \cdot)$ be a regular conditional P_i-distribution for S given $\mathscr{F}(\tau+)$. Remember that $\mathbf{P}$ is the set of probabilities on $\bar{\Omega}$ which satisfy (43). Let Δ_P be the set of $\omega \in \Delta$ such that $Q(\omega, \cdot) \in \mathbf{P}$.

   (c) *$\Delta_P \in \mathscr{F}(\tau+)$, and $P_i\{\Delta_P \,|\, \Delta\} = 1$.*

PROOF. *Claim (a)* repeats (44c–d).
Claim (b) follows from the definitions.
Claim (c) follows from (44b) and (34), as in (7.41). $\star$

ASIDE. Let $\mathscr{F}_Y(0+) = \bigcap_{\varepsilon>0} \mathscr{F}_Y(\varepsilon)$, where $\mathscr{F}_Y(\varepsilon)$ is the σ-field spanned by $Y(t)$ for $0 \leq t \leq \varepsilon$ and all measurable subsets of $\Omega_q \backslash \Delta$. Given $\mathscr{F}_Y(0+)$, on Δ the process Y and the σ-field $\mathscr{F}(\tau+)$ are conditionally P_i-independent. This generalizes (41d).

NOTE. Strong Markov (41, 45) holds with $\mu \in \mathbf{P}$ in place of P_i: review the proofs.

For another treatment of strong Markov, see (Chung, 1960, II.9), and for a complete strong Markov property on $\{X(\tau) = \varphi\}$, see (Doob, 1968).

5. THE POST-EXIT PROCESS

For $\omega \in \Omega_q$, let

$$\tau(\omega) = \inf \{t : X(t, \omega) \neq X(0, \omega)\}.$$

Plainly, τ is Markov. Let

$$Y(t, \omega) = X[\tau(\omega) + t, \omega] \quad \text{if } \tau(\omega) < \infty,$$

the *post-exit processes*. Following Section 7.3, let $i \in I$ and $0 < q(i) < \infty$. Let Ω_i be the set of $\omega \in \Omega_q$ with $X(0, \omega) = i$ and $\tau(\omega) < \infty$. Let

$$\Gamma(i, j) = Q(i, j)/q(i) \quad \text{for } j \neq i,$$
$$\Gamma(i, j) = 0 \quad \text{for } j = i.$$

(46) Theorem.   **(a)** $P_i\{\tau < t\} = e^{-q(i)t}$, so $P_i\{\Omega_i\} = 1$.

(b) τ and Y are independent.

(c) $P_i\{Y(0) = j\} = \Gamma(i,j)$.

(d) Y is Markov and finitary on $(0, \infty)$, with stationary transitions P and almost all sample functions quasiregular on $(0, \infty)$.

PROOF.   *Claim (a)* follows from (18).

Claim (d) follows from (44c–d).

Claims (b) and (c) are proved as in (7.21), with a new difficulty for instantaneous j. Let τ_n be the least $m/2^n$ greater than τ. Let σ_n be the least $m/2^n$ with $X(m/2^n) \neq i$. Now $\sigma_n \downarrow \tau$, but $X(\sigma_n)$ need not converge to $X(\tau)$, if $X(\tau)$ is not a stable state. To overcome this difficulty, suppose $\omega \in \Omega(i)$, as defined in (19). Check that $\tau_n(\omega) = \sigma_n(\omega)$ for all large enough n, using (20). So, $P_i\{\tau_n = \sigma_n\} \to 1$. Consequently, (33) implies

$$P_i\{X(\tau + t) = j \text{ and } X(\sigma_n + t) \neq j\} \to 0. \qquad \bigstar$$

(47) Remark.   (7.46) continues to hold, supposing $q(i_1), \ldots, q(i_n)$ positive and finite. The proof still works.

Similarly, (7.33) holds with the proper convention. Let $\{\xi_n, \tau_n\}$ be the successive jumps and holding times in X, so far as they are defined. That is, $\xi_0 = X(0)$ and

$$\tau_0 = \inf \{t : X(t) \neq \xi_0\}.$$

Put $\sigma_{-1} = -\infty$ and $\sigma_0 = 0$ and

$$\sigma_n = \tau_0 + \cdots + \tau_{n-1} \quad \text{for} \quad n \geq 1.$$

Suppose $\xi_0, \ldots, \xi_n$ and $\tau_0, \ldots, \tau_n$ are all defined. If τ_n is 0 or ∞, then $\xi_{n+1}, \xi_{n+2}, \ldots$ as well as $\tau_{n+1}, \tau_{n+2}, \ldots$ are left undefined. If $0 < \tau_n < \infty$ and $X(\sigma_{n+1}) = \varphi$, then $\xi_{n+1} = \varphi$; but $\xi_{n+2}, \xi_{n+3}, \ldots$ and $\tau_{n+1}, \tau_{n+2}, \ldots$ are left undefined. If $0 < \tau_n < \infty$ and $X(\sigma_{n+1}) \in I$, then

$$\xi_{n+1} = X(\sigma_{n+1})$$

$$\tau_{n+1} + \sigma_{n+1} = \inf \{t : t > \sigma_{n+1} \text{ and } X(t) \neq \xi_{n+1}\}.$$

Here is an inductive measurability check. First, $\sigma_0 = 0$ and $\xi_0 = X(0)$ are measurable. Next, $\sigma_{n+1} < t$ iff $\xi_n \in I$ and there is a binary rational r with

$$\sigma_{n-1} < \sigma_n < r < t \quad \text{and} \quad X(r) \neq \xi_n.$$

Finally, $\xi_{n+1} = j \in I$ iff: $\sigma_{n+1} < \infty$, and for any binary rational r bigger than σ_{n+1}, there is a binary rational s with

$$\sigma_n < \sigma_{n+1} < s < r \quad \text{and} \quad X(s) = j.$$

Recall $Q = P'(0)$ and $q(i) = -Q(i, i)$. Let

$$\Gamma(i, j) = Q(i, j)/q(i) \quad \text{for } 0 < q(i) < \infty \text{ and } i \neq j$$

$$= 0 \qquad\qquad \text{elsewhere.}$$

(48) Theorem. *Let $i_0 = i, i_1, \ldots, i_N$ be in I, and let $t_0, \ldots, t_N$ be non-negative numbers. Then*

$$P_i\{\xi_n = i_n \text{ and } \tau_n > t_n \text{ for } n = 0, \ldots, N\} = pe^{-t},$$

where

$$p = \Pi_{n=0}^{N-1} \Gamma(i_n, i_{n+1})$$

and

$$t = \Sigma_{n=0}^{N} q(i_n)t_n.$$

Here, $\infty \cdot 0 = 0$ and $\infty + 0 = \infty$; while $\infty \cdot x = \infty + x = \infty$ for $x > 0$; and $e^{-\infty} = 0$. In particular, if ξ_n is defined and absorbing, then $\tau_n = \infty$ a.e.; if ξ_n is defined and instantaneous, then $\tau_n = 0$ a.e.; in either case, further ξ's or τ's are defined almost nowhere. If ξ_n is defined and stable but not absorbing, then $0 < \tau_n < \infty$ a.e. The proof of (7.33) still works.

The backward and forward equations

The results of Section 7.6 can be extended to the general case. For (7.52–58) on the backward equation, assume i is stable. The argument is about the same, because (7.21) works in general (46). For (7.60–68) on the forward equation, assume j is stable. To handle (7.66), let D be the complement of

$$X(s) = X(t) \quad \text{for } 0 \leqq s \leqq t.$$

On D, let γ be the sup of s with $X(s) \neq X(t)$, and let

$$X(\gamma-) = q\text{-}\lim_{s \uparrow t} X(s).$$

Rescue the argument by using the idea of (33) on the reversed process

$$\{X(t - s){:}0 \leqq s \leqq t\},$$

which is Markov with nonstationary transitions. Details for this maneuver appear in (Chung, 1967, pp. 198–199). You will have to check that given $X(t) = j$, the time t is almost surely interior to a j-interval, by adapting the proof of (7.6). For (7.68), say X jumps from φ to j at time τ iff

$$X(\tau) = j \quad \text{and} \quad X(\tau-) = q\text{-}\lim_{s \uparrow t} X(s) = \varphi.$$

6. THE ABSTRACT CASE

The results of this chapter, notably (41–48), can be applied to abstract Markov chains. Let $(\mathcal{X}, \Sigma, \mu)$ be an abstract probability triple, and $\{Z(t):0 \leq t < \infty\}$ an I-valued process on $(\mathcal{X}, \Sigma)$. Suppose Z is a Markov chain with stationary transitions P and starting state i, relative to μ. Suppose that all the sample functions of Z are quasiregular: if not, Z can be modified using (17). There is no difficulty in transcribing (46–48) to this situation: see (5.46–48) for the style. Strong Markov is something else. Probably the easiest thing to do is to review the proof and make sure it still works. For ideological reasons, I will use the Chapter 5 approach.

Let $\Sigma(t)$ be the σ-field in $\mathcal{X}$ spanned by $Z(s)$ for $0 \leq s \leq t$. Let σ be a random variable on $\mathcal{X}$, with values in $[0, \infty]$. Suppose σ is *Markov for Z*, namely $\{\sigma < t\} \in \Sigma(t)$ for all t. Let $\Sigma(\sigma+)$ be the σ-field of all $A \in \Sigma$ such that $A \cap \{\sigma < t\} \in \Sigma(t)$ for all t. Let W be the post-σ process:

$$W(t, x) = Z[\sigma(x) + t, x] \quad \text{when } \sigma(x) < \infty.$$

Let $Tx \in \bar{\Omega}$ be the function $W(\cdot, x)$ retracted to the binary rationals R in $[0, \infty)$. Define $\mathbf{P}$ by (43). The object is to prove:

(49) Theorem. *Let $Q(\cdot, \cdot)$ be a regular conditional μ-distribution for T given $\Sigma(\sigma+)$. Let $\bar{\mathcal{X}}_q$ be the set of $x \in \{\sigma < \infty\}$ such that $W(\cdot, x)$ is quasiregular on $(0, \infty)$.*

(a) $\bar{\mathcal{X}}_q \in \Sigma$ *and* $\mu\{\bar{\mathcal{X}}_q \mid \sigma < \infty\} = 1$.

(b) *If* $x \in \bar{\mathcal{X}}_q$ *then* $W = \bar{X} \circ T$; *that is,* $W(t, x) = \bar{X}(t, Tx)$ *for all* $t \geq 0$: *where $\bar{X}$ was defined by (42).*

(c) $\mu\{Q \in \mathbf{P} \mid \sigma < \infty\} = 1$.

(d) *Given* $\{\sigma < \infty\}$ *and* $Z(\sigma) = j \in I$, *the pre-σ sigma field $\Sigma(\sigma+)$ is conditionally μ-independent of the shift T, and the conditional μ-distribution of T is P_j.*

(e) *Let $\mathcal{F}$ be the product σ-field in Ω relativized to Ω_q. Let F be a nonnegative, measurable function on the cartesian product*

$$(\mathcal{X}, \Sigma(t)) \times (\Omega_q, \mathcal{F}).$$

Let $j \in I$ and

$$D = \{\bar{\mathcal{X}}_q \text{ and } Z(\sigma) = j\}.$$

Then

$$\int_D F(x, Tx)\,\mu(dx) = \int_D F^*(x)\,\mu(dx),$$

where

$$F^*(x) = \int_{\Omega_q} F(x, \omega) \, P_j(d\omega).$$

PROOF.   Use (41), (45), and (50) below.   ★

(50) Proposition.   *Let M be this mapping from $\mathcal{X}$ to Ω_q:*

$$X(r, Mx) = Z(r, x) \quad \text{for all } r \in R.$$

There is a Markov time τ on Ω_q such that $\sigma = \tau \circ M$. Then

$$\{\sigma < \infty\} \cap \Sigma(\sigma+) = M^{-1}[\{\tau < \infty\} \cap \mathcal{F}(\tau+)].$$

Let Y be the post-τ process on Ω_q, and let S be Y with time domain retracted to R. Then

$$W(t, x) = Y(t, Mx)$$

and

$$Tx = SMx$$

for all $x \in \mathcal{X}$.

PROOF.   The first problem is to find a Markov time τ on Ω_q such that $\sigma = \tau \circ M$. Let $A \in \Sigma(\sigma+)$ with $A \subset \{\sigma < \infty\}$. The second problem is to find $B \in \mathcal{F}(\tau+)$ with $B \subset \{\tau < \infty\}$ and $A = M^{-1}B$. The rest is easy. To start the constructions, let $\Sigma(\infty)$ be the σ-field spanned by Z, and let $\mathcal{F}(\infty)$ be the full σ-field in Ω_q, namely the σ-field spanned by X. Check

$$\Sigma(t) = M^{-1}\mathcal{F}(t) \quad \text{for } 0 \leq t \leq \infty.$$

I remind you that M^{-1} commutes with set operations. Start work on τ. Confine r and s to R. Now $\{\sigma < r\} \in \Sigma(r)$, so $\{\sigma < r\} = M^{-1}F_r$ for some $F_r \in \mathcal{F}(r)$. Let

$$G_r = \bigcup_{s < r} F_s.$$

Then $G_r \in \mathcal{F}(r)$ and $\{\sigma < r\} = M^{-1}G_r$. Moreover,

$$G_r = \bigcup_{s < r} G_s.$$

Let $G = \bigcup_r G_r$. Off G, let $\tau = \infty$. For $\omega \in G$, let $\tau(\omega)$ be the sup of r with $\omega \notin G_r$. You should check that $\{\tau < r\} = G_r$. So τ is Markov, because $\{\tau < t\} = \bigcup_{r < t} \{\tau < r\}$. And $\tau \circ M = \sigma$, because $\tau \circ M < r$ iff $\sigma < r$.

Turn to the second problem. Let $A \in \Sigma(\sigma+)$ and $A \subset \{\sigma < \infty\}$. Now $A \cap \{\sigma < r\} \in \Sigma(r)$, so $A \cap \{\sigma < r\} = M^{-1}H_r$ for some $H_r \in \mathcal{F}(r)$. Let

$$J_r = H_r \cap \{\tau < r\}$$
$$K_r = \bigcup_{s < r} \{\tau < s\} \backslash J_s$$
$$B_r = J_r \backslash K_r$$
$$B = \bigcup_r B_r.$$

Check $A = M^{-1}B$. I claim $B \in \mathscr{F}(\tau +)$. I only have to prove

$$B \cap \{\tau < t\} = \bigcup_{r < t} B_r.$$

Clearly, $r < t$ makes $B_r \subset \{\tau < t\}$, so

$$B \cap \{\tau < t\} \supset \bigcup_{r < t} B_r.$$

For the opposite inclusion, fix $\omega \in B \cap \{\tau < t\}$. Choose s with $\tau(\omega) < s < t$. Now $\omega \in B_r$ for some r; if $r < t$, you're done: suppose $r \geq t$. But $\omega \notin K_r$ and $\tau(\omega) < s$, so $\omega \in J_s$. And K is nondecreasing, so $\omega \notin K_s$. That is, $\omega \in B_s$.
    The rest is haggling.   ★

    The final version of this proof is due to Mike Orkin.

    QUESTION.   Can you lift a strict Markov time this way?

10

APPENDIX

1. NOTATION

★ is the end of a proof, or a discussion.

iff is if and only if.

$A \setminus B$ is the set of points in A but not in B.

$A \triangle B = (A \setminus B) \cup (B \setminus A)$.

$\emptyset$ is the empty set.

∘ is composition.

If $\mathscr{F}$ is a σ-field and $A \in \mathscr{F}$, then $A\mathscr{F}$ is the σ-field of subsets of A of the form $A \cap F$, with $F \in \mathscr{F}$.

X is measurable on Y means that X is measurable relative to any σ-field over which Y is measurable. Usually, this means you can compute X in a measurable way from Y.

If π is a statement about points x, then

$$\{\pi\} = \{x : \pi(x)\} = [\pi] = [x : \pi(x)]$$

is the set of x for which $\pi(x)$ is true. And

$$\Sigma_x \{a_x : \pi(x)\} = \Sigma \{a_x : \pi(x)\}$$

is the sum of a_x over all x for which $\pi(x)$ is true.

An empty sum is 0; an empty product is 1.

$a_n = 0(b_n)$ means: there are finite K and N such that

$$|a_n| \leq K|b_n| \quad \text{for all } n \geq N.$$

$a_n = o(b_n)$ means: for any positive ε, there is a finite N_ε such that

$$|a_n| \leq \varepsilon|b_n| \quad \text{for all } n \geq N_\varepsilon.$$

I want to thank Allan Izenman for checking the final draft of this chapter.

$a_n \sim b_n$ means: there are finite, positive K and N such that

$$\frac{1}{K} b_n \leq a_n \leq K b_n \quad \text{for all } n \geq N.$$

$a_n \approx b_n$ means: for any ε with $0 < \varepsilon < 1$, there is a finite N_ε such that

$$(1 - \varepsilon) b_n \leq a_n \leq (1 + \varepsilon) b_n \quad \text{for all } n \geq N_\varepsilon.$$

$[0, 1) = \{x : 0 \leq x < 1\}$.

$\langle x \rangle$ means the greatest integer $n \leq x$.

x is *positive* means $x > 0$, while x is *nonnegative* means $x \geq 0$. When it seems desirable, the redundancy "x is *strictly positive*" is employed. Similarly for *increasing* and *nondecreasing*.

Real-valued means in $(-\infty, \infty)$, while *extended real-valued* means in $[-\infty, \infty]$. Random variables are allowed to take infinite values without explicit warning.

Clearly usually means that the assertion which follows is clear. Sometimes, by force of habit, it means that I didn't feel like writing out the argument.

Let f be a real-valued function on $S \times T$. Then $f(s) = f(s, \cdot)$ is the real-valued function $t \to f(s, t)$ on T, while $f(t) = f(\cdot, t)$ is the real-valued function $s \to f(s, t)$ on S. Furthermore, f is used indifferently for the real-valued function $(s, t) \to f(s, t)$ on $S \times T$, the function-valued mapping $s \to f(s)$ on S, and the function-valued mapping $f \to f(t)$ on T. Whenever this threatens to get out of hand, some explanation is provided.

2. NUMBERING

In each chapter, all important formulas, definitions, theorems and so on are treated as displays and numbered consecutively from 1 on. Inside chapter a: display (b) means the display numbered b in chapter a; for $a' \neq a$, display $(a' \cdot b)$ means the display number b in chapter a'. Section $a \cdot b$ is section b of chapter a. And $(MC, a \cdot b)$ is the display numbered b in chapter a of MC; this kind of reference is used in $B \& D$ and *Approx*.

3. BIBLIOGRAPHY

(Blackwell, 1958) and Blackwell (1958) refer to the work of Blackwell listed in the bibliography with year of publication 1958. The obvious problem is settled by this device: (Lévy, 1954a). Each entry in the bibliography gives the edition I used when writing the book. In certain cases, notably (Chung, 1960), a more recent edition is now available. When this is known to me, the newer edition is referred to in parentheses following the main entry. Journals are abbreviated following *Math. Rev.* practice. I do not give references to my own articles.

This book is part of a triology, published by Holden-Day at San Francisco in 1970. The titles, and their abbreviations, are:

> *Markov Chains (MC)*
> *Brownian Motion and Diffusion (B & D)*
> *Approximating Countable Markov Chains by Finite Ones (Approx.)*

4. THE ABSTRACT LEBESGUE INTEGRAL†

As usual, Ω is a set, and $\mathscr{F}$ is a σ-field of subsets of Ω; that is, $\Omega \in \mathscr{F}$ and $\mathscr{F}$ is closed under complementation and the formation of countable unions. A *probability* $\mathscr{P}$ on $\mathscr{F}$ is a countably additive, nonnegative function, with $\mathscr{P}(\Omega) = 1$. Then $(\Omega, \mathscr{F}, \mathscr{P})$ is called a *probability triple*. If $\mathscr{P}(\Omega) \leq 1$, then $\mathscr{P}$ is a *subprobability*. On occasion, reference will be made to the *inner measure* $\mathscr{P}_*(A)$ of $A \subset \Omega$, which is sup $\{\mathscr{P}(F): F \in \mathscr{F}$ and $F \subset A\}$. Similarly, the *outer measure* $\mathscr{P}^*(A) = \inf \{\mathscr{P}(F): F \in \mathscr{F}$ and $A \subset F\}$.

The *indicator function* 1_A is 1 on A and 0 on $\Omega \setminus A$, the set of points in Ω but not in A. A *random variable* X is an $\mathscr{F}$-*measurable* function from Ω to the real line: that is,

$$[X \leq x] = \{X \leq x\} = \{\omega : \omega \in \Omega \text{ and } X(\omega) \leq x\} \in \mathscr{F}$$

for all x. If X takes infinite values, it will sometimes, but not always, be described as *extended real-valued*. A *partially defined* random variable is a random variable on $(A, A\mathscr{F})$, where $A \in \mathscr{F}$ and $A\mathscr{F} = \{A \cap B : B \in \mathscr{F}\}$. For $X \geq 0$, the *expectation* (or $\mathscr{P}$-expectation, if several probabilities are in sight) of X is a nonnegative, extended real number $E(X)$:

$$E(1_A) = \mathscr{P}(A); \quad \text{and} \quad E(\alpha X + \beta Y) = \alpha E(X) + \beta E(Y);$$

and

(1) Monotone convergence theorem.   *If* $0 \leq X_n \uparrow X$, *then* $E(X_n) \uparrow E(X)$.

Here $\alpha \geq 0$ and $\beta \geq 0$; and $(\alpha X + \beta Y)(\omega) = \alpha X(\omega) + \beta Y(\omega)$; and $X_n \uparrow X$ means that $X_n(\omega)$ is nondecreasing with n and tends to $X(\omega)$ as $n \to \infty$ for each $\omega \in \Omega$. Moreover,

(2) Fatou's Lemma.   *If* $X_n \geq 0$, *then* $E(\liminf X_n) \leq \liminf E(X_n)$.

If $\{A_n\}$ is a sequence of sets, then $\limsup A_n$ is the set of ω which are elements of infinitely many A_n; and $\liminf A_n$ is the set of ω which are

† References: (Loève, 1963, Part 1); (Neveu, 1965, Chapters 1 and 2).

elements of all but finitely many A_n. Thus,

$$1_{\liminf A_n} = \liminf 1_{A_n} \quad \text{and} \quad 1_{\limsup A_n} = \limsup 1_{A_n}.$$

Furthermore, $\Omega \setminus (\limsup A_n) = \liminf (\Omega \setminus A_n)$. By definition $A_n \to A$ iff $A = \limsup A_n = \liminf A_n$. As (2) implies,

(3) Corollary. *If $A_n \to A$, then $\mathscr{P}(A_n) \to \mathscr{P}(A)$.*

In general, $X = X^+ - X^-$, where $X^+ = X \cup 0 = \max\{X, 0\}$, and

$$E(X) = \int X \, d\mathscr{P} = \int X(\omega) \, \mathscr{P}(d\omega) = E(X^+) - E(X^-),$$

except $\infty - \infty$ is taboo. Write $E(1_A X) = \int_A X = \int_A X \, d\mathscr{P}$.

(4) Dominated convergence theorem. *If $X_n \to X$ and $|X_n| \leq Y$ for all n and $E(Y) < \infty$, then $E(X_n) \to E(X)$.*

If $\lim_{k \to \infty} \int_{\{|X_n| \geq k\}} |X_n| \, d\mathscr{P} = 0$ uniformly in n, the X_n are *uniformly integrable*.

(5) Criterion. *$\{X_n\}$ is uniformly integrable if either (a) or (b) holds:*

(a) $E(|X_n|^p) \leq K$ for all n, for some $p > 1$ and $K < \infty$.
(b) $\mathscr{P}\{|X_n| \geq x\} \leq \mathscr{P}\{Y \geq x\}$ for all n and all $x > K$, where $0 \leq K < \infty$ and $0 \leq Y$ and $E(Y) < \infty$.

(6) Theorem. *If $X_n \to X$ and the X_n are uniformly integrable, $E(X_n) \to E(X)$; in fact, $E(|X - X_n|) \to 0$.*

Easy, useful estimates:

(7) Chebychev inequality. $\mathscr{P}\{X \geq k\} \leq k^{-1} E(X)$ for $X \geq 0$ and $k > 0$.

(8) Schwarz inequality. $[E(XY)]^2 \leq E(X^2) \cdot E(Y^2)$. *Equality holds iff* $\mathscr{P}\{Y = aX\} = 1$ *for some constant a.*

(9) Jensen's inequality. *Suppose $E(|X|) < \infty$ and f is convex. Then $E[f(X)] \geq f[E(X)]$. Equality holds iff f is linear on an interval $[a,b]$ with $\mathscr{P}\{a \leq X \leq b\} = 1$.*

PROOF. Use (76) below.   ★

Useful miscellany:

(10) Theorem. (a) *Let X and Y be random variables. Suppose*

$$\int_A X \, d\mathscr{P} = \int_A Y \, d\mathscr{P}$$

for all measurable A. Then $\mathscr{P}\{X = Y\} = 1$.

(b) $E(|X|) < \infty$ *implies* $\mathcal{P}\{|X| < \infty\} = 1.$
(c) $E(|X|) = 0$ *implies* $\mathcal{P}\{X = 0\} = 1.$

Absolute continuity

Suppose P and Q are two probabilities on $\mathcal{F}$. Then $P \ll Q$, or P is *absolutely continuous* with respect to Q, iff $Q(A) = 0$ implies $P(A) = 0$.

(11) Radon-Nikodym Theorem. $P \ll Q$ *iff there is a nonnegative measurable finite f with*

$$P(A) = \int_A f \, dQ.$$

This f is unique up to changes on null sets, by (10a), *and is the* Radon-Nikodym *derivative of P with respect to Q. For any nonnegative measurable g,*

$$\int g \, dP = \int gf \, dQ.$$

Say P is *orthogonal* or *singular* with respect to Q, or $P \perp Q$, iff there is a set A with $P(A) = 0$ and $Q(\Omega \setminus A) = 0$.

(12) Lebesgue decomposition. $P = P_0 + P_1$, *where* $P_0 \ll Q$ *and* $P_1 \perp Q$ *and the P_i are subprobabilities.*
Say P is *equivalent* to Q, or $P \equiv Q$, iff $P \ll Q$ and $Q \ll P$.

Convergence

Suppose X_n and X are finite almost surely. Say X_n *converges to X, or $X_n \to X$, in $\mathcal{P}$-probability iff*

$$\mathcal{P}\{|X_n - X| \geq \varepsilon\} \to 0$$

for any $\varepsilon > 0$. Say $\{X_n\}$ is *fundamental in probability* iff

$$\lim_{n,m \to \infty} \mathcal{P}\{|X_n - X_m| \geq \varepsilon\} = 0$$

for any $\varepsilon > 0$. For extended real-valued X_n and X, say $X_n \to X$ in $\mathcal{P}$-probability iff for any positive, finite ε and K:

$$\mathcal{P}\{|X_n - X| \geq \varepsilon \text{ and } |X| < \infty\} \to 0$$

$$\mathcal{P}\{X_n \leq K \text{ and } X = \infty\} \to 0$$

$$\mathcal{P}\{X_n \geq -K \text{ and } X = -\infty\} \to 0.$$

(13) Theorem.   *Suppose $\{X_n\}$ are finite almost surely and fundamental in* *probability. Then there is a random variable X, also finite almost surely, such that $X_n \to X$ in probability. Conversely, if $\{X_n\}$ and X are finite almost surely, and $X_n \to X$ in probability, then $\{X_n\}$ is fundamental in probability.*

Say X_n *converges to* X, or $X_n \to X$, *almost surely* iff

$$\mathscr{P}\{X_n \to X\} = 1.$$

(14) Theorem. *If $X_n \to X$ almost surely, then $X_n \to X$ in probability. If $X_n \to X$ in probability, there is a nonrandom subsequence n* such that $X_{n*} \to X$ almost surely.*

If $\pi(\omega)$ is a statement about ω, then π *almost everywhere*, or π $\mathscr{P}$-a.e. or π *almost surely* means $\mathscr{P}(\omega:\pi(\omega)$ is false$\} = 0$. Similarly, π a.e. *on A* means $\mathscr{P}\{\omega:\pi(\omega)$ is false and $\omega \in A\} = 0$. Finally π *almost nowhere* means that $\mathscr{P}\{\omega:\pi(\omega)$ is true$\} = 0$.

The L^p-spaces

A random variable X is in L^p relative to $\mathscr{P}$ iff $\int |X|^p \, d\mathscr{P} < \infty$. The p-th root of this number is the L^p-norm of X. A sequence $X_n \to X$ in L^p if the norm of $X - X_n$ tends to 0. After identifying functions which are equal a.e., L^p is a Banach space for $p \geqq 1$. This popular fact is not used in the book.

The results of this section (except that uniform integrability gets more complicated) are usually true, and sometimes used, for measures $\mathscr{P}$ which are not probabilities; a measure on $\mathscr{F}$ is nonnegative and countably additive. In places like (11), you have to assume that $\mathscr{P}$ is σ-finite:

$$\Omega = \cup_{i=1}^{\infty} \Omega_i \quad \text{with} \quad \mathscr{P}(\Omega_i) < \infty.$$

For the rest, suppose $\mathscr{P}$ is a probability; although I occasionally use converse Fubini (22) for σ-finite measures.

5. ATOMS*

If Σ is a σ-field of subsets of Ω and $\omega \in \Omega$, the Σ-*atom* containing ω is $\hat{\Sigma}(\omega)$, the intersection of all the Σ-sets containing ω. Say Σ is *separable* or *countably generated* iff it is the smallest σ-field which includes some countable collection $\mathscr{C}$ of sets. In this case, let $\mathscr{A}$ be the smallest *field* containing $\mathscr{C}$. Namely, $\Omega \in \mathscr{A}$; and $\mathscr{A}$ is closed under the formation of complements and finite unions. Then $\mathscr{A}$ is countable and generates Σ. That is, Σ is the smallest σ-field which includes $\mathscr{A}$. Let $\mathscr{A}(\omega)$ be the intersection of all the $\mathscr{A}$-sets

---

* References: (Blackwell, 1954); (Loève, 1963, Secs. 1.6, 25.3 and 26.2).

containing ω, which by definition is the $\mathscr{A}$-atom containing ω. Then $\mathscr{A}(\omega) \in \Sigma$, and $\mathscr{A}(\omega) = \Sigma(\omega)$. Indeed, $\mathscr{A}(\omega)$ is wholly included in or wholly disjoint from each $\mathscr{A}$-set. By the *monotone class argument*, which I will make in a second, this goes for Σ as well.

Call M a *monotone class* iff:

(a) $A_n \in M$ and $A_1 \subset A_2 \subset \cdots$ imply $\bigcup_n A_n \in M$; and

(b) $A_n \in M$ and $A_1 \supset A_2 \supset \cdots$ imply $\bigcap_n A_n \in M$.

The monotone class argument. Let M be the set of $A \in \Sigma$ such that: $\mathscr{A}(\omega)$ is wholly included in or wholly disjoint from A. Then M is monotone and includes $\mathscr{A}$. Now (15) below implies $M \supset \Sigma$.

(15) Lemma. *The smallest monotone class which includes a field $\mathscr{A}$ coincides with the smallest σ-field which includes $\mathscr{A}$.*

(16) Theorem. *Let $\mathscr{C}$ be a collection of sets which is closed under intersection and generates the σ-field $\mathscr{F}$. Let P and Q be two subprobabilities on $\mathscr{F}$, which agree on $\mathscr{C}$. Suppose Ω is a countable union of pairwise disjoint elements of $\mathscr{C}$, or more generally that $P(\Omega) = Q(\Omega)$. Then $P = Q$ on $\mathscr{F}$.*

PROOF.   Let $\mathscr{E}$ be the class of $A \in \mathscr{F}$ with $P(A) = Q(A)$. Clearly, $\Omega \in \mathscr{E}$ and $\mathscr{C} \subset \mathscr{E}$. If $A \in \mathscr{E}$ and $B \in \mathscr{E}$ and $A \supset B$, then $A \setminus B \in \mathscr{E}$ because $R(A \setminus B) = R(A) - R(B)$ for $R = P$ or Q. If $A \in \mathscr{E}$ and $B \in \mathscr{E}$ and $A \cap B = \emptyset$, then $A \cup B \in \mathscr{E}$, for a similar reason.

If $A_i \in \mathscr{C}$ for $i = 1, \ldots, n$, then $B_n = \bigcup_{i=1}^n A_i \in \mathscr{E}$ by induction on n. The case $n = 1$ is clear. And

$$B_{n+1} = B_n \cup A_{n+1}$$
$$= B_n \cup (A_{n+1} \setminus B_n)$$
$$= B_n \cup (A_{n+1} \setminus C_n)$$

where

$$C_n = A_{n+1} \cap B_n = \bigcup_{i=1}^n (A_{n+1} \cap A_i).$$

Now $A_{n+1} \cap A_i \in \mathscr{C}$, because $\mathscr{C}$ was assumed closed under intersection. So C_n, being the union of n sets in $\mathscr{C}$, is in $\mathscr{E}$ by inductive assumption. But $C_n \subset A_{n+1}$, so $A_{n+1} \setminus C_n \in \mathscr{E}$. Finally, $A_{n+1} \setminus C_n$ is disjoint from B_n, so its union B_{n+1} with B_n is in $\mathscr{E}$. This completes the induction.

Let $A^* = A$ or $\Omega \setminus A$. If $A_i \in \mathscr{C}$ for $i = 1, \ldots, n$. I will get

$$B = \bigcap_{i=1}^n A_i^* \in \mathscr{E}.$$

Using the assumption that $\mathscr{C}$ is closed under intersection, you can rewrite B as

$$B = \left[\bigcap_{i=1}^m (\Omega \setminus C_i) \right] \cap D,$$

with $C_1, \ldots, C_m$ and D in $\mathscr{C}$. Let $C = \bigcup_{i=1}^{m} C_i$. Then

$$B = D \setminus C$$

$$= D \setminus (C \cap D).$$

Now $C \cap D \subset D$ and $C \cap D = \bigcup_{i=1}^{m} (C_i \cap D) \in \mathscr{E}$, because $C_i \cap D \in \mathscr{C}$. This forces $B \in \mathscr{E}$.

If $A_i \in \mathscr{C}$ for $i = 1, \ldots, n$, the field generated by $A_1, \ldots, A_n$ is included in $\mathscr{E}$; the typical atom was displayed above as $B \in \mathscr{E}$, and any nonempty set in the field is a disjoint union of some atoms. Consequently, the field $\mathscr{A}$ generated by $\mathscr{C}$ is included in $\mathscr{E}$. Of course, $\mathscr{A}$ generates $\mathscr{F}$. Now use the monotone class argument: $\mathscr{E}$ is a monotone class, and includes $\mathscr{A}$; so $\mathscr{E}$ includes the smallest monotone class which includes $\mathscr{A}$, namely $\mathscr{F}$.   ★

Let Σ be a σ-field of subsets of Ω. The set of probabilities on Σ is *convex*: the convex combination of two probabilities is again a probability. A probability is *extreme* iff it cannot be represented as the convex combination of two distinct probabilities. A probability is 0–1 iff it only assumes the values 0 and 1; sometimes, Σ is called *trivial* relative to such a probability.

(17) Theorem.   *Let m be a probability on Σ.*

   (a)   *m is extreme iff m is $0 - 1$.*

   (b)   *Suppose Σ is countably generated. Then m is $0 - 1$ iff $m(B) = 1$ for some atom B of Σ.*

PROOF.   *Claim (a).*   Suppose m is not $0 - 1$. Then $0 < m(A) < 1$ for some $A \in \Sigma$. And

$$m = m(A) \cdot m(\cdot \mid A) + [1 - m(A)] \cdot m(\cdot \mid \Omega \setminus A)$$

is not extreme. Conversely, suppose m is not extreme. Then

$$m = pm' + (1 - p)m''$$

for $0 < p < 1$ and $m' \neq m''$. Find $A \in \Sigma$ with $m'(A) \neq m''(A)$. Conclude $0 < m(A) < 1$ and m is not $0 - 1$.

Claim (b).   The *if* part is easy. For *only if*, let $\mathscr{A}$ be a countable generating field for Σ. Let $\mathscr{A}_i$ be the set of $A \in \mathscr{A}$ such that $m(A) = i$, for $i = 0$ or 1. Now $\mathscr{A} = \mathscr{A}_0 \cup \mathscr{A}_1$, and $A \in \mathscr{A}_0$ iff $\Omega \setminus A \in \mathscr{A}_1$. Let B be the intersection of all $A \in \mathscr{A}_1$. Then $B \in \Sigma$ has m-probability 1, and in particular is nonempty. Fix $\omega \in B$. If $A \in \mathscr{A}_1$, then $A \supset B$. If $A \in \mathscr{A}_0$, then $\Omega \setminus A \in \mathscr{A}_1$ and $A \cap B = \emptyset$. Thus, $\omega \in A \in \mathscr{A}$ iff $A \in \mathscr{A}_1$, and B is an atom.   ★

Say X is a *measurable mapping* from $(\Omega, \mathscr{F})$ to $(\Omega', \mathscr{F}')$ iff X is a function from Ω to Ω' and

$$X^{-1} \mathscr{F}' = \{ X^{-1} A : A \in \mathscr{F}' \} \subset \mathscr{F}.$$

(18) Theorem. *Let X be a measurable mapping from $(\Omega, \mathscr{F})$ to $(\Omega', \mathscr{F}')$. If $\mathscr{F}'$ is countably generated, so is $X^{-1}\mathscr{F}'$. The atoms of $X^{-1}\mathscr{F}'$ are precisely the X-inverse images of the atoms of $\mathscr{F}'$.*

6. INDEPENDENCE†

Let $(\Omega, \mathscr{F}, \mathscr{P})$ be a probability triple. Sub σ-fields $\mathscr{F}_1, \mathscr{F}_2, \ldots$ are *independent* (with respect to $\mathscr{P}$) iff $A_i \in \mathscr{F}_i$ implies

$$\mathscr{P}(A_1 \cap A_2 \cap \cdots) = \mathscr{P}(A_1) \cdot \mathscr{P}(A_2) \cdots.$$

Random variables $X_1, X_2, \ldots$ are *independent* iff the σ-fields they *span* are independent; the σ-field *spanned* or *generated* by X_i is the smallest σ-field with respect to which X_i is measurable. Sets $A, B, \ldots$ are *independent* iff $1_A, 1_B, \ldots$ are independent.

(19) Borel-Cantelli Lemma. (a) $\Sigma \mathscr{P}\{A_n\} < \infty$ *implies* $\mathscr{P}\{\limsup A_n\} = 0$;

(b) $\Sigma \mathscr{P}\{A_n\} = \infty$ *and* $A_1, A_2, \ldots$ *independent implies* $\mathscr{P}\{\limsup A_n\} = 1$.

Suppose X_i is a *measurable mapping* from $(\Omega, \mathscr{F})$ to $(\Omega_i, \mathscr{F}_i)$ for $i = 1, 2$.

The *distribution* or $\mathscr{P}$-*distribution* $\mathscr{P}X_1^{-1}$ of X_1 is a probability on $\mathscr{F}_1$: namely,

$$(\mathscr{P}X_1^{-1})(A) = \mathscr{P}(X_1^{-1}A) \quad \text{for } A \in \mathscr{F}_1.$$

(20) Change of variables formula. *If f is a random variable on $(\Omega_1, \mathscr{F}_1)$, then*

$$E[f(X_1)] = \int_{x_1 \in \Omega_1} f(x_1)(\mathscr{P}X_1^{-1})(dx_1).$$

Let X_1 and X_2 be independent: that is, $X_1^{-1}\mathscr{F}_1$ and $X_2^{-1}\mathscr{F}_2$ are. Let $\mathscr{F}_1 \times \mathscr{F}_2$ be the smallest σ-field of $\Omega_1 \times \Omega_2$ containing all sets $A_1 \times A_2$ with $A_i \in \mathscr{F}_i$. Let f be a random variable on $(\Omega_1 \times \Omega_2, \mathscr{F}_1 \times \mathscr{F}_2)$ such that $E[f(X_1, X_2)]$ exists.

(21) Fubini's theorem. *If X_1 and X_2 are independent,*

$$E[f(X_1, X_2)] = \int_{x_1 \in \Omega_1} E[f(x_1, X_2)] \mathscr{P}X_1^{-1}(dx_1)$$

$$= \int_{x_2 \in \Omega_2} E[f(X_1, x_2)] \mathscr{P}X_2^{-1}(dx_2).$$

In particular,

(21a) $\quad E(X_1 X_2) = E(X_1) \cdot E(X_2).$

† References: (Loève, 1963, Sections 8.2 and 15); (Neveu, 1965, Section IV.4).

Conversely, suppose $(\Omega_i, \mathscr{F}_i, \mathscr{P}_i)$ are probability triples for $i = 1, 2$. Let $\Omega = \Omega_1 \times \Omega_2$ and $\mathscr{F} = \mathscr{F}_1 \times \mathscr{F}_2$.

(22) Converse Fubini. *There is a unique probability $\mathscr{P} = \mathscr{P}_1 \times \mathscr{P}_2$ on $\mathscr{F}_1 \times \mathscr{F}_2$ such that*

$$\mathscr{P}(A_1 \times A_2) = \mathscr{P}_1(A_1) \cdot \mathscr{P}_2(A_2)$$

for $A_i \in \mathscr{F}_i$. Then

$$\int f \, d\mathscr{P} = \int \int f(x_1, x_2) \, \mathscr{P}_1(dx_1) \, \mathscr{P}_2(dx_2)$$

for nonnegative $(\mathscr{F}_1 \times \mathscr{F}_2)$-measurable f.

PROOF.   The uniqueness comes from (16). For existence, if $A \in \mathscr{F}_1 \times \mathscr{F}_2$, let $A(\omega_1)$ be the ω_1-*section* of A, namely, the set of $\omega_2 \in \Omega_2$ with $(\omega_1, \omega_2) \in A$. Let

$$\mathscr{P}\{A\} = \int \mathscr{P}_2\{A(\omega_1)\} \, \mathscr{P}_1(d\omega_1). \qquad\qquad ★$$

Let $X_1, X_2, \ldots$ be independent and identically distributed. Suppose X_1 has finite mean μ.

(23a) Weak law of large numbers. $n^{-1}(X_1 + \cdots + X_n)$ *converges to μ in probability.*

(23b) Strong law of large numbers. $n^{-1}(X_1 + \cdots + X_n)$ *converges to μ with probability 1.*

Let $\mathscr{F}_1, \mathscr{F}_2, \ldots$ be independent sub σ-fields in $(\Omega, \mathscr{F}, \mathscr{P})$. Let $\mathscr{F}^{(n)}$ be the σ-field generated by $\mathscr{F}_n, \mathscr{F}_{n+1}, \ldots$. The *tail σ-field* $\mathscr{F}^{(\infty)}$ is $\cap_{n=1}^{\infty} \mathscr{F}^{(n)}$.

(24) Kolmogorov 0-1 Law. *Each $\mathscr{F}^{(\infty)}$-set has $\mathscr{P}$-probability 0 or 1.*

7. CONDITIONING†

Let $\mathscr{A}$ be a sub-σ-field of $\mathscr{F}$, and let X be a random variable with expectation. The *conditional expectation* or $\mathscr{P}$-*expectation of X given $\mathscr{A}$* is $E\{X|\mathscr{A}\} = Y$, the $\mathscr{A}$-measurable random variable Y such that $\int_A Y \, d\mathscr{P} = \int_A X \, d\mathscr{P}$ for all $A \in \mathscr{A}$. For $B \in \mathscr{F}$, the *conditional probability* or $\mathscr{P}$-*probability of B given $\mathscr{A}$* is $\mathscr{P}\{B|\mathscr{A}\} = E\{1_B|\mathscr{A}\}$. If Z is a measurable mapping, $E\{X|Z\} = E\{X|\mathscr{A}\}$, where $\mathscr{A}$ is the σ-field spanned by Z. According to convenience, $\{ \ \}$ changes to $[\ ]$ or $(\ )$. Conditional expectations are unique up to changes on sets of measure 0, by (10a), and exist by Radon–Nikodym (11).

† References: (Loève, 1963, Sections 24 and 25); (Neveu, 1965, Chapter IV).

Let $\mathscr{B}$ be a sub-σ-field of $\mathscr{A}$. These facts about conditional expectations are used rather casually: equality and inequality are only a.e.

(25) Facts. **(a)** $X \geq 0$ *implies* $E\{X|\mathscr{A}\} \geq 0$.

    (b) $E\{X|\mathscr{A}\}$ *depends linearly on* X.

    (c) $E\{X|\mathscr{A}\} = X$ *if* X *is* $\mathscr{A}$-*measurable.*

    (d) $E\{XY|\mathscr{A}\} = XE\{Y|\mathscr{A}\}$ *if* X *is* $\mathscr{A}$-*measurable.*

    (e) $E\{X\} = E\{E(X|\mathscr{A})\}$.

    (f) $E\{E(X|\mathscr{A})|\mathscr{B}\} = E\{X|\mathscr{B}\}$.

    (g) $E\{X|\mathscr{A}\} = E\{X|\mathscr{B}\}$ *if* $E\{X|\mathscr{A}\}$ *is* $\mathscr{B}$-*measurable.*

    (h) *If* X *is independent of* $\mathscr{A}$, *then* $E\{X|\mathscr{A}\} = E\{X\}$.

Say $\mathscr{A}$ is trivial *iff* $\mathscr{P}(A) = 0$ *or* 1 *for any* $A \in \mathscr{A}$.

    (i) *If* $\mathscr{A}$ *is trivial, then* $\mathscr{A}$ *is independent of* X, *and* $E\{X|\mathscr{A}\} = E\{X\}$.

8. MARTINGALES†

Let T be a subset of the line. For $t \in T$, let $\mathscr{F}_t$ be a sub-σ-field of $\mathscr{F}$, and let X_t be an $\mathscr{F}_t$-measurable function on Ω, with finite expectation. Suppose that $s < t$ in T implies: $\mathscr{F}_s \subset \mathscr{F}_t$, and for $A \in \mathscr{F}_s$,

$$\int_A X_s \, d\mathscr{P} = \int_A X_t \, d\mathscr{P}.$$

Then $\{X_t, \mathscr{F}_t : t \in T\}$ is a *martingale*, or $\{X_t\}$ is a *martingale relative* to $\{\mathscr{F}_t\}$. If under similar circumstances,

$$\int_A X_s \, d\mathscr{P} \geq \int_A X_t \, d\mathscr{P},$$

then $\{X_t\}$ is an *expectation-decreasing* martingale relative to $\{\mathscr{F}_t\}$. If $\{\mathscr{F}_t\}$ is not specified, then $\mathscr{F}_t$ is the σ-field generated by X_s for $s \in T$ with $s \leq t$.

(26) Example. Suppose $\mathscr{F}_t$ is a sub-σ-field of $\mathscr{F}$ for each $t \in T$, and $\mathscr{F}_s \subset \mathscr{F}_t$ for $s < t$. Let X be a random variable on $(\Omega, \mathscr{F})$ with finite expectation, and $X_t = E\{X|\mathscr{F}_t\}$. Then $\{X_t, \mathscr{F}_t : t \in T\}$ is a martingale.

(27) Lemma. *Let* $\{X_t\}$ *be an expectation-decreasing martingale, and let* f *be concave and nondecreasing. Then* $\{f(X_t)\}$ *is an expectation-decreasing martingale.*

---

† References: (Doob, 1953, Chapter VII, Sections 1–4); (Loève, 1963, Section 29); (Neveu, 1965, Section IV.5).

Let τ be a random variable on $(\Omega, \mathscr{F}, \mathscr{P})$ which is ∞ or in T. Suppose $\mathscr{P}\{\tau < \infty\} = 1$ and $\{\tau \leq t\} \in \mathscr{F}_t$. Then τ is *admissible* or a *stopping time*.

(28) Theorem. *Suppose $T = \{0, 1, \ldots\}$. Suppose τ is admissible. Suppose $\{X_t, \mathscr{F}_t : t \in T\}$ is a martingale, and*

  (a) $E(|X_\tau|) < \infty$

and

  (b) $\liminf_{n \to \infty} \int_{\{\tau > n\}} |X_n| \, d\mathscr{P} = 0.$

Then (X_0, X_τ) is a martingale, so $E(X_0) = E(X_\tau)$.

PROOF. Let $A \in \mathscr{F}_0$. Let $A_m = \{A \text{ and } \tau = m\}$. Then $A_m \in \mathscr{F}_m$. If $m \leq n$,

$$\int_{A_m} X_\tau \, d\mathscr{P} = \int_{A_m} X_m \, d\mathscr{P} = \int_{A_m} X_n \, d\mathscr{P}.$$

Sum out $m = 0, \ldots, n$:

$$\int_{\{A \text{ and } \tau \leq n\}} X_\tau \, d\mathscr{P} = \int_{\{A \text{ and } \tau \leq n\}} X_n \, d\mathscr{P}.$$

Now

$$\int_A X_0 \, d\mathscr{P} = \int_A X_n \, d\mathscr{P}$$

$$= \int_{\{A \text{ and } \tau \leq n\}} X_n \, d\mathscr{P} + \int_{\{A \text{ and } \tau > n\}} X_n \, d\mathscr{P}$$

$$= \int_{\{A \text{ and } \tau \leq n\}} X_\tau \, d\mathscr{P} + \int_{\{A \text{ and } \tau > n\}} X_n \, d\mathscr{P}.$$

This doesn't use (a–b). By (a),

$$\lim_{n \to \infty} \int_{\{A \text{ and } \tau \leq n\}} X_\tau \, d\mathscr{P} = \int_A X_\tau \, d\mathscr{P}.$$

By (b), if n increases through the right subsequence,

$$\int_{\{A \text{ and } \tau > n\}} X_n \, d\mathscr{P} \to 0. \qquad\qquad \bigstar$$

Example. Let the Y_n be independent and identically distributed, each being 0 or 2 with probability $\frac{1}{2}$ each. Let $X_0 = 1$ and let $X_n = Y_1 \cdots Y_n$ for $n \geq 1$. Let τ be the least n if any with $X_n = 0$, and $\tau = \infty$ if none. Then $\{X_n\}$ is a nonnegative martingale, τ is a stopping time, $E(\tau) < \infty$, and $X_\tau = 0$ almost surely. This martingale was proposed by David Gilat. $\qquad \bigstar$

Example.   There is a martingale, and a stopping time with finite mean, which satisfy (28b) but not (28a).

DISCUSSION.   For $n \geq 1$, let

$$a_n = \frac{(n-1)^2}{\log(3+n)} \quad \text{and} \quad b_{n+1} = a_{n+1}^2 - a_n^2.$$

Let $b_1 = 1$. Check $b_n > 0$ for all n. Let Y_n be $N(0, b_n)$, and let $Y_1, Y_2, \ldots$ be independent. Let $X_n = Y_1 + \cdots + Y_n$, so X_n is a martingale. For $n \geq 1$, let

$$p_n = \left(\frac{1}{n}\right)^2 - \left(\frac{1}{n+1}\right)^2,$$

so

$$\Sigma_{m=n}^{\infty} p_m = \left(\frac{1}{n}\right)^2.$$

Let f be a measurable function from $(-\infty, \infty)$ to $\{1, 2, \ldots\}$, with

$$\mathscr{P}\{f(Y_1) = n\} = p_n \quad \text{for } n = 1, 2, \ldots.$$

So $\tau = f(Y_1)$ is a stopping time, with finite mean.

Let θ be the distribution of Y_1, and let $S_n = Y_2 + \cdots + Y_n$, so $S_1 = 0$. Now Y_1 is independent of $S_1, S_2, \ldots$, and $X_n = Y_1 + S_n$, and S_n is $N(0, a_n^2)$. Compute.

$$E\{|X_\tau|\} = \int E\{|y + S_{f(y)}|\} \, \theta(dy)$$

$$\geq \int E\{|S_{f(y)}|\} \, \theta(dy)$$

$$= c \int a_{f(y)} \, \theta(dy) \qquad \text{where } c = E\{|Y_1|\}$$

$$= c \, \Sigma_{n=1}^{\infty} p_n a_n$$

$$= \infty.$$

Continuing,

$$\int_{\{\tau > n\}} |X_n| \, d\mathscr{P} = \int_{\{f > n\}} E\{|y + S_n|\} \, \theta(dy)$$

$$\leq \int_{\{f > n\}} [|y| + E\{|S_n|\}] \, \theta(dy)$$

$$= \int_{\{f > n\}} |y| \, \theta(dy) + c a_n \cdot \theta\{f > n\}$$

$$\to 0.$$

★

(29) Theorem.   *Suppose $T = [0, \infty)$. Suppose $\{X_t, \mathscr{F}_t : t \in T\}$ is a martingale, and τ is admissible. Suppose*

(a)   *$t \to X_t(\omega)$ is continuous from the right, for each ω*

(b)   *$E\{\sup_{0 \leq s \leq t} |X_s|\} < \infty$,   for each t*

(c)   *$E(|X_\tau|) < \infty$*

(d)   *$\liminf_{t \to \infty} \int_{\{\tau > t\}} |X_t| \, d\mathscr{P} = 0$.*

Then (X_0, X_τ) is a martingale, so $E(X_0) = E(X_\tau)$.

PROOF.   Fix t. Let τ_n be the least $(j/n)t \geq \tau$. Fix $A \in \mathscr{F}_0$. As in (28),

$$\int_{\{A \text{ and } \tau \leq t\}} X_{\tau_n} \, d\mathscr{P} = \int_{\{A \text{ and } \tau \leq t\}} X_t \, d\mathscr{P}.$$

Let $n \to \infty$; use (a), (b), and dominated convergence:

$$\int_{\{A \text{ and } \tau \leq t\}} X_\tau \, d\mathscr{P} = \int_{\{A \text{ and } \tau \leq t\}} X_t \, d\mathscr{P}.$$

Now

$$\int_A X_0 \, d\mathscr{P} = \int_A X_t \, d\mathscr{P}$$

$$= \int_{\{A \text{ and } \tau \leq t\}} X_t \, d\mathscr{P} + \int_{\{A \text{ and } \tau > t\}} X_t \, d\mathscr{P}$$

$$= \int_{\{A \text{ and } \tau \leq t\}} X_\tau \, d\mathscr{P} + \int_{\{A \text{ and } \tau > t\}} X_t \, d\mathscr{P}.$$

Let $t \to \infty$; use (c) and (d).   ★

Theorem (30) partially extends (28). To state it, let

$$\{X_t, \mathscr{F}_t : t = 0, 1, \ldots\}$$

be a nonnegative, expectation-decreasing martingale. Let $\tau_0 \leq \tau_1 \leq \cdots$ be stopping times. Let

$$Y_n = X_{\tau_n}.$$

Let $\mathscr{G}_n$ be the σ-field of measurable A such that

$$\{A \text{ and } \tau_n \leq t\} \in \mathscr{F}_t$$

for all $t = 0, 1, \ldots$.

(30) Theorem.   *$\{Y_n, \mathscr{G}_n : n = 0, 1, \ldots\}$ is an expectation-decreasing martingale.*

PROOF.   You should check that Y_n is $\mathscr{G}_n$-measurable, and $\mathscr{G}_n \subset \mathscr{G}_{n+1}$. Let $A \in \mathscr{G}_n$. I have to prove

$$\int_A Y_n \, d\mathscr{P} \geqq \int_A Y_{n+1} \, d\mathscr{P}.$$

It is enough to do this with $A \cap \{\tau_n = m\}$ in place of A: afterwards, sum over m. Abbreviate $\sigma = \tau_{n+1}$, so $Y_{n+1} = X_\sigma$. On $A \cap \{\tau_n = m\}$,

$$\sigma \geqq m \quad \text{and} \quad Y_n = X_m.$$

Check $\{\tau_n = m\}$ and $A \cap \{\tau_n = m\}$ are both in $\mathscr{F}_m$. My problem is reduced to showing that

(31) $$\int_A X_m \, d\mathscr{P} \geqq \int_A Y_{n+1} \, d\mathscr{P},$$

for a typical $A \in \mathscr{F}_m$ with $A \subset \{\tau_n = m\}$. I will argue inductively that for $M \geqq m$,

(32) $$\int_A X_m \, d\mathscr{P} \geqq \int_{A \cap \{\sigma \leqq M\}} X_\sigma \, d\mathscr{P} + \int_{A \cap \{\sigma > M\}} X_m \, d\mathscr{P}.$$

This is clear for $M = m$. Suppose it true for some $M \geqq m$. Now

$$\{\sigma > M\} = \Omega \setminus \{\sigma \leqq M\} \in \mathscr{F}_M, \quad \text{and} \quad A \in \mathscr{F}_m \subset \mathscr{F}_M.$$

The computation rolls on:

$$\int_{A \cap \{\sigma > M\}} X_M \, d\mathscr{P} \geqq \int_{A \cap \{\sigma > M\}} X_{M+1} \, d\mathscr{P}$$

$$= \int_{A \cap \{\sigma = M+1\}} X_{M+1} \, d\mathscr{P} + \int_{A \cap \{\sigma > M+1\}} X_{M+1} \, d\mathscr{P}$$

$$= \int_{A \cap \{\sigma = M+1\}} X_\sigma \, d\mathscr{P} + \int_{A \cap \{\sigma > M+1\}} X_{M+1} \, d\mathscr{P}.$$

This proves (32). Now (32) is even truer without the rightmost term. Drop it, and let M increase to ∞ to get (31).   ★

If $s_1, \ldots, s_N$ is a sequence of real numbers, and $a < b$ are real numbers, the number of *downcrossings* of $[a, b]$ by $s_1, \ldots, s_N$ is the largest positive integer k for which there exists integers $1 \leqq n_1 < n_2 < \cdots < n_{2k} \leqq N$ with

$$s_{n_1} \geqq b, s_{n_2} \leqq a, \ldots, s_{n_{2k-1}} \geqq b, s_{n_{2k}} \leqq a.$$

If no such k exists, the number of downcrossings is 0.

(33) Downcrossings inequality.   *Let $X_0, X_1, \ldots$ be a nonnegative, expectation-decreasing martingale. Let $0 \leqq a < b < \infty$. The mean number of downcrossings of $[a, b]$ by $X_0, X_1, \ldots$ is at most $(b - a)^{-1}$ times the mean of X_0.*

This differs only in detail from the upcrossings inequality (Doob, 1953, Theorem 3.3 on p. 316).

PROOF.   Introduce β_n for the number of downcrossings of $[a, b]$ by $X_0, \ldots, X_n$. It is enough to prove

$$E(\beta_n) \leqq E(X_0)/(b - a)$$

in the case $X_i \leqq b$ for $i = 0, \ldots, n$; use (27) on the function $x \to \max\{x, b\}$. Let σ_0 be the least $m = 0, \ldots, n$ if any with $X_m = b$: if none, let $\sigma_0 = n$. Let σ_1 be the least $m = 0, \ldots, n$ if any with $m > \sigma_0$ and $X_m \leqq a$; if none, let $\sigma_1 = n$. And so on, up to σ_n. Now $X_0, X_{\sigma_0}, \ldots, X_{\sigma_n}$ is an expectation-decreasing martingale by (30). Check,

$$\beta_n(b - a) \leqq \Sigma_m \{X_{\sigma_m} - X_{\sigma_{m+1}} : m = 0, \ldots, n - 1 \text{ and } m \text{ is even}\}.$$

Therefore,

$$E(X_0) \geqq E(X_{\sigma_0})$$
$$\geqq E(X_{\sigma_0}) - E(X_{\sigma_n})$$
$$= \Sigma_{m=0}^{n-1} \{E(X_{\sigma_m}) - E(X_{\sigma_{m+1}})\}$$
$$\geqq \Sigma_m \{E(X_{\sigma_m}) - E(X_{\sigma_{m+1}}) : m = 0, \ldots, n - 1 \text{ and } m \text{ is even}\}$$
$$= E[\Sigma_m \{X_{\sigma_m} - X_{\sigma_{m+1}} : m = 0, \ldots, n - 1 \text{ and } m \text{ is even}\}]$$
$$\geqq E[\beta_n(b - a)]. \qquad \qquad \bigstar$$

Martingale convergence theorem

(34) Theorem. Forward martingales.   *Let $\{X_n : n = 0, 1 \ldots\}$ be a martingale. If $\sup_n E(|X_n|) < \infty$, then X_n converges a.e. as $n \to \infty$; the convergence is L^1 iff the X_n are uniformly integrable. If $p > 1$ and $\sup_n E(|X_n|^p) < \infty$, then X_n converges in L^p as $n \to \infty$. Suppose $\{\mathscr{F}_n : n = 0, 1, \ldots\}$ are nondecreasing σ-fields and $X \in L^p$ for $p \geqq 1$ and $X_n = E\{X | \mathscr{F}_n\}$. Then $\{X_n\}$ is a martingale, and $E(|X_n|^p) \leqq E(|X|^p)$, so the previous assertions apply: The X_n are automatically uniformly integrable, even for $p = 1$. The limit of X_n is the conditional expectation of X given the σ-field generated by all the $\mathscr{F}_n$.*

Backward martingales.   *Let $\{X_n : n = \ldots, -3, -2, -1\}$ be a martingale. Then X_n converges a.e. as $n \to -\infty$. The X_n are automatically uniformly integrable, and the convergence is also L^1. If $X_{-1} \in L^p$ for $p > 1$, so are all the X_n, and the convergence is also L^p. Suppose $\{\mathscr{F}_n : = \ldots, -3, -2, -1\}$ are* *nondecreasing σ-fields and $X \in L^p$ for $p \geq 1$ and $X_n = E\{X|\mathscr{F}_n\}$. Then $\{X_n\}$ is a martingale and $E(|X_n|^p) \leq E(|X|^p)$, so the previous assertions apply. The limit of X_n is the conditional expectation of X given the intersection of the $\mathscr{F}_n$.*

PROOF. If the X_n are nonnegative, the a.e. convergence follows from (33). General X_n follow the same route, with minor changes. The L^p convergence follows from (6).   ★

Differentiation

For (35–36), let P and Q be two probabilities on $\mathscr{F}$. Then Ω divides up into three $\mathscr{F}$-sets, Ω_P and Ω_e and Ω_Q, such that $P(\Omega_Q) = Q(\Omega_P) = 0$ and P is equivalent to Q on Ω_e. This partition is unique up to $(P + Q)$-null sets. Let

$$\frac{dP}{dQ} = 0 \quad \text{on } \Omega_Q$$

$$= \infty \quad \text{on } \Omega_P$$

and let dP/dQ be the Radon-Nikodym derivative of P with respect to Q on Ω_e. This function is $\mathscr{F}$-measurable and unique up to changes on $(P + Q)$-null sets. Let $\{\mathscr{A}_n\}$ be a nondecreasing or nonincreasing sequence of σ-fields. In the former case, let $\mathscr{A}_\infty$ be the σ-field generated by the union of the $\mathscr{A}_n$. In the latter case, let $\mathscr{A}_\infty$ be the intersection of the $\mathscr{A}_n$. For any measure R, let R_n be the retraction of R to $\mathscr{A}_n$. Define dP_n/dQ_n like dP/dQ above, with $\mathscr{A}_n$ replacing $\mathscr{F}$. Thus, dP_n/dQ_n is an $\mathscr{A}_n$-measurable function to $[0, \infty]$, unique up to changes on $(P_n + Q_n)$-null sets.

(35) Theorem. $\lim_{n\to\infty} \dfrac{dP_n}{dQ_n} = \dfrac{dP_\infty}{dQ_\infty}$, *except on a $(P + Q)$-null set.*

PROOF. Introduce the probability $A = \frac{1}{2}(P + Q)$. Then $P \leq 2A$, so $P_n \ll A_n$ and $0 \leq dP_n/dA_n \leq 2$. Let E_A be expectation relative to A. Abbreviate $r_n = dP_n/dQ_n$. Suppose the $\mathscr{A}_n$ are nondecreasing. Make the convention $\infty/\infty = 1$, and use (11) to check that

$$\frac{2r_n}{1 + r_n} = \frac{dP_n}{dA_n} = E_A\left\{\frac{dP_\infty}{dA_\infty}\Big|\mathscr{A}_n\right\};$$

and use (34). Suppose the $\mathscr{A}_n$ are non-increasing. Check that

$$\frac{2r_n}{1 + r_n} = \frac{dP_n}{dA_n} = E_A\left\{\frac{dP_1}{dA_1}\Big|\mathscr{A}_n\right\},$$

and use (34).   ★

(36) Example. Suppose the $\mathscr{A}_n$ are nondecreasing. If $P_n \ll Q_n$ for $n = 1, 2, \ldots$, then $\{dP_n/dQ_n : n = 1, 2, \ldots\}$ is a martingale.

9. METRIC SPACES†

A *metric* ρ on Ω is a nonnegative real-valued function on $\Omega \times \Omega$ such that: $\rho(x, y) = 0$ iff $x = y$; and

$$\rho(x, y) + \rho(y, z) \geq \rho(x, z).$$

Say x_n *converges to* x or $x_n \to x$ iff $\rho(x_n, x) \to 0$. Say $V \subset \Omega$ is *closed* iff $x_n \in V$ and $x_n \to x$ implies $x \in V$. Say $U \subset \Omega$ is *open* iff $\Omega \setminus U$ is closed. The *topology* or ρ-*topology* of Ω is the set of open U. A sequence $\{x_n\}$ in Ω is *Cauchy* iff $\rho(x_n, x_m) \to 0$ as $n, m \to \infty$. Say Ω is *complete* iff each Cauchy sequence converges. Say Ω is *separable* if there is a countable subset C of Ω *dense* in Ω: for each $x \in \Omega$, there are $x_n \in C$ with $x_n \to x$.

If (Ω, ρ) is complete and separable, the *Borel-σ-field* of (Ω, ρ) is the smallest σ-field containing the ρ-topology. Then $(\Omega_1, \mathscr{F}_1)$ is *Borel* iff there is a complete, separable metric space (Ω, ρ), such that Ω_1 is in the Borel σ-field $\mathscr{F}$ of (Ω, ρ), and $\mathscr{F}_1 = \Omega_1\mathscr{F}$. In this case also, $\mathscr{F}_1$ is called the *Borel-σ-field* of Ω_1. If $(\Omega_i, \mathscr{F}_i)$ are Borel, so is $(\Omega_1 \times \Omega_2 \times \cdots, \mathscr{F}_1 \times \mathscr{F}_2 \times \cdots)$. Borel σ-fields are countably generated.

If I is a countably infinite set, $\rho(i, j) = 1$ or 0 according as $i \neq j$ or $i = j$ is a perfectly good metric; I is complete and separable. The corresponding topology and σ-field are called *discrete*. The *one-point compactification* $\bar{I} = I \cup \{\varphi\}$ is obtained as follows. Let $I = \{i_1, i_2, \ldots\}$; let $\varphi \notin I$; and let $\bar{I} = \{i_1, i_2, \ldots, i_\infty\}$ with $i_\infty = \varphi$; let

$$\bar{\rho}(i_m, i_n) = \left| \frac{1}{m} - \frac{1}{n} \right|,$$

where $1/\infty = 0$. Of course, $\bar{\rho}$ retracted to I produces the same topology as ρ.

(37) Example. Let Ω be Euclidean n-space R^n. Let ρ be the usual distance: $\rho(x, y) = \|x - y\|$ and $\|u\|^2 = \sum_{i=1}^n u_i^2$. Then Ω is complete and separable.

(38) Example. Let Ω be the rationals. There is no way to metrize the usual topology so that Ω is complete.

† References: (Dunford and Schwartz, 1958), (Hausdorff, 1957), (Kuratowski, 1958), (Loève, 1963, Section 2).

10.   REGULAR CONDITIONAL DISTRIBUTIONS†

COMFORT.   The material in Sections 10–15 is fairly exotic, and is used only on special occasions.

Let X_i be a measurable mapping from Ω to $(\Omega_i, \mathscr{F}_i)$ for $i = 1, 2$. A *regular conditional $\mathscr{P}$-distribution* for X_2 given X_1 is a function $Q(\cdot, \cdot)$ on $\Omega_1 \times \mathscr{F}_2$ with the following properties:

(39a) $Q(x_1, \cdot)$ is a probability on $\mathscr{F}_2$ for each $x_1 \in \Omega_1$;

(39b) $Q(\cdot, A_2)$ is $\mathscr{F}_1$-measurable for each $A_2 \in \mathscr{F}_2$;

(39c)   for each $A_2 \in \mathscr{F}_2$, the function $\omega \to Q[X_1(\omega), A_2]$ is a version of

$$\mathscr{P}\{X_2 \in A_2 | X_1^{-1}\mathscr{F}_1\}.$$

Condition (c) can be rephrased as follows: if $A_i \in \mathscr{F}_i$, then

(39d) $$\int_{A_1} Q(x_1, A_2)(\mathscr{P}X_1^{-1})(dx_1) = \mathscr{P}\{X_1 \in A_1 \text{ and } X_2 \in A_2\}.$$

You only have to check this for generating classes of A_1's and A_2's closed under intersection, by (16). Sometimes $Q(x_1, \cdot)$ is called a regular conditional $\mathscr{P}$-distribution for X_2 given $X_1 = x_1$.

Suppose Q is a regular conditional $\mathscr{P}$-distribution for X_2 given X_1. Let ϕ be a measurable mapping from $(\Omega_2, \mathscr{F}_2)$ to $(\Omega_\phi, \mathscr{F}_\phi)$. Let $Q_\phi(x_1, \cdot)$ be the $Q(x_1, \cdot)$-distribution of ϕ. Make sure that $Q_\phi(x_1, \cdot)$ is a probability on Ω_ϕ.

(40) Lemma. Q_ϕ *is a regular conditional $\mathscr{P}$-distribution for $\phi(X_2)$ given X_1.*

EXAMPLE.   Let $\Omega_1 = \Omega_2 = \Omega = [0, 1]$. Let $X_1(\omega) = X_2(\omega) = \omega$. Let $\mathscr{F}_1$ be the Borel σ-field of $[0, 1]$. Let λ be Lebesgue measure on $\mathscr{F}_1$. Let B be a non-Lebesgue-measurable set, namely $\lambda_*(B) < \lambda^*(B)$. Let $\mathscr{F}_2 = \mathscr{F}$ be the σ-field generated by $\mathscr{F}_1$ and B. Extend λ to a probability $\mathscr{P}$ on $\mathscr{F}$. You can do this so $\mathscr{P}(B)$ is any number in the interval $[\lambda_*(B), \lambda^*(B)]$. There is no regular conditional $\mathscr{P}$-distribution for X_2 given X_1. For suppose Q were such an object. Theorem (51) below produces a $\mathscr{P}$-null set $N \in \mathscr{F}_1$, such that $Q(\omega, \cdot)$ is point mass at ω for $\omega \notin N$. In particular,

$$Q(\omega, B) = 1_B(\omega) \quad \text{for } \omega \notin N.$$

The left side is an $\mathscr{F}_1$-measurable function of ω. So B differs by a subset of the null set N from an $\mathscr{F}_1$-set, a contradiction.   ★

(41) Theorem.   *If $(\Omega_2, \mathscr{F}_2)$ is Borel, then a regular conditional $\mathscr{P}$-distribution for X_2 given X_1 exists.*

† References: (Blackwell, 1954); (Loève, 1963, Sections 26 and 27).

Theorem (41) is hard. One of its virtues (although this does not materially increase the difficulty) is the absence of conditions on $\mathscr{F}$ or $\mathscr{F}_1$.

(42) Theorem. *Suppose $\mathscr{F}_2$ is countably generated. Suppose Q and Q^* are two regular conditional $\mathscr{P}$-distributions for X_2 given X_1. Then*

$$\{x : x \in \Omega_1 \text{ and } Q(x, \cdot) = Q^*(x, \cdot)\}$$

is in $\mathscr{F}_1$ and has $\mathscr{P}X_1^{-1}$-probability 1.

PROOF.   Let $\mathscr{A}$ be a countable, generating algebra for $\mathscr{F}_2$. Then

$$\{x : x \in \Omega_1 \text{ and } Q(x, A) = Q^*(x, A)\}$$

is an $\mathscr{F}_1$-set of $\mathscr{P}X_1^{-1}$-probability 1, for each $A \in \mathscr{A}$. The intersection over all $A \in \mathscr{A}$ is the set described in the theorem.   ★

The next result generalizes converse Fubini (22). Suppose $\Omega = \Omega_1 \times \Omega_2$ and $\mathscr{F} = \mathscr{F}_1 \times \mathscr{F}_2$. Suppose $\mathscr{P}_1$ is a probability on $\mathscr{F}_1$, and Q satisfies (39a, b). Let

$$X_1(x_1, x_2) = x_1 \quad \text{and} \quad X_2(x_1, x_2) = x_2.$$

(43) Theorem. *There is a unique probability $\mathscr{P}$ on $\mathscr{F}$ satisfying the two conditions:*

(a) $\mathscr{P}X_1^{-1} = \mathscr{P}_1$, *and*

(b)   *Q is a regular conditional $\mathscr{P}$-distribution for X_2 given X_1.*

If f is a nonnegative, measurable function on $(\Omega_1 \times \Omega_2, \mathscr{F}_1 \times \mathscr{F}_2)$, then

$$\int_\Omega f \, d\mathscr{P} = \int_{\Omega_1} \int_{\Omega_2} f(x_1, x_2) \, Q(x_1, dx_2) \mathscr{P}_1(dx_1).$$

PROOF.   The uniqueness follows from (16). For existence, define

$$\mathscr{P}(A) = \int_{\Omega_1} Q(x_1, A(x_1)) \, \mathscr{P}_1(dx_1)$$

for $A \in \mathscr{F}_1 \times \mathscr{F}_2$: as before, $A(x_1)$ is the x_1-section of A, namely the set of x_2 with $(x_1, x_2) \in A$. Check that $\mathscr{P}$ is a probability satisfying (a) and (b). The integration formula now holds with $f = 1_A$; both sides are linear and continuous under increasing passages to the limit.   ★

Regular conditional distributions given Σ

In the book, the usual case is: $\Omega_1 = \Omega$ and $\mathscr{F}_1 \subset \mathscr{F}$ and $X_1(\omega) = \omega$. Then, a regular conditional $\mathscr{P}$-distribution for X_2 given X_1 is called a regular conditional $\mathscr{P}$-distribution for X_2 given $\mathscr{F}_1$. The next theorem (43) embodies the main advantage of regular distributions. It is easy to prove, and intuitive: it says that when you condition on a σ-field Σ, you can put any Σ-measurable function U equal to a typical value u, and then substitute U for u when you're through conditioning. That is, U is truly constant given Σ. However, example (48) shows that something a little delicate happened.

I will state (44) in its most popular form. The notation will be used through (50). Let $(\Omega, \mathscr{F}, \mathscr{P})$ be the basic probability triple. Let Σ be a sub-σ-field of $\mathscr{F}$. Let U be a measurable mapping from (Ω, Σ) to a new space $(\Omega_U, \mathscr{F}_U)$. Let V be a measurable mapping from $(\Omega, \mathscr{F})$ to a new space $(\Omega_V, \mathscr{F}_V)$. Thus, U is Σ-measurable and V is $\mathscr{F}$-measurable. The situation is summarized in Figure 1. Let Q be a regular conditional $\mathscr{P}$-distribution for V given Σ, so Q is a function of pairs (ω, C) with $\omega \in \Omega$ and $C \in \mathscr{F}_V$. Let F be a nonnegative, measurable function on $(\Omega_U \times \Omega_V, \mathscr{F}_U \times \mathscr{F}_V)$.

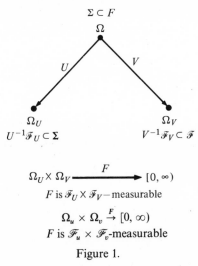

Figure 1.

(44) Theorem.   *Let* $F^*(u, \omega) = \int_{\Omega_V} F(u, v)\, Q(\omega, dv)$. *Then* F^* *is* $\mathscr{F}_U \times \Sigma$-*measurable. And* $F^*(U, \cdot)$, *namely*

$$\omega \to F^*(U(\omega), \omega),$$

is a version of $E\{F(U, V)|\Sigma\}$.

PROOF.   You can check the measurability of F^* by using (47) below. Now $\omega \to (U(\omega), \omega)$ is a measurable mapping from (Ω, Σ) to $(\Omega_U \times \Omega, \mathscr{F}_U \times \Sigma)$. But $F^*(U, \cdot)$ is the composition of F^* with this mapping, and is Σ-measurable. Fix $A \in \Sigma$. I have to show

$$(45) \qquad \int_A \int_{\Omega_V} F(U(\omega), v)\, Q(\omega, dv)\, \mathscr{P}(d\omega) = \int_A F(U(\omega), V(\omega))\mathscr{P}(d\omega).$$

I know

$$\int_S Q(\omega, C)\, \mathscr{P}(d\omega) = \mathscr{P}\{S \text{ and } V \in C\}$$

for $S \in \Sigma$ and $C \in \mathscr{F}_V$. Rewrite this with $\{A \text{ and } U \in B\}$ in place of S, where B is a variable element of $\mathscr{F}_U$. This is legitimate because U is Σ-measurable. I now have (45) for a special F:

$$F(u, v) = 1_B(u) \cdot 1_C(v).$$

Both sides of (45) are linear in F, and continuous under increasing passages to the limit. Use (47) below.   ★

(46) Corollary. $E[F(U, V)] = \int_\Omega \int_{\Omega_V} F(U(\omega), v)\, Q(\omega, dv)\, \mathscr{P}(d\omega).$

(47) Lemma. *Let* F *be a family of nonnegative, $(\mathscr{F}_U \times \mathscr{F}_V)$-measurable functions on $\Omega_U \times \Omega_V$. Suppose $af + bg \in$ F when $f, g \in$ F and a, b are nonnegative constants. Suppose $f - g \in$ F when $f, g \in$ F and $1 \geq f \geq g$. Suppose $f \in$ F when $f_n \in$ F and $f_n \uparrow f$. Finally, suppose*

$$(u, v) \rightarrow 1_B(u) \cdot 1_C(v)$$

is in F *when $B \in \mathscr{F}_U$ and $C \in F_V$. Then* F *consists of all the nonnegative measurable functions on $(\Omega_U \times \Omega_V, \mathscr{F}_U \times \mathscr{F}_V)$.*

(48) Example. Suppose $U = V$ is uniform on $[0, 1]$. Let $F(u, v)$ be 1 or 0 according as $u = v$ or $u \neq v$. Then $F(U, V) = 1$ almost surely, so

$$E\{F(U, V)|U\} = 1$$

almost surely. But $F(u, V) = 0$ almost surely for any particular u, forcing $E\{F(u, V)|U\} = 0$ almost surely. Theorem (43) rescues this example by defining

$$E\{F(u, V)|U = u\} = 1.$$ ★

Theorem (49) sharpens (44). To state it and (50), let ϕ be a measurable mapping from $(\Omega_U \times \Omega_V, \mathscr{F}_V \times \mathscr{F}_V)$ to some new space $(\Omega_\phi, \mathscr{F}_\phi)$. Temporarily, fix $\omega \in \Omega$ and $u \in \Omega_U$. Then $\phi(u, \cdot)$ is a measurable mapping from $(\Omega_V, \mathscr{F}_V)$ to $(\Omega_\phi, \mathscr{F}_\phi)$. And $Q(\omega, \cdot)$ is a probability on $\mathscr{F}_V$. So I am entitled to define $D(\omega, u, \cdot)$ as the $Q(\omega, \cdot)$-distribution of $\phi(u, \cdot)$; this comes out to a probability on $\mathscr{F}_\phi$. Let $R(\omega, \cdot) = D(\omega, U(\omega), \cdot)$, again a probability on $\mathscr{F}_\phi$.

(49) Theorem. *$R(\cdot, \cdot)$ is a regular conditional $\mathscr{P}$-distribution for $\phi(U, V)$ given Σ.*

PROOF.   Let $A \in \Sigma$ and $B \in \mathscr{F}_\phi$. I have to check that

$$\mathscr{P}\{A \text{ and } \phi(U, V) \in B\} = \int_A R(\omega, B)\, \mathscr{P}(d\omega).$$

Let $F(u, v) = 1$ or 0, according as $\phi(u, v) \in B$ or $\notin B$. Use (44):

$$\mathscr{P}\{A \text{ and } \phi(U, V) \in B\} = \int_A F(U, V)\, d\mathscr{P}$$

$$= \int_A E\{F(U, V)|\Sigma\}\, d\mathscr{P}$$

$$= \int_A \int_{\Omega_V} F[U(\omega), v]\, Q(\omega, dv)\, \mathscr{P}(d\omega).$$

Recognize

$$\int_{\Omega_V} F[U(\omega), v]\, Q(\omega, dv) = Q(\omega, \{v : \phi[U(\omega), v] \in B\})$$

$$= R(\omega, B). \qquad \bigstar$$

The situation is more tractable when Σ and V are independent, which will be assumed in (50). Let $D(u, \cdot)$ be the $\mathscr{P}$-distribution of $\phi(u, \cdot)$, a probability on $\mathscr{F}_\phi$ for each $u \in \Omega_U$.

(50) Theorem. *Suppose Σ and V are independent. Then $D(U, \cdot)$ is a regular conditional $\mathscr{P}$-distribution for $\phi(U, V)$ given Σ.*

PROOF.   Let $A \in \Sigma$ and $B \in \mathscr{F}_\phi$. I have to check that

$$\mathscr{P}\{A \text{ and } \phi(U, V) \in B\} = \int_A D[U(\omega), B]\, \mathscr{P}(d\omega).$$

Use Fubini (21) to evaluate the left side. Keep $(\Omega, \mathscr{F}, \mathscr{P})$ for the basic probability triple. Put (Ω, Σ) for $(\Omega_1, \mathscr{F}_1)$, with $X_1(\omega) = \omega$. Put $(\Omega_V, \mathscr{F}_V)$ for $(\Omega_2, \mathscr{F}_2)$, with $X_2(\omega) = V(\omega)$. Let $f(\omega, v) = 1$ if $\omega \in A$ and $\phi[U(\omega), v] \in B$; otherwise, let $f(\omega, v) = 0$. Let $\hat{\mathscr{P}}$ be $\mathscr{P}$ retracted to Σ. Then

$$\mathscr{P}\{A \text{ and } \phi(U, V) \in B\} = \int_\Omega f[X_1(\omega), V(\omega)]\, \mathscr{P}(d\omega)$$

$$= \int_\Omega \int_\Omega f[\omega, V(\omega')]\, \mathscr{P}(d\omega')\, \hat{\mathscr{P}}(d\omega)$$

$$= \int_A \int_\Omega f[\omega, V(\omega')]\, \mathscr{P}(d\omega')\, \hat{\mathscr{P}}(d\omega),$$

because $f(\omega, \cdot) \equiv 0$ for $\omega \notin A$. Recognize

$$\int_\Omega f[\omega, V(\omega')]\, \mathscr{P}(d\omega') = \mathscr{P}\{\omega' : \phi[U(\omega), V(\omega')] \in B\}$$

$$= D[U(\omega), B]$$

for $\omega \in A$. $\qquad \bigstar$

Regular conditional probabilities

If $\Omega_V = \Omega$ and $\mathscr{F}_V = \mathscr{F}$ and $V(\omega) = \omega$, then a regular conditional $\mathscr{P}$-distribution for V given Σ is called a regular conditional $\mathscr{P}$-probability given Σ. For (51) and (52), let Q be a regular conditional $\mathscr{P}$-probability given Σ. That is, $(\Omega, \mathscr{F}, \mathscr{P})$ is the basic probability triple, and Σ is a sub-σ-field of $\mathscr{F}$. Moreover Q is a function of pairs (ω, B), with $\omega \in \Omega$ and $B \in \mathscr{F}$. The function $Q(\cdot, B)$ is a version of $\mathscr{P}(B|\Sigma)$, and the function $Q(\omega, \cdot)$ is a probability. Recall that $\Sigma(\omega)$ is the Σ-atom containing ω.

(51) Theorem. *Let Σ be countably generated. Then the set of ω such that $Q(\omega, \Sigma(\omega)) = 1$ is a Σ-set of $\mathscr{P}$-probability 1.*

Proof. Let $\mathscr{A}$ be a countable generating algebra for Σ. For each $A \in \mathscr{A}$, let A^* be the set of ω such that $Q(\omega, A) = 1_A(\omega)$. Then A^* is a Σ-set of $\mathscr{P}$-probability 1, and the intersection of A^* as A varies over $\mathscr{A}$ is the set described in the theorem.   ★

For (52), do not assume that Σ is countably generated. Let $\mathscr{E}$ be the smallest σ-field over which $\omega \to Q(\omega, A)$ is measurable, for all $A \in \mathscr{F}$. Thus, $\mathscr{E} \subset \Sigma$. Let E be the set of ω such that

$$Q(\omega, \{\omega' : Q(\omega', \cdot) = Q(\omega, \cdot)\}) = 1.$$

(52) Theorem. *Suppose $\mathscr{F}$ is countably generated.*

(a) *$\mathscr{E}$ is countably generated.*

(b) *$E \in \mathscr{E}$.*

(c) *$\mathscr{P}(E) = 1$.*

Proof. Let $\mathscr{A}$ be a countable generating algebra for $\mathscr{F}$. Then $\mathscr{E}$ is also the smallest σ-field over which $\omega \to Q(\omega, A)$ is measurable, for all $A \in \mathscr{A}$, by the monotone class argument (Section 5). This proves (a). As (18) now implies,

$$\mathscr{E}(\omega) = \{\omega' : Q(\omega', A) = Q(\omega, A) \text{ for all } A \in \mathscr{A}\}.$$

Of course, Q is a regular conditional $\mathscr{P}$-probability given $\mathscr{E}$. Finally, (51) proves (b) and (c).   ★

Regular conditional distributions for partially defined random variables

Let $(\Omega, \mathscr{F}, \mathscr{P})$ be the basic probability triple, and let Σ be a sub-σ-field of $\mathscr{F}$. Let $D \in \Sigma$. Let V be a measurable mapping from $(D, D\Sigma)$ to a new space $(\Omega_V, \mathscr{F}_V)$. As usual, $D\Sigma$ is the σ-field of subsets of D of the form $D \cap S$ with $S \in \Sigma$. A *regular conditional $\mathscr{P}$-distribution for V given Σ on D* is a function

Q of pairs (ω, B) with $\omega \in D$ and $B \in \mathcal{F}_V$, such that:

$Q(\omega, \cdot)$ is a probability on $\mathcal{F}_V$ for each $\omega \in D$;

$Q(\cdot, B)$ is $D\Sigma$-measurable for each $B \in \mathcal{F}_V$; and

$$\int_A Q(\omega, B)\, \mathcal{P}(d\omega) = \mathcal{P}\{A \text{ and } V \in B\}$$

for all $A \in \Sigma$ with $A \subset D$ and all $B \in \mathcal{F}_V$. Of course, A and B can be confined to generating classes in the sense of (16). The partially defined situation is isomorphic to a fully defined one. Replace Ω by D, and $\mathcal{F}$ by $D\mathcal{F}$, and Σ by $D\Sigma$, and $\mathcal{P}$ by $\mathcal{P}\{\cdot | D\}$. Theorems like (44) can therefore be used in partially defined situations.

Conditional independence

Let $(\Omega, \mathcal{F})$ and $(\Omega_i, \mathcal{F}_i)$ be Borel. Let $\mathcal{P}$ be a probability on $\mathcal{F}$, and X_i a measurable mapping from $(\Omega, \mathcal{F})$ to $(\Omega_i, \mathcal{F}_i)$. Let Σ be a sub-σ-field of $\mathcal{F}$. What does it mean to say X_1 and X_2 are *conditionally $\mathcal{P}$-independent given* Σ? The easiest criterion is

$$\mathcal{P}\{X_1 \in A_1 \text{ and } X_2 \in A_2 | \Sigma\} = \mathcal{P}\{X_1 \in A_1 | \Sigma\} \cdot \mathcal{P}\{X_2 \in A_2 | \Sigma\} \quad \text{a.e.}$$

for all $A_i \in \mathcal{F}_i$. Nothing is changed if A_i is confined to a generating class for $\mathcal{F}_i$ in the sense of (16). Here is an equivalent criterion. Let $Q(\cdot, \cdot)$ be a regular conditional $\mathcal{P}$-distribution for (X_1, X_2) given Σ. Then $Q(\omega, \cdot)$ is a probability on $\mathcal{F}_1 \times \mathcal{F}_2$. The variables X_1, X_2 are conditionally $\mathcal{P}$-independent given Σ iff for $\mathcal{P}$-almost all ω,

$$Q(\omega, \cdot) = Q_1(\omega, \cdot) \times Q_2(\omega, \cdot)$$

where $Q_i(\omega, \cdot)$ is the projection of $Q(\omega, \cdot)$ onto $\mathcal{F}_i$. Necessarily, Q_i is a regular conditional $\mathcal{P}$-distribution for X_i given Σ. The equivalence of these conditions is easy, using (10a).

11.   THE KOLMOGOROV CONSISTENCY THEOREM†

Let $(\Omega_i, \mathcal{F}_i)$ be Borel for $i = 1, 2, \ldots$. Let

$$\Omega = \Omega_1 \times \Omega_2 \times \cdots \quad \text{and} \quad \mathcal{F} = \mathcal{F}_1 \times \mathcal{F}_2 \times \cdots.$$

Let π_n *project* $\Omega_1 \times \cdots \times \Omega_{n+1}$ onto $\Omega_1 \times \cdots \times \Omega_n$: namely,

$$\pi_n(\omega_1, \ldots, \omega_n, \omega_{n+1}) = (\omega_1, \ldots, \omega_n).$$

† References: (Loève, 1963, Section 4.3); (Neveu, 1965, Section III.3).

Let π_n *project* Ω onto $\Omega_1 \times \cdots \times \Omega_n$: namely,

$$\pi_n(\omega_1, \ldots, \omega_n, \omega_{n+1}, \ldots) = (\omega_1, \ldots, \omega_n).$$

For $n = 1, 2, \ldots$, let $\mathscr{P}_n$ be a probability on $(\Omega_1 \times \cdots \times \Omega_n, \mathscr{F}_1 \times \cdots \times \mathscr{F}_n)$. Suppose the $\mathscr{P}_n$ are *consistent*, namely, $\mathscr{P}_{n+1}\pi_n^{-1} = \mathscr{P}$ for all n.

(53) Theorem. *There is a unique probability $\mathscr{P}$ on $(\Omega, \mathscr{F})$ with $\mathscr{P}\pi_n^{-1} = \mathscr{P}_n$ for all n.*

12. THE DIAGONAL ARGUMENT

Let Z be the positive integers. Let S be the set of strictly increasing functions from Z to Z. Call $s \in S$ a *subsequence* of Z. For $s \in S$, the *range* of s is the s-image $s(Z)$ of Z; and $s(n) \geq n$. Say s is a *subsequence* or on special occasions a *sub-subsequence* of $t \in S$ iff $s \in S$ and for each $n \in Z$, there is a $\sigma(n) \in Z$ with $s(n) = t[\sigma(n)]$. This well-defines σ, and forces $\sigma \in S$. Further, $s(n) \geq t(n)$, because $\sigma(n) \geq n$. Geometrically, $s \in S$ is a subsequence of $t \in S$ iff the range of s is a subset of the range of t. Thus, if s is a subsequence of t, and t is a subsequence of $u \in S$, then s is a subsequence of u. If $s \in S$, and $m = 0, 1, \ldots$, define $s(m + \cdot) \in S$ as follows:

$$s(m + \cdot)(n) = s(m + n) \quad \text{for } n \in Z.$$

Of course, $s(m + \cdot)$ is a subsequence of s.

Here is a related notion. Say s is *eventually* a subsequence of $t \in S$ iff $s \in S$ and $s(m + \cdot)$ is a subsequence of t for some $m = 0, 1, \ldots$. Geometrically, $s \in S$ is eventually a subsequence of $t \in S$ iff the range of s differs by a finite set from a subset of the range of t. In particular, if s is eventually a subsequence of t, and t is eventually a subsequence of $u \in S$, then s is eventually a subsequence of u.

To state the *first diagonal principle*, let $s_1 \in S$ and let s_{n+1} be a subsequence of s_n for $n = 1, 2, \ldots$. Let d be the *diagonal sequence*:

$$d(n) = s_n(n) \quad \text{for } n = 1, 2, \ldots.$$

(54) First diagonal principle. *The diagonal sequence d is a subsequence of s_1, and is eventually a subsequence of s_n for all n.*

PROOF. I claim $d \in S$:

$$d(n + 1) = s_{n+1}(n + 1) \geq s_n(n + 1) > s_n(n) = d(n).$$

Fix $n = 1, 2, \ldots$. I claim $d(n - 1 + \cdot)$ is a subsequence of s_n. Indeed, fix $m = 1, 2, \ldots$. Then $m - 1 \geqq 0$, so

$$d(n - 1 + m) = s_{n-1+m}(n - 1 + m) \in s_{n-1+m}(Z) \subset s_n(Z).$$ ★

(55) Illustration.   Let $s_n(m) = n + m$. So $d(n) = 2n$, as in Figure 2.

$\overset{\displaystyle m}{s}$	1	2	3	4	•	•	•	•	•
s_1	2	3	4	5	•	•	•	•	•
s_2	3	4	5	6	•	•	•	•	•
s_3	4	5	6	7	•	•	•	•	•
s_4	5	6	7	8	•	•	•	•	•
•	•	•	•	•	•	•	•	•	•
•	•	•	•	•	•	•	•	•	•
•	•	•	•	•	•	•			
•	•	•	•	•	•	•			
•	•	•	•	•	•				

Figure 2.

To make this a little more interesting, introduce a metric space (Ω, ρ). Let f be a function from Z to Ω. Let $t \in S$. Suppose

$$\lim_{n \to \infty} f[t(n)] = y \in \Omega.$$

If s is eventually a subsequence of t, you should check

$$\lim_{n \to \infty} f[s(n)] = y.$$

For the *second diagonal principle*, let C be a countable set. For $c \in C$, let f_c be a function from Z to Ω. Suppose that for each $c \in C$ and $t \in S$, there is a subsequence s_c of t, such that

$$\lim_{n \to \infty} f_c[s_c(n)]$$

exists. This s_c depends on c and t.

(56) Second diagonal principle.   *For each* $t \in S$, *there is a subsequence* d *of* t *such that*

$$\lim_{n \to \infty} f_c[d(n)]$$

exists for all $c \in C$. *This* d *depends on* t, *but not on* c.

PROOF.   Enumerate C as $\{c_1, c_2, \dots\}$. Abbreviate $f_n = f_{c_n}$. Inductively, construct $s_n \in S$ such that $s_0 = t$ and s_n is a subsequence of s_{n-1} and

$$\lim_{m \to \infty} f_n[s_n(m)]$$

exists. Call this limit y_n; of course, y_n may depend on $s_1, \dots, s_n$. Using the first diagonal principle, construct the diagonal subsequence d, which is a subsequence of t and eventually a subsequence of each s_n. So,

$$\lim_{m \to \infty} f_n[d(m)] = y_n.   \bigstar$$

13.   CLASSICAL LEBESGUE MEASURE

Euclidean n-space R^n comes equipped with a metric ρ_n, and is complete and separable (37). A real-valued random variable on $(\Omega, \mathscr{F})$ is now a measurable mapping to Borel R^1. The classical n-dimensional Lebesgue measure λ_n is the countably additive, nonnegative set function on Borel R^n, whose value at an n-dimensional cube is its n-dimensional volume.

(57) Theorem.   *Suppose* f *is a bounded function on* R^n *which vanishes outside a cube. Then* f *is Riemann integrable iff* f *is continuous* λ_n-*almost everywhere, and its Riemann integral coincides with* $\int f \, d\lambda_n$.

This theorem will be used only to evaluate Lebesgue integrals, and the Riemann integrability of f will be obvious.

(58) Theorem.   *Let* f *be a measurable function on* R^n, *with finite Lebesgue integral. Then*

$$\lim_{h \to 0} \int_{R^n} |f(x + h) - f(x)| \lambda_n(dx) = 0.$$

PROOF.   If f is continuous and vanishes outside a large cube, the result is clear. General f can be approximated by these special f in L^1-norm. $\bigstar$

If A is a Borel subset of R^1, the *metric density of A at x* is

$$\lim_{\varepsilon, \delta \downarrow 0} \frac{\lambda_1\{(x - \varepsilon, x + \delta) \cap A\}}{\varepsilon + \delta}.$$

(59) Metric density theorem.[1] *Let A be a Borel subset of R^1. There is a Borel set B with $\lambda_1(B) = 0$, such that A has metric density 1 at all $x \in A \setminus B$.*

14. REAL VARIABLES

Let f be a real-valued function on $[0, 1]$. Let $S = \{s_0, s_1, \ldots, s_n\}$ be a finite subset of $[0, 1]$ with $0 = s_0 < s_1 < \cdots < s_n = 1$. Let

$$\delta S = \max \{(s_{j+1} - s) : j = 0, \ldots, n - 1\},$$

and

$$Sf = \Sigma_{j=0}^{n-1} |f(s_{j+1}) - f(s_j)|.$$

Let $W(S, f) = \Sigma_{j=0}^{n-1} (M_j - m_j)$, where

$$M_j = \max \{f(t) : s_j \leqq t \leqq s_{j+1}\} \quad \text{and} \quad m_j = \min \{f(t) : s_j \leqq t \leqq s_{j+1}\}.$$

The *variation of f* is $\sup_S Sf$; if this number is finite, f is of *bounded variation*. If S_n is nondecreasing and $\delta S_n \downarrow 0$, then $S_n f$ tends to the variation of f; so $W(S_n, f)$ must tend to the variation of f also.

(60) Lebesgue's theorem.[2] *If f is of bounded variation, then f has a finite derivative Lebesgue almost everywhere.*

Theorem (60) can be sharpened as follows.

(61) Theorem.[3] *Suppose f is of bounded variation. The pointwise derivative of f is a version of the Radon–Nikodym derivative of the absolutely continuous part of f, with respect to Lebesgue measure.*

Even more is true.

(62) de la Vallée Poussin's theorem.[4] *Suppose f is of bounded variation. The positive, continuous, singular part of f is concentrated on $\{x : f'(x) = \infty\}$.*

ASSUMPTION. For the rest of this section, assume f is a continuous function on $[0, 1]$.

[1] Reference: (Saks, 1964, Theorem 6.1 on p. 117. Theorem 10.2 on p. 129 is the n-dimensional generalization, which is harder.)
[2] References: (Saks, 1964, Theorem 5.4 on p. 115); (Riesz-Nagy, 1955, Chapter 1) has a proof from first principles. This theorem is hard.
[3] References: (Dunford and Schwartz, 1958, III.12); (Saks, 1964, Theorem 7.4 on p. 119). It's hard.
[4] Reference: (Saks, 1964, Theorem 9.6 on p. 127). Theorems (60–62) are hard.

Let $s(y)$ be the number of x with $f(x) = y$, so $s(y) = 0, 1, \ldots, \infty$.

(63) Banach's theorem.[1]   *The variation of f is $\int_{-\infty}^{\infty} s(y)\, dy$.*

Proof.   Let $S_n = \{0, 1/2^n, 2/2^n, \ldots, 1\}$. Let $s_{n,0}$ be the indicator function of the f-image of the interval $[0, 1/2^n]$. For $j = 1, \ldots, 2^n - 1$, let $s_{n,j}$ be the indicator function of the f-image of $(j/2^n, (j + 1)/2^n]$. Let $s_n = \Sigma_{j=0}^{2^n-1} s_{n,j}$. Verify that $s_n \uparrow s$, so s is Borel and

$$\int_{-\infty}^{\infty} s(y)\, dy = \lim_n \int_{-\infty}^{\infty} s_n(y)\, dy = \lim_n W(S_n, f):$$

because

$$\int_{-\infty}^{\infty} s_{n,j}(y)\, dy = M_{n,j} - m_{n,j},$$

where

$$M_{n,j} = \max \{f(t) : j/2^n \leq t \leq (j + 1)/2^n\}$$

$$m_{n,j} = \min \{f(t) : j/2^n \leq t \leq (j + 1)/2^n\}. \qquad \bigstar$$

The *upper right Dini derivative $D*f$* is defined by:

$$D*f \cdot x = \lim \sup_{\varepsilon \downarrow 0} [f(x + \varepsilon) - f(x)]/\varepsilon,$$

for $0 \leq x < 1$.

(64) Zygmund's theorem.[2]   *If the set of values assumed by f on the set of x with $D*f \cdot x \leq 0$ includes no proper interval, then f is nondecreasing.*

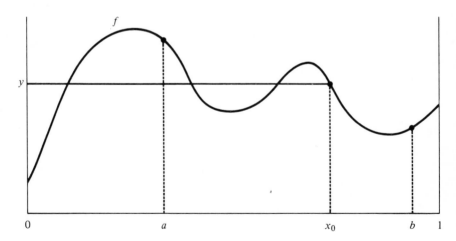

Figure 3.

[1] Reference: (Saks, 1964, Theorem 6.4 on p. 280).
[2] Reference: (Saks, 1964, Theorem 7.1 on p. 203).

PROOF. Suppose by way of contradiction that there are a and b with $0 < a < b < 1$ and $f(a) > f(b)$. Find one y with $f(a) > y > f(b)$, such that $y \notin f\{D*f \leq 0\}$; that is, $y = f(x)$ entails $D*f{\cdot}x > 0$. Let x_0 be the largest $x \in [a, b]$ with $f(x) = y$, so $x_0 < b$. But $f < y$ on $(x_0, b]$, so $D*f{\cdot}x_0 \leq 0$. See Figure 3.   ★

(65) Corollary. *If the set of x with $D*f{\cdot}x < 0$ is at most countable, then f is nondecreasing.*

PROOF. Let $\varepsilon > 0$ and $f_\varepsilon(x) = f(x) + \varepsilon x$. Now $D*f_\varepsilon{\cdot}x = D*f{\cdot}x + \varepsilon$, so $\{D*f_\varepsilon \leq 0\}$ is at most countable. By Zygmund's theorem, f_ε is nondecreasing. Let $\varepsilon \to 0$.   ★

(66) Dini's theorem.[1]

(a) $\sup_{0 \leq x < 1} D*f{\cdot}x = \sup_{0 \leq x < y \leq 1} \dfrac{f(y) - f(x)}{y - x}$

and

(b) $\inf_{0 \leq x < 1} D*f{\cdot}x = \inf_{0 \leq x < y \leq 1} \dfrac{f(y) - f(x)}{y - x}.$

PROOF. Fix a finite m. Then $D*f{\cdot}x \geq m$ for $0 \leq x < 1$ implies $f(x) - mx$ is nondecreasing with x, by (65). By algebra, for $0 \leq x < y \leq 1$,

$$\frac{f(y) - f(x)}{y - x} \geq m.$$

Consequently, $\inf [f(y) - f(x)]/(y - x) \geq \inf D*f \cdot x$. The opposite inequality is clear, proving (b). Assertion (a) is easy.   ★

(67) Corollary.[2] *If $D*f$ is continuous at x, then f is differentiable at x.*

PROOF. Use (66).   ★

(68) Corollary. *If $D*f \equiv 0$ on $[0, 1)$, then f is constant on $[0, 1]$.*

PROOF. Use (67).   ★

(69) Theorem. *Suppose f has a finite, right continuous, right derivative f^+ on $(0, 1)$, which has a finite integral over $(\varepsilon, 1 - \varepsilon)$ for any $\varepsilon > 0$. If $0 < x < y < 1$, then*

$$f(y) = f(x) + \int_x^y f^+(t)\, dt.$$

[1] Reference: (Saks, 1964, p. 204).
[2] Reference: Saks, 1964, p. 204).

PROOF.   Let

$$g(y) = f(y) - f(x) - \int_x^y f^+(t)\, dt.$$

Then g is continuous and $D*g = 0$ on $[x, 1)$, while $g(x) = 0$. Use (68).   ★

Miscellany

Let μ be a probability on the Borel subsets of $[0, \infty)$. Its *Laplace transform* φ is this function of nonnegative λ:

$$\varphi(\lambda) = \int_{[0, \infty)} e^{-\lambda x}\, \mu(dx).$$

(70) Theorem.[1] φ *determines* μ.

As usual, $0! = 1$, and $1! = 1$, and $(n + 1)! = (n + 1)n!$. Write

$$\binom{n}{m} = \frac{n!}{m!(n - m)!},$$

the number of subsets with m elements which can be chosen from a set with n elements. Temporarily, let

$$s(n) = \sqrt{2\pi n} \left(\frac{n}{e}\right)^n.$$

(71) Stirling's formula.[2] $n!/s(n) \to 1$ as $n \to \infty$.

15.   ABSOLUTE CONTINUITY

A function f on $[0, 1]$ is *absolutely continuous* iff for every positive ε there is a positive δ such that:

$$0 \leq t_1 < t_2 < t_3 < \cdots < t_{2n-1} < t_{2n} \leq 1$$

and

$$\Sigma_{m=1}^n (t_{2m} - t_{2m-1}) < \delta$$

imply

$$\Sigma_{m=1}^n |f(t_{2m}) - f(t_{2m-1})| < \varepsilon.$$

In other words, f is absolutely continuous iff f is of bounded variation,

[1] Reference: (Feller, 1966, Theorem 1 on p. 408).
[2] Reference: (Feller, 1968, II.9).

and the measure induced by f is absolutely continuous with respect to Lebesgue measure. The pointwise derivative of f is then a version of the Radon–Nikodym derivative by (61). Suppose g is another function on $[0, 1]$. If f and g are absolutely continuous, and $f' = g'$ a.e. it follows that $f - g$ is constant. Suppose f and g are absolutely continuous, and g is nondecreasing, and $0 \leq g(0) \leq g(1) \leq 1$. Then $f \circ g$ is absolutely continuous, as is immediate from the definition; and

THE CHAIN RULE.[1] $(f \circ g)' = (f' \circ g)g'$ a.e.

PROOF. For simplicity, suppose $f(0) = g(0) = 0$ and $g(1) = 1$. Let $0 \leq t \leq 1$. Here is a computation.

$$f \circ g(t) = \int_0^{g(t)} f'(u)\, du$$

$$= \int_0^t f'[g(t)]\, g(dt)$$

$$= \int_0^t f'[g(t)]g'(t)\, dt.$$

The first line holds by (61); the second by (20), for the g-distribution of g is uniform on $[0, 1]$; and the third by (61, 11). Now use (61, 11) to differentiate.
★

(72) Theorem. *Suppose the upper or lower right Dini derivative of f is finite at all but a countable number of points, and f is of bounded variation. Then f is absolutely continuous.*

PROOF. Use (62).

16. CONVEX FUNCTIONS

Let f be a real-valued function on the open interval (a, b). Abbreviate $\bar{p} = 1 - p$. Then f is *convex* iff

$$a < x < y < b \quad \text{and} \quad 0 < p < 1$$

imply

$$f(px + \bar{p}y) \leq pf(x) + \bar{p}f(y).$$

Geometrically, each chord of f is nowhere below f, as in Figure 4. Say f is *strictly convex* iff inequality holds. Say f is *concave* or *strictly concave* iff

[1] Reference: (Serrin and Varberg, 1969).

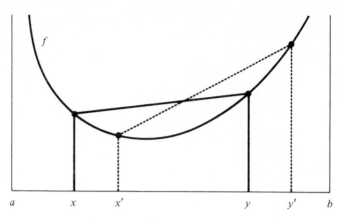

Figure 4.

$-f$ is convex or strictly convex. For (73), suppose f is convex and

$$a < x < y < b \quad \text{and} \quad a < x' < y' < b$$

$$x \leqq x' \quad \text{and} \quad y \leqq y'.$$

As in Figure 4,

(73)
$$\frac{f(y') - f(x')}{y' - x'} \geqq \frac{f(y) - f(x)}{y - x}.$$

Indeed, the case $x = x'$ restates the definition of convexity, as does the case $y = y'$. General (73) follows by combining the two cases: the slope of f over (x, y) is at most the slope over (x, y'), which does not exceed the slope over (x', y').

(74) Theorem. *Suppose f is convex on (a, b).*

    (a) *f is continuous.*

    (b) *f has a finite right derivative f^+, which is nondecreasing and continuous from the right.*

    (c) *f has a finite left derivative f^-, which is nondecreasing and continuous from the left.*

    (d) *$f^+ \geqq f^-$.*

    (e) *The discontinuity sets of f^+ and f^- coincide, and are countable. Off this set, $f^+ = f^-$.*

    (f) *For $a < x < y < b$,*

$$f(y) - f(x) = \int_x^y f^+(t)\, dt = \int_x^y f^-(t)\, dt.$$

(g) *If f is strictly convex, then f^+ and f^- are strictly increasing.*

(h) *Suppose a and $f(a+)$ are finite. Define $f(a) = f(a+)$. Then f has a right derivative $f^+(a)$ at a, and $-\infty \le f^+(a) < \infty$. And $f^+(a+) = f^+(a)$. The situation at b is symmetric.*

PROOF.   *Claim (b).*   Let y decrease to $x > w$. The slope of f over (x, y) nonincreases and is at least the slope over (w, x) by (73), proving that

(75) $f^+(x)$ exists and is at most the slope of f over (x, y), and at least the slope over (w, x).

You can use (73) to show that f^+ is nondecreasing. I will argue that f^+ is continuous from the right at x. Let

$$a < x < y < z < b,$$

as in Figure 5.

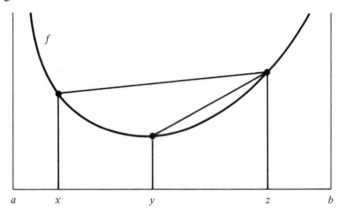

Figure 5.

Then $f^+(x) \le f^+(y)$. But $f^+(y)$ is at most the slope of f over (y, z) by (75), which tends to the slope of f over (x, z) when y tends to x. That is,

$$\lim_{y \to x} f^+(y) \le \frac{f(z) - f(x)}{z - x} \downarrow f^+(x) \quad \text{as } z \downarrow x.$$

Claim (c) is symmetric.

Claim (d) follows from (73).

Claim (a). Get $f(x) = f(x+)$ from the finitude of f^+, and $f(x) = f(x-)$ from symmetry.

Claim (e). Let $h > 0$. Then $f^-(x + h) \ge f^+(x)$ by (73). And (d) makes $f^-(x + h) \le f^+(x + h)$. Let $h \to 0$ and use (b):

$$f^-(x+) = f^+(x+) = f^+(x).$$

Similarly

$$f^-(x-) = f^-(x-) = f^-(x).$$

Claim (f).   Use (a, b) and (69), then (e).

Claim (g).   Improve (73) to strict inequality, when $x < x'$ or $y < y'$. Let

$$a < x < y < z < w < b.$$

Use (75) and improved (73):

$$f(x+) \leqq \frac{f(y) - f(x)}{y - x} < \frac{f(z) - f(y)}{z - y} < \frac{f(w) - f(z)}{w - z} \downarrow f^+(z)$$

as $w \downarrow z$.

Claim (h) is like (b).   ★

(76) Theorem.   *Suppose f is convex on (a, b), and $x \in (a, b)$. Using (74d), choose s with*

$$f^-(x) \leqq s \leqq f^+(x).$$

Let

$$\lambda(t) = s(t - x) + f(x),$$

the linear function with slope s which agrees with f at x. Then

$$\lambda \leqq f \quad on \; (a, b).$$

In particular, a convex function on a finite interval is bounded below.

PROOF.   Let $x < t < b$: the other case is symmetric. Then

$$s \leqq f^+(x) \leqq \frac{f(t) - f(x)}{t - x}$$

by (75); look at Figure 6. So

$$f(t) \geqq s(t - x) + f(x) = \lambda(t).$$ ★

NOTE.   Suppose f is a function on (a, b). If either

(a) f has a finite, right continuous derivative f^+ which is nondecreasing,

or

(b)   there is a nondecreasing g with

$$f(y) = f(x) + \int_x^y g(t) \, dt \quad when \; a < x < y < b,$$

then f is convex.

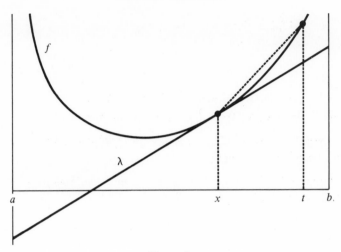

Figure 6.

PROOF.   Condition (a) implies (b) with $g = f^+$, by (69). Suppose (b). Fix x, y, z with

$$a < x < y < z < b.$$

Abbreviate c for the slope of f over (x, y), and d for the slope of f over (y, z). Then

$$c = \frac{1}{y - x} \int_x^y g(t)\, dt \leq \frac{1}{z - y} \int_y^z g(t)\, dt = d.$$

So $(y, f(y))$ cannot be above the chord joining $(x, f(x))$ and $(z, f(z))$.   ★

17.   COMPLEX VARIABLES

If $z = x + iy$, where x and y are real and $i^2 = -1$, then $\operatorname{Re} z = x$ and $\operatorname{Im} z = y$ and $|z| = (x^2 + y^2)^{\frac{1}{2}}$. Moreover, $e^z = e^x(\cos y + i \sin y)$, so[1]

(77) $|e^z| = e^x.$

A complex function f of a complex variable is *analytic* on an open half-plane H iff it is differentiable at every point of H.

(78) Theorem.[2] *Suppose f is analytic on an open half-plane H. Then f has derivatives $f^{(n)}$ of all orders n at all points of H. If $z_0 \in H$, choose a positive*

[1] Reference: (Ahlfors, 1953, p. 47).
[2] Reference: (Ahlfors, 1953, pp. 96 and 142). Hard.

real number r so small that

$$D_r = \{z : |z - z_0| \leqq r\} \subset H.$$

Then

$$f(z) = \Sigma_{n=0}^{\infty} f^{(n)}(z_0) \cdot (z - z_0)^n/n!.$$

The series converges absolutely and uniformly on D_r.

(79) Theorem.[1] *If f_n is an analytic function on a half-plane and $f_n \to f$ uniformly on compact sets, then f is analytic.*

(80) Theorem.[2] *If f is an analytic function on a half-plane, and f vanishes on a set which has a point of accumulation, then f is identically 0.*

---

[1] Reference: (Ahlfors, 1953, Theorem 1 on p. 138). Hard.
[2] Reference: (Ahlfors, 1953, p. 102). Hard.

BIBLIOGRAPHY

LARS V. AHLFORS (1953; 2nd ed., 1965). *Complex Analysis*, McGraw-Hill, New York.

DAVID BLACKWELL (1954). On a class of probability spaces, *Proc. 3rd Berk. Symp.*, Vol. 2, pp. 1–6.

DAVID BLACKWELL (1955). On transient Markov processes with a countable number of states and stationary transition probabilities, *Ann. Math. Statist.*, Vol. 26, pp. 654–658.

DAVID BLACKWELL (1958). Another countable Markov process with only instantaneous states, *Ann. Math. Statist.*, Vol. 29, pp. 313–316.

DAVID BLACKWELL (1962). Representation of nonnegative martingales on transient Markov chains, Mimeograph, Statistics Department, University of California at Berkeley.

DAVID BLACKWELL and LESTER DUBINS (1963). A converse to the dominated convergence theorem, *Illinois J. Math.*, Vol. 7, pp. 508–514.

DAVID BLACKWELL and DAVID A. FREEDMAN (1964). The tail σ-field of a Markov chain and a theorem of Orey, *Ann. Math. Statist.*, Vol. 35, pp. 1291–1295.

DAVID BLACKWELL and DAVID FREEDMAN (1968). On the local behavior of Markov transition probabilities, *Ann. Math. Statist.*, Vol. 39, pp. 2123–2127.

DAVID BLACKWELL and DAVID KENDALL (1964). The Martin boundary for Pólya's urn scheme and an application to stochastic population growth, *J. Appl. Probability*, Vol. 1, pp. 284–296.

R. M. BLUMENTHAL (1957). An extended Markov property, *Trans. Amer. Math. Soc.*, Vol. 85, pp. 52–72.

R. M. BLUMENTHAL and R. K. GETOOR (1968). *Markov Processes and Potential Theory*, Academic Press, New York.

R. M. BLUMENTHAL, R. GETOOR, and H. P. MCKEAN, Jr. (1962). Markov processes with identical hitting distributions, *Illinois J. Math.*, Vol. 6, pp. 402–420.

LEO BREIMAN (1968). *Probability*, Addison-Wesley, Reading.

D. BURKHOLDER (1962). Transient processes and a problem of Blackwell, Mimeograph, Statistics Department, University of California at Berkeley.

D. BURKHOLDER (1962). Successive conditional expectations of an integrable function, *Ann. Math. Statist.*, Vol. 33, pp. 887–893.

KAI LAI CHUNG (1960; 2nd ed., 1967). *Markov Chains with Stationary Transition Probabilities*, Springer, Berlin.

KAI LAI CHUNG (1963). On the boundary theory for Markov chains, I, *Acta. Math.*, Vol. 110, pp. 19–77.

KAI LAI CHUNG (1966). On the boundary theory for Markov chains, II, *Acta. Math.*, Vol. 115, pp. 111–163.

KAI LAI CHUNG and W. H. J. FUCHS (1951). On the distribution of values of sums of random variables, *Mem. Amer. Math. Soc.*, no. 6.

R. COGBURN and H. G. TUCKER (1961). A limit theorem for a function of the increments of a decomposable process, *Trans. Amer. Math. Soc.*, Vol. 99, pp. 278–284.

HARALD CRAMÉR (1957). *Mathematical Methods of Statistics*, Princeton University Press.

ABRAHAM DE MOIVRE (1718). *The Doctrine of Chances*, Pearson, London, Chelsea, New York (1967).

C. DERMAN (1954). A solution to a set of fundamental equations in Markov chains, *Proc. Amer. Math. Soc.*, Vol. 5, pp. 332–334.

W. DOEBLIN (1938). Sur deux problèmes de M. Kolmogorov concernant les chaînes dénombrables, *Bull. Soc. Math. France*, Vol. 66, pp. 210–220.

W. DOEBLIN (1939). Sur certains mouvements aléatoires discontinus, *Skand. Akt.*, Vol. 22, pp. 211–222.

MONROE D. DONSKER (1951). An invariance principle for certain probability limit theorems, *Mem. Amer. Math. Soc.*, no. 6.

J. L. DOOB (1942). Topics in the theory of Markoff chains, *Trans. Amer. Math. Soc.*, Vol. 52, pp. 37–64.

J. L. DOOB (1945). Markoff chains-denumerable case, *Trans. Amer. Math. Soc.*, Vol. 58, pp. 455–473.

J. L. DOOB (1953). *Stochastic Processes*, Wiley, New York.

J. L. DOOB (1959). Discrete potential theory and boundaries, *J. Math. Mech.*, Vol. 8, pp. 433–458, 993.

J. L. DOOB (1968). Compactification of the discrete state space of a Markov process, *Z. Wahrscheinlichkeitstheorie*, Vol. 10, pp. 236–251.

LESTER E. DUBINS and DAVID A. FREEDMAN (1964). Measurable sets of measures, *Pac. J. Math.*, Vol. 14, pp. 1211–1222.

LESTER E. DUBINS and DAVID A. FREEDMAN (1965). A sharper form of the Borel-Cantelli lemma and the strong law, *Ann. Math. Statist.*, Vol. 36, pp. 800–807.

LESTER E. DUBINS and GIDEON SCHWARZ (1965). On continuous martingales, *Proc. Nat. Acad. Sci. USA*, Vol. 53, pp. 913–916.

NELSON DUNFORD and JACOB T. SCHWARTZ (1958). *Linear operators, Part I*, Wiley, New York.

A. DVORETZKY, P. ERDÖS, and S. KAKUTANI (1960). Nonincrease everywhere of the Brownian motion process, *Proc. 4th Berk. Symp.*, Vol. 2, pp. 103–116.

E. B. DYNKIN (1965). *Markov Processes*, Springer, Berlin.

P. ERDÖS and M. KAC (1946). On certain limit theorems of the theory of probability, *Bull. Amer. Math. Soc.*, Vol. 52, pp. 292–302.

WILLIAM FELLER (1945). On the integro-differential equations of purely discontinuous Markoff processes, *Trans. Amer. Math. Soc.*, Vol. 48, pp. 488–515.

WILLIAM FELLER (1956). Boundaries induced by nonnegative matrices, *Trans. Amer. Math. Soc.*, Vol. 83, pp. 19–54.

WILLIAM FELLER (1957). On boundaries and lateral conditions for the Kolmogoroff differential equations, *Ann. of Math.*, Vol. 65, pp. 527–570.

WILLIAM FELLER (1959). Non-Markovian processes with the semigroup property, *Ann. Math. Statist.*, Vol. 30, pp. 1252–1253.

WILLIAM FELLER (1961). A simple proof for renewal theorems, *Comm. Pure Appl. Math.*, Vol. 14, pp. 285–293.

WILLIAM FELLER (1966). *An introduction to probability theory and its applications*, Vol. 2, Wiley, New York.

WILLIAM FELLER (1968). *An introduction to probability theory and its applications*, Vol. 1, 3rd ed., Wiley, New York.

WILLIAM FELLER and H. P. McKEAN, Jr. (1956). A diffusion equivalent to a countable Markov chain, *Proc. Nat. Acad. Sci. USA*, Vol. 42, pp. 351–354.

R. GETOOR (1965). Additive functionals and excessive functions, *Ann. Math. Statist.*, Vol. 36, pp. 409–423.

G. H. HARDY, J. E. LITTLEWOOD, and G. POLYA (1934). *Inequalities*, Cambridge University Press.

T. E. HARRIS (1952). First passage and recurrence distributions, *Trans. Amer. Math. Soc.*, Vol. 73, pp. 471–486.

T. E. HARRIS and H. ROBBINS (1953). Ergodic theory of Markov chains admitting an infinite invariant measure, *Proc. Nat. Acad. Sci. USA*, Vol. 39, pp. 860–864.

P. HARTMAN and A. WINTNER (1941). On the law of the iterated logarithm, *Amer. J. Math.*, Vol. 63, pp. 169–176.

FELIX HAUSDORFF (1957). *Set Theory*, Chelsea, New York.

EDWIN HEWITT and L. J. SAVAGE (1955). Symmetric measures on Cartesian products, *Trans. Amer. Math. Soc.*, Vol. 80, pp. 470–501.

E. HEWITT and K. STROMBERG (1965). *Real and Abstract Analysis*, Springer, Berlin.

G. A. HUNT (1956). Some theorems concerning Brownian motion, *Trans. Amer. Math. Soc.*, Vol. 81, pp. 294–319.

G. A. HUNT (1957). Markoff processes and potentials, 1, 2, 3, *Illinois J. Math.*, Vol. 1, pp. 44–93; Vol. 1, pp. 316–369; Vol. 2, pp. 151–213 (1958).

G. A. HUNT (1960). Markoff chains and Martin boundaries, *Illinois J. Math.*, Vol. 4, pp. 313–340.

K. ITÔ and H. P. McKEAN, Jr. (1965). *Diffusion Processes and Their Sample Paths*, Springer, Berlin.

W. B. JURKAT (1960). On the analytic structure of semigroups of positive matrices, *Math. Zeit.*, Vol. 73, pp. 346–365.

A. A. JUSKEVIC (1959). Differentiability of transition probabilities of a homogeneous Markov process with countably many states, *Moskov. Gos. Univ. Učenye Zapiski*, No. 186, pp. 141–159; in Russian. Reviewed in *Math. Rev.* No. 3124 (1963).

M. KAC (1947). On the notion of recurrence in discrete stochastic processes, *Bull. Amer. Math. Soc.*, Vol. 53, pp. 1002–1010.

S. KAKUTANI (1943). Induced measure preserving transformations, *Proc. Impl. Acad. Tokyo*, Vol. 19, pp. 635–641.

J. G. KEMENY and J. L. SNELL (1960). *Finite Markov Chains*, Van Nostrand, Princeton.

J. G. KEMENY, J. SNELL, and A. W. KNAPP (1966). *Denumerable Markov Chains*, Van Nostrand, Princeton.

A. KHINTCHINE (1924). Ein Satz der Wahrscheinlichkeitsrechnung, *Fund. Math.*, Vol. 6, pp. 9–20.

J. F. C. KINGMAN (1962). The imbedding problem for finite Markov chains, *Z. Wahrscheinlichkeitstheorie*, Vol. 1, pp. 14–24.

J. F. C. KINGMAN (1964). The stochastic theory of regenerative events, *Z. Wahrscheinlichkeitstheorie*, Vol. 2, pp. 180–224.

J. F. C. KINGMAN (1968). On measurable *p*-functions, *Z. Wahrscheinlichkeitstheorie*, Vol. 11, pp. 1–8.

J. F. C. KINGMAN and STEVEN OREY (1964). Ratio limit theorems for Markov chains, *Proc. Amer. Math. Soc.*, Vol. 15, pp. 907–910.

FRANK KNIGHT and STEVEN OREY (1964). Construction of a Markov process from hitting probabilities, *J. Math. Mech.*, Vol. 13, pp. 857–873.

A. KOLMOGOROV (1931). Über die analytischen Methoden in der Wahrscheinlich-keitsrechnung, *Math. Ann.*, Vol. 104, pp. 415–458.

A. KOLMOGOROV (1936). Anfangsgründe der Theorie der Markoffschen Ketten mit unendlichen vielen möglichen Züstanden, *Mat. Sb.*, pp. 607–610.

A. KOLMOGOROV (1951). On the differentiability of the transition probabilities in stationary Markov processes with a denumerable number of states, *Moskov. Gos. Univ. Učenye Zapiski*, Vol. 148, Mat. 4, pp. 53–59; in Russian. Reviewed on p. 295 of *Math. Rev.* (1953).

ULRICH KRENGEL (1966). On the global limit behavior of Markov chains and of general nonsingular Markov processes, *Z. Wahrscheinlichkeitstheorie*, Vol. 4, pp. 302–316.

CASIMIR KURATOWSKI (1958). *Topologie I*, 4th ed. Warsaw.

PAUL LÉVY (1951). Systèmes markoviens et stationnaires. Cas dénombrable, *Ann. Sci. École Norm. Sup.*, (3), Vol. 68, pp. 327–381.

PAUL LÉVY (1952). Complément à l'étude des processus de Markoff, *Ann. Sci. École Norm. Sup.*, (3) Vol. 69, pp. 203–212.

PAUL LÉVY (1953). Processus markoviens et stationnaires du cinquième type, *C. R. Acad. Sci. Paris*, Vol. 236, pp. 1630–1632.

PAUL LÉVY (1954). *Le Mouvement Brownien*, Gauthier Villars, Paris.

PAUL LÉVY (1954a). *Théorie de l'Addition des Variables Aléatoires*, Gauthier Villars, Paris.

PAUL LÉVY (1958). Processus markoviens et stationnaires. Cas dénombrable, *Ann. Inst. H. Poincaré*, Vol. 16, pp. 7–25.

PAUL LÉVY (1965). *Processus Stochastiques et Mouvement Brownien*, Gauthier Villars, Paris.

MICHEL LOÈVE (1963). *Probability Theory*, 3rd ed., Van Nostrand, Princeton.

A. A. MARKOV (1906). Extension of the law of large numbers to dependent events, *Bull. Soc. Phys. Math. Kazan.*, (2), Vol. 15, pp. 135–156; in Russian.

JACQUES NEVEU (1965). *Mathematical Foundations of the Calculus of Probability*, Holden-Day, San Francisco.

STEVEN OREY (1962). An ergodic theorem for Markov chains, *Z. Wahrscheinlich-keitstheorie*, Vol. 1, pp. 174–176.

DONALD ORNSTEIN (1960). The differentiability of transition functions, *Bull. Amer. Math. Soc.*, Vol. 66, pp. 36–39.

DANIEL RAY (1956). Stationary Markov processes with continuous paths, *Trans. Amer. Math. Soc.*, Vol. 82, pp. 452–493.

DANIEL RAY (1967). Some local properties of Markov processes, *Proc. 5th Berk. Symp.*, Vol. 2, part 2, pp. 201–212.

G. E. H. REUTER (1957). Denumerable Markov processes and the associated contraction semigroups on ρ, *Acta Math.*, Vol. 97, pp. 1–46.

G. E. H. REUTER (1959). Denumerable Markov processes. *J. London Math. Soc.*, Vol. 34, pp. 81–91.

G. E. H. REUTER (1969). Remarks on a Markov chain example of Kolmogorov, *Z. Wahrscheinlichkeitstheorie*, Vol. 13, pp. 315–320.

F. RIESZ and B. SZ. NAGY (1955). *Functional Analysis*, Ungar, New York.

B. A. ROGOZIN (1961). On an estimate of the concentration function, *Theor. Probability Appl.*, Vol. 6, pp. 94–96.

H. L. ROYDEN (1963). *Real Analysis*, Macmillan, New York.

STANISLAW SAKS (1964). *Theory of the Integral*, 2nd rev. ed., Dover, New York.

JAMES SERRIN and D. E. VARBERG (1969). A general chain rule for derivatives and the change of variables formula for the Lebesgue integral, *Amer. Math. Monthly*, Vol. 76, pp. 514–520.

A. SKOROKHOD (1965). *Studies in the Theory of Random Processes*, Addison-Wesley, Reading.

GERALD SMITH (1964). Instantaneous states of Markov processes, *Trans. Amer. Math. Soc.*, Vol. 110, pp. 185–195.

J. M. O. SPEAKMAN (1967). Two Markov chains with a common skeleton, *Z. Wahrscheinlichkeitstheorie*, Vol. 7, p. 224.

FRANK SPITZER (1956). A combinatorial lemma and its applications to probability theory, *Trans. Amer. Math. Soc.*, Vol. 82, pp. 323–339.

FRANK SPITZER (1964). *Principles of Random Walk*, Van Nostrand, Princeton.

VOLKER STRASSEN (1964). An invariance principle for the law of the iterated logarithm, *Z. Wahrscheinlichkeitstheorie*, Vol. 3, pp. 211–226.

VOLKER STRASSEN (1966). A converse to the law of the interated logarithm, *Z. Wahrscheinlichkeitstheorie*, Vol. 4, pp. 265–268.

VOLKER STRASSEN (1966a). Almost sure behavior of sums of independent random variables and martingales, *Proc. 5th Berk. Symp.*, Vol. 2, part 1, pp. 315–343.

H. F. TROTTER (1958). A property of Brownian motion paths, *Illinois J. Math.*, Vol. 2, pp. 425–433.

A. WALD (1944). On cumulative sums of random variables, *Ann. Math. Statist.*, Vol. 15, pp. 283–296.

N. WIENER (1923). Differential space, *J. Math. and Phys.*, Vol. 2, pp. 131–174.

DAVID WILLIAMS (1964). On the construction problem for Markov chains, *Z. Wahrscheinlichkeitstheorie*, Vol. 3, pp. 227–246.

DAVID WILLIAMS (1966). A new method of approximation in Markov chain theory and its application to some problems in the theory of random time substitution, *Proc. Lond. Math. Soc. (3)*, Vol. 16, pp. 213–240.

DAVID WILLIAMS (1967). Local time at fictitious states, *Bull. Amer. Math. Soc.*, Vol. 73, pp. 542–544.

DAVID WILLIAMS (1967a). A note on the Q-matrices of Markov chains, *Z. Wahrscheinlichkeitstheorie*, Vol. 7, pp. 116–121.

DAVID WILLIAMS (1967b).   On local time for Markov chains, *Bull. Amer. Math. Soc.*, Vol. 73, pp. 432–433.

HELEN WITTENBERG (1964).   Limiting distributions of random sums of independent random variables, *Z. Wahrscheinlichkeitstheorie*, Vol. 1, pp. 7–18.

A. ZYGMUND (1959).   *Trigonometric Series*, Cambridge University Press.

INDEX

SYMBOL FINDER

DESCRIPTION

I've listed here the symbols with some degree of permanence; the list is not complete, and local usage is sometimes different. The listing is alphabetical, first English then Greek; I give a quick definition, if possible, and a page reference for the complete definition.

Sections 10.1–3 discuss notation and references.

ENGLISH

C, C_d: index sets, Chapter 6 only, page 184
$C_f[\omega, w] = C_f[w]$: set, Chapter 6 only, page 184
$C(i)$: communicating class containing i, page 17
$C_r(i)$: cyclically moving class, page 18
C_X: concentration function of X, page 99
eP: expected number of visits, page 19
$e^n P$: expected number of visits, page 49
ePH: expected number of visits, page 34
$eP\{i\}$: expected number of visits, page 47
E is expectation
E_i is P_i-expectation
E: set, Chapter 4 only, page 118
$\mathscr{E}$: exchangeable σ-field, Chapter 1 only, page 39
$\mathscr{E}$: equivalent to invariant σ-field, Chapter 4 only, page 118
fP: hitting probability, page 19
$f^n P$: hitting probability, page 19
fPH: hitting probability, page 34
$f \times v$: measure, Chapter 2 only, page 51
$\mathscr{F}_\tau$: pre-τ sigma-field, page 11
$\mathscr{F}(\tau)$: pre-τ sigma-field, page 203
$\mathscr{F}(\tau+)$: pre-τ sigma-field

379

GREEK

$\phi(i, j)$: hitting probability, Chapter 4 only, page 124
φ: infinite state, pages 174, 184, 218, 299
$\omega \in I^\infty$: I-valued function on the nonnegative integers, pages 2, 7, 155
$\omega \in \Omega$: I-valued function, pages 112, 185, 218, 304
Ω^*: set, pages 112, 185
Ω^∞: set, Chapter 4 only, page 112
Ω_g: good sample functions, Chapter 7 only, page 219
Ω_m: metrically perfect sample functions, Chapter 9 only, page 311
Ω_n: hit I_n, Chapter 4 only, page 113
Ω_q: quasiregular sample functions, Chapter 9 only, page 307
Ω_v: very good sample functions, Chapter 7 only, page 219